Minds Behind the Brain

Minds Behind the Brain

A History of the Pioneers and Their Discoveries

Stanley Finger, Ph.D.

OXFORD
UNIVERSITY PRESS

2000

OXFORD
UNIVERSITY PRESS

Oxford   New York
Athens   Auckland   Bangkok   Bogotá   Buenos Aires
Calcutta   Cape Town   Chennai   Dar es Salaam   Delhi
Florence   Hong Kong   Istanbul   Karachi   Kuala Lumpur
Madrid   Melbourne   Mexico City   Mumbai   Nairobi
Paris   São Paulo   Singapore   Taipei   Tokyo   Toronto   Warsaw

and associated companies in

Berlin   Ibadan

Copyright © 2000 by Stanley Finger

Published by Oxford University Press, Inc.
198 Madison Avenue, New York, New York 10016

Oxford is a registered trademark of Oxford University Press

Library of Congress Cataloging-in-Publication Data
Finger, Stanley.
Minds Behind the Brain : a history of
the pioneers and their discoveries / Stanley Finger.
p.   cm.
Includes bibliographical references and index.
ISBN 0-19-508571-X
1. Neurosciences—History. 2. Brain—Research—History.
I. Title.
QP353.F548 1999 612.8'2'09—dc21 99-17110

1 3 5 7 9 8 6 4 2

Printed in the United States of America
on acid-free paper

To my wife Wendy and our two sons, Robert and Bradley

Aside from increased feelings of self-esteem and the approval of our own conscience, the conquest of a new truth is undoubtedly the greatest adventure to which a man can aspire.

—Santiago Ramón y Cajal
Advice for a Young Investigator

Contents

Preface

The purpose of this book is to look at the lives and discoveries of some of the major Western thinkers who contemplated how the brain may work. The stimulus for this project came from the many lectures I give each year to undergraduate and graduate students at Washington University in St. Louis, where I have long held a faculty position. My students wanted to know more about these pioneers as real people, what led to their discoveries, and the ramifications of their insights. For these more inquisitive minds, knowing the year of a landmark and a "beard" was not enough.

I realized that what my better students were asking for would benefit all concerned. I personally would gain perspective by looking at the scientific literature in social context. In addition, by saying more about the men, women, and cultures behind the great discoveries, my lectures would become more interesting and stimulating. And with clearer images of the people involved, my students would find the important facts easier to retain and associate with other events worth remembering.

For example, all of my students knew that French philosopher René Descartes wrote about animals being "beast machines" without souls. But they became even more fascinated by Descartes and his ideas when I told them that the man who denied higher thought and conscious feelings to animals had a pet dog of his own. The dog was suitably named Monsieur Grat ("Mr. Scratch"), and Descartes adored his pet and treated him with great affection. They were also intrigued by several other biographical facts about Descartes, such as how he was influenced by the mechanical revolution taking place in Western Europe the 1600s, why he chose to leave France to write in Holland, and how Galileo's tribunal in Italy led him to withhold publication of his most important work on the mind and brain.

To cite a second example of how an aspect of personal history can make a name and a discovery even more memorable, we can turn to David Ferrier, who stimulated or removed different parts of the roof brain in monkeys and other animals during the 1870s and 1880s. Ferrier's efforts led to a better understanding of the sensory and motor areas of the cerebrum. He became much more of a "modern," however, when my students learned how the militant animal-rights activists in London did everything in their power to make life hell for him. They even brought him to trial for allegedly experimenting on animals without an appropriate license.

With examples such as these implanted in my mind, I began to plan my book about the scientific elite and their insights about the nervous system. My first step was to decide who my major figures would be. Guiding me were two thoughts. First, I hoped to cover a wide expanse of time in order to show how physicians and scientists from different eras and cultures perceived the brain. Second, I did not want to deal with men and women whose discoveries and theories were so new that they had not undergone the test of time or could easily be found in other places.

Actually, I was also influenced by a third thought, although it was probably more subconscious in nature. There was no reason for my chosen few to have been on target when seen through today's eyes. The important thing was only that these dominating individuals dramatically changed the scientific or medical landscape; that is, what they said and wrote had great impact.

I now tried to think of twelve individuals associated with changes in thinking about the nervous system, each of whom would become a main subject for a chapter. Finding a dozen highly influential players proved to be fairly easy. I mailed my tentative list to some historically oriented neuroscientists for suggestions and comments and also whipped it out at professional meetings. These acts of insecurity did not lead to any deletions. Instead, my list took on a life of its own and swelled to twice its original size. In the end, I wound up with sixteen chapters dealing with nineteen key figures, as well as with a sizable supporting cast, all of whom, I was convinced, deserved mention.

While engaged in the research for this book, I became obsessed with several thoughts. For example: What traits did these heroes of the nervous system have in common? Did unexpected events and chance observations play significant roles in their great insights and discoveries? And what benefits could we accrue from knowing how the great minds behind the brain thought, worked, and approached life in general? These issues will be addressed in the final chapter.

Before we embark on what I have come to think of as a voyage across time, some words of appreciation are in order. First, I would like to thank Joan Bossert at Oxford University Press for encouraging me to follow up on my 1994 *Origins of Neuroscience* with this more selective, in-depth, and personal sort of book. I am also grateful to the McDonnell Foundation for providing the financial support for the photography, to Lilla Vekerdy and Nada Vaughn in the rare books and interlibrary loan departments at Washington University for their assistance, and to "Roddy" Roediger for giving me the sabbatical I needed to work on this project.

Equally important are those individuals who have taken the time to review specific chapters or the entire book. In the whole-book category, I am especially indebted to an international quintet of multitalented scholars and good friends: Francis Schiller (USA), C. U. M. Smith (UK), Peter Koehler (Netherlands), Axel Karenberg (Germany), and Ulf Norrsell (Sweden). Each had perceptive comments and helpful suggestions that guided me deeper into fertile domains of science, culture, and history. Finally, Amy Dogette and Susan Polinski helped shrink my text to its present size, after which my wife, Wendy, reinserted passages to make sure it still had color.

Without question, I could not have undertaken this project without the support of Wendy and our two sons, Robert and Bradley. All adapted tolerably well to my work schedule and to my so-called creative spurts. In addition, they did the right thing by having me swear a blood oath to join the family for meals, to go on vacations without computers, and to remain my enthusiastic self with regard to my many "side" interests: collecting antique clocks, old barometers, and Post-Impressionist drawings, as well as fishing, golfing, and Scrabble. Wendy, Rob, and Brad, thank you for making my life so interesting and so much fun, and for teaching me how to pace myself to still smell the flowers.

Minds Behind the Brain

Introduction

A Voyage Across Time

We need only view a Dissection of that large Mass, the Brain, to have ground to bewail our Ignorance. . . . We admire . . . the Fibres of every Muscle, and ought still more to admire their disposition in the Brain, where an infinite number of them contained in a very small Space, do each execute their particular Offices without confusion or disorder.

—Nicolaus Steno, 1668

Some Aesthetics and Numbers

The brain, from which our grandest ideas and aspirations originate, is not large or especially heavy. In adult humans, it weighs about 1,400 grams (3 pounds) and accounts for less than 3 percent of our body weight. Neither is it a beautiful organ. Its exterior is covered with dull fibrous material (the meninges) and its own surface has an unappealing buff or beige color that even scientists with excellent color vision have opted, for some obscure reason, to call gray or *substantia cinerea,* meaning ash-like in appearance.[1]

Today we know that this undignified, odd-looking mass contains several billion very fragile neurons; put end to end, they would form a cable several hundred thousand miles long. These nerve cells, some long and others short, communicate with each other by releasing chemicals or neurotransmitters across gaps 0.02 to 0.05 microns, or millionths of a meter, wide. The number of synaptic "connections" is staggering scientists give estimates as high as a hundred trillion. Small wonder that the brain requires almost 20 percent of the body's energy to perform its functions.

The number of different routes that can be used to shuttle sensory messages to the brain, to process them, and to respond to them voluntarily is almost limitless. This was recognized late in the nineteenth century by Frederic Myers.[2] He asked his readers to "picture the human brain as a vast manufactory, in which thousands of looms, of complex and differing patterns, are habitually at work."[3] The same metaphor was later adopted by Sir Charles Scott Sherrington, who was awarded the 1932 Nobel Prize in Physiology or Medicine. Writing in 1941, when considerably more was known about the synapse (his term) than in Myers' day, he described the brain as "an enchanted loom where millions of flashing shuttles weave a dissolving pattern."[4]

In terms of when he lived and what he thought about the brain with its multitude of circuits, Sherrington was decidedly "modern." Yet some would argue that much more has been learned about the brain in the last two decades than in all of the rest of time. Still, what newer scientists have inherited from Sherrington and other figures of the past deserves recognition. The ways in which the pieces of the brain puzzle have come together over time is, in fact, like a play with many acts. In this case, however, the play is still being written, and even those working on the latest act cannot envision how their efforts will end.

The Organ of Mind

The "loom" envisioned by Frederic Myers and Sir Charles Sherrington is now being described with newer terminology from the world of computers. Although analogies and models may change, it can be said that the more we learn about the machinery of the brain, the more we are amazed at how it works and what it can do. It is the brain that generates perception, cognition, and memory, and it is the brain that houses the specialized circuitry that allows us to make a cup of coffee, catch a ball, or write a note to a friend.

Roger W. Sperry, the 1981 Nobel Prize winner whose life and discoveries will be described toward the end of this book, looked upon consciousness as the greatest but least understandable achievement of the brain. He wrote: "Before brains there was no color and no sound in the universe, nor was there any flavor or aroma and probably little sense and no feeling or emotion."[5] Before brains the play still took place, but the seats in the theater were empty; there were no appreciative audiences, no discussions, and later no reminiscences.

To think that the brain just manages "higher" cognitive functions or conscious awareness, however, would be to overlook its more basic roles. The brain also tells us when we are hungry, determines our patterns of sleep and dreaming, and allows us to express and control our emotions. Even gender differences in play and fighting are in part a function of the brain.

When we are young and the brain is still developing, it does not yet possess masterly control over each of its duties. Some functions, such as memory for names or events, may be poor. Later, the ability of the brain to work at peak efficiency may be diminished by diseases, injuries, and the inevitable loss of connections that accompanies the steady march into advanced age.

Initially by looking at the effects of brain damage, we have begun to appreciate the functional organization of this miraculous, odd-shaped organ. More than just a device that can sustain life and ensure the survival of the species, we see that the brain grants us our humanity and make us different from each other. Our "great raveled knot," to use more of Sherrington's colorful terminology, allows us to portray ourselves as thinking animals—creatures that can make choices in the present, be appreciative of the past, and become excited about the future.[6]

A Voyage Across Time

Although it is still the least understood and most mysterious of all bodily organs, we have come a long way in understanding how brain structure and events determine be-

FIGURE 1.1.

Nicolaus Steno (1638–1686) thought scientists should try harder to understand the functional organization of the brain. He admired the ability of this organ to govern thoughts, movements, and the senses without confusion.

havior. This book will examine how Western ideas about the brain emerged and developed. It will look at neurological, anatomical, and physiological discoveries. It will also delve into theories to account for the ability of the brain to perform its various functions "without confusion or disorder," to repeat the phrase used by Steno, the seventeenth-century anatomist quoted at the beginning of this chapter (Fig. 1.1).[7]

Our journey will involve stepping back in time and embarking on a tour across the ages, one that will culminate two-thirds of the way through the twentieth century. At each stop, our tour will allow us to meet some remarkable individuals who promoted new ways of thinking about the machinery of the mind—men and women whose discoveries and insights have led us into the present era.

We shall see that new ideas have sometimes split scientists into opposing camps, each looking at the world of the brain in a different way. One such division will pit the "holists" against the "reductionists." The former will argue that the hemispheres of the brain function as an indivisible unit—as a whole. Looking upon this orientation as archaic, the reductionists will envision the brain as a collection of highly specialized centers. Interestingly, Camillo Golgi and Santiago Ramón y Cajal, the two brain scientists who shared the 1906 Nobel Prize in Physiology or Medicine, were flag bearers for these two different groups—Golgi for the holists and Cajal for the reductionists—even though both were histologists dependent upon the microscope.

In looking at the lives of these and other outstanding individuals who changed the course of thinking about the brain, we shall also find that some people whose names are rarely mentioned today were really the first to make an important discovery. One is Emanuel Swedenborg, an eighteenth-century man of many interests. He might well have been the first to propose that the cerebral cortex, the roof of the brain, can be subdivided into distinct functional units—an area for moving a limb, a region for hearing, and so forth. A second case involves Marc Dax, an early-nineteenth-century French physician. He was probably the first person to recognize that the left hemisphere of the cerebrum is more involved with speaking than is the right hemisphere.

During the 1860s, Paul Broca championed both cortical localization of function and what we now call "cerebral dominance," citing his own clinical cases. In part because Broca worked out of Paris, better disseminated his findings to the larger scientific community through lectures and articles, and was highly regarded by other scientists, he received the lion's share of the credit for both Swedenborg's and Dax's extremely important insights.

The Forces of Change

When we meet Paul Broca, as well as Ramón y Cajal and other pioneers of the brain, a number of conditions that affected how they thought about the brain will become apparent. Without doubt, one of the most significant forces, especially early on, was religion. Religious beliefs led civilized people to ask how the heart and later the brain can act as the "seat of the soul," meaning the physical organ of the mind itself. Religion also played a role when it came to defining acceptable methods of study. During Greco-Roman times, for example, there were strong religious and cultural sentiments against performing human autopsies. Civil and religious laws, as well as powerful emotions relating to how the dead should be treated, have clearly affected the advancement of science throughout history.

Another important factor in the history of ideas about the brain relates to the use of animals. The idea of dissecting animals for anatomical and physiological study was never very popular among the pet-loving masses, who equated such acts with barbaric practices and satanic rituals. Even among some of the most enlightened ancient physicians of the past, including Hippocrates and his followers, there were cultural prohibitions against vivisecting live animals or even dissecting dead ones. In contrast, Greek scientists in Alexandria, Egypt, as well as their brethren in imperial Rome, were able to do the unmentionable to animals. And, by moving in this direction, they revolutionized the study of the brain. Even so, those dedicated scientists of the past who conducted experiments on animals were often put on the defensive—a situation similar to that encountered by even the most well-meaning laboratory researchers of today.

Technology has also played a crucial role in how scientists have viewed the brain. The advent of powerful microscopes, the discovery of new histological stains, and the development of devices that permitted researchers to record the electrical activity of the brain have all allowed scientists to see the organ of mind in exciting new ways. In the language of science, these technological advances helped to trigger significant "paradigm shifts."

Also worth noting is the role played by chance in many of the discoveries that will be described. Although we tend to think that great discoveries are based on hard logic, formal hypotheses, and well-planned experiments, we shall see that this is not always the case. Many great discoveries about the nervous system have involved "fortunate accidents."

The term often used to describe such a stroke of good luck is *serendipity*. This word was coined in 1754 by British statesman Horace Walpole in a letter to his friend, American educator Horace Mann. It was derived from Serendip, the old name for the large Asian island of Sri Lanka (Ceylon). In the story called "The Three Princes of Serendip," the fairy-tale characters had an unusual knack for making accidental discoveries as they traveled about. Some scientists and practitioners have also been blessed with this kind of luck or good fortune, making discoveries more significant than those they actually sought.[8]

Henry Dale, who won the Nobel Prize in 1936 for his research on chemical transmission in the nervous system, once wrote an article entitled "Accident and Opportunism in Medical Research."[9] Dale maintained that serendipity guided his own successful career. But, he went on to say, accidents would probably mean little in science unless the recipient has a well-trained mind and a willingness to follow through on accidental encounters. Throughout his delightful autobiography, Walter Cannon, still another Nobel Prize–winning physiologist, also emphasized the importance of having a mind flexible enough to deal with unexpected findings and trained enough to put them in proper context.[10]

In the last chapter of this book, much more will be said about chance and prepared minds. Special attention will also be paid to how the roads to a discovery are sometimes indirect and laced with blind alleys. But first, let us start our lengthy journey with a physician who lived almost five thousand years ago in the land of the Nile.

2

An Ancient Egyptian Physician
The Dawn of Neurology

If thou examinest a man having a gaping wound in his head, penetrating to the bone [and] splitting his skull, thou shouldst palpate his wound. Shouldst thou find something disturbing therein under thy fingers, [and] he shudders exceedingly . . . Thou shouldst say regarding him: . . . "An ailment with which I will contend."

—*Anonymous, Edwin Smith Surgical Papyrus, c. 1550 B.C.*

The Time of Imhotep

During the Third Dynasty of Egypt's Old Kingdom, there lived a high priest named Imhotep (Fig. 2.1). Later depictions of Imhotep are common, but the exact dates of his life are uncertain. In fact, scholars are not even sure of the dates of the Third Dynasty. Dating is estimated by working backward from well-defined archeological landmarks. Early in the twentieth century, the Third Dynasty was thought to have existed about 3000 B.C., whereas contemporary archeologists are more likely to suggest the twenty-seventh century B.C.

Imhotep, whose name means "he who comes in peace," was born a few hundred years after the kingdoms of Upper and Lower Egypt were united. He lived under the rule of King Djoser (Zoser), who resided near Memphis, the ancient city located near the top of the fertile Nile delta. Egypt was experiencing tremendous cultural changes at this time. If we close our eyes and think about the past glories of Egypt, we are likely to envision the great stone pyramids of Giza. These limestone block pyramids were erected during the Fourth Dynasty of the Old Kingdom, no more than a few generations after Djoser ruled and Imhotep served him.

Whether Imhotep was related by blood to his king, a noble by birth, or a commoner may never be known.[1] Whatever his lineage, he captured the king's favor with his creative and administrative abilities. He served as grand vizier, a position similar to a high minister of state. His responsibilities included overseeing the king's records, the seal of state, the judiciary, the treasury, and the war and agriculture departments. Imhotep was also reputed to be one of Egypt's greatest sages, a "patron of scribes," a venerated astronomer, a magician, and the architect of the step pyramid of Sakkara, one of the earliest cut-stone structures remaining (Fig. 2.2).

After his death, Imhotep received divine honors, a rare event for a man who was

FIGURE 2.1.

*Statue of Imhotep,
vizier and
physician to King
Djoser, pharaoh of
the Third Dynasty.
(Courtesy of the
Egyptian Museum,
Cairo.)*

FIGURE 2.2.

*The step pyramid
of Sakkara,
designed by
Imhotep for
King Djoser
around 2700 B.C.*

not a pharaoh. And with his death, legends about him grew. At first just a remarkably talented man, he rose after death to be perceived as a demigod, a being with mixed human and divine attributes. By the time of the Ptolemies, the Greeks who began their rule of Egypt in 332 B.C., the Egyptians worshiped him as a full deity.

Myths, legends, and a large healing cult came to surround the name of Imhotep. At the height of the cult, his powers overshadowed those of the other Egyptian healing gods. Healing temples were erected in his honor at Memphis, Sakkara, Philae, and other locations. It was to these sites that the sick came to seek counsel and medical advice.

The Egyptians believed that if a person made appropriate offerings and prayers, and spent the night in one of these sanctuaries, Imhotep would provide helpful insights and needed aid. The assistance would come through the person's dreams, and the directives would be interpreted by priests at the temple sites. It was by deciphering the often cryptic messages given to a person's soul that ways were found to help the sick and needy.

Unfortunately, we know of no papyruses written during the Third Dynasty that have withstood the ravages of time. This is not really surprising, since papyruses were made from the thin, cut stems of the reeds that grew along the Nile. The Egyptians pasted the cut stems together and wrote on them with a reed dipped in a mixture of gum, soot, and water. The technique may not seem very sophisticated to us today, but the end product, although fragile, represented a major advance over the bulky clay tablets that had been used in Babylonia.

Nevertheless, the earliest known document in the field of medicine, a papyrus known as the Edwin Smith Surgical Papyrus, can trace its roots to the Third Dynasty. Although this papyrus was written approximately one thousand years after Imhotep's death, the cases described in it are believed to have taken place during King Djoser's reign. The papyrus is filled with descriptions of injuries that can be associated with large-scale building projects, such as those Imhotep oversaw in his role as an architect. In addition, many of the wounds could have been caused by the weapons of war used at the time that Imhotep was responsible for keeping such records. Thus, the injuries described in the Edwin Smith Surgical Papyrus have led some historians to speculate that the legendary Imhotep could have compiled the material found in this medical papyrus, even though the surviving copy of the original document is too new to have been written by his hand.[2]

Whether associating Imhotep with the Edwin Smith Surgical Papyrus is a correct attribution or just loose conjecture about a charismatic figure in Egyptian history and mythology is uncertain. Sadly, the celebrated papyrus is missing its beginning and ending lines, and the author's name is nowhere revealed. Even the original title of the document is unknown.

It has been suggested that the papyrus may be a copy of the venerated *Secret Book of the Physician,* a title referred to in later papyruses. The Egyptians believed that all medical knowledge was revealed by the gods. Specifically, they thought that Thoth, the birdlike god resembling an ibis, recorded cures in secret books that only great priest-magicians, like the revered Imhotep, could read.

Thus, there are many things that are not known about the Edwin Smith Surgical Papyrus. Fortunately, this lack of information pales in comparison to what we have learned from the document, especially about brain injuries and their treatment during this murky and elusive stage in antiquity. The picture that has been painted, like the discovery of the papyrus itself, stirs the imagination. It tells us how the early healers worked and makes us wonder just how we would have been treated had we suffered an unexpected head injury in ancient Egypt.

The History of the Edwin Smith Surgical Papyrus

In 1862 the archeologist Edwin Smith bought the papyrus that now bears his name in the Egyptian city of Luxor, once a part of the ancient city of Thebes.[3] Actually, the papyrus was purchased in two separate transactions. The first purchase encompassed the bulk of the papyrus. At the time, Smith was told that it had been found in an ancient Egyptian coffin along with a mummy, presumably the owner of the document.

The second purchase occurred a few months later and involved a patched-up "dummy" papyrus. Smith astutely recognized that the pasted assemblage contained eight missing fragments from the first papyrus, along with much unrelated material. Having studied Egyptology in London and Paris, and having resided in Luxor for several years, Smith was able to insert some of the new fragments into their correct places on his first scroll. Although he was hardly an expert in the early Egyptian language, he knew enough to recognize that the subject of the papyrus was medicine.

The door for Smith or another Egyptologist to attempt a translation of the papyrus had been opened in 1799. In that year a captain of artillery in Napoleon's expeditionary corps in Lower Egypt, a man named Boussard, discovered the Rosetta Stone (Fig. 2.3) This black basalt slab was given its name because it was found near Rosetta, a town at the western mouth of the Nile. It dates from 195 B.C. and is inscribed with three types of writing: hieroglyphs, late Egyptian demotic script, and Greek.

FIGURE 2.3.

The Rosetta Stone, found in Egypt in 1799. (Courtesy of the British Museum, London.)

FIGURE 2.4.

James Breasted, translator of the Edwin Smith Surgical Papyrus, shown with his wife and son at the Amada Temple, 1906. (Courtesy of the Oriental Institute of the University of Chicago.)

In 1822 Jean-François Champollion deciphered the Egyptian writing on the stone by matching it with the Greek inscriptions. His work set the stage for translations of a wealth of ancient material into modern languages. By midcentury, when Smith lived, early Egyptian writing was no longer quite as mysterious as it had been. Amateur and professional archeologists were now able to do some of their own translating and these men also knew where to find skilled linguists who would work for fees.

But for some unknown reason or reasons, Smith, who was born in the very year of Champollion's translation, failed to seize the moment. He did not hire scholars to help him translate his prized possession, nor did he sit down to do the job himself. He never published a book or even a paper describing the contents of the papyrus. While he was alive, the public knew nothing about this document.

Smith passed away in 1906, but fortunately he had an altruistic daughter who survived him. She donated his medical papyrus to the New-York Historical Society several months later. There it gathered dust until 1920, when James Breasted (Fig. 2.4), a highly respected Egyptologist from the newly formed Oriental Institute of the University of Chicago, was asked to translate the document.

Breasted had studied the Egyptian language and culture with leading scholars in Germany, and had spent considerable time digging and excavating in Egypt. He held the first American chair of Egyptology and had a strong interest in translating everything Egyptian. But he was so inundated with work that he was not inclined to take on yet another job. After looking at the ancient document and recognizing its importance, however, he accepted the challenge. Slowly the arduous chore of precise translation turned into a labor of love.

In 1930, a decade after he began, Breasted's complete translation, with notes and commentaries, rolled off the press. Medical historians who read his work immediately recognized the need to revise their history books. The long-held theory that the Greeks of the fifth century B.C. were the first to write about the brain was no longer tenable. The head-injured men described in the Edwin Smith Surgical Papyrus lived well over two thousand years before Greek scientists and naturalists had anything meaningful to say about the brain.

The Contributors to the Papyrus

Breasted suggested that there were at least three contributors to the papyrus. The first was the compiler of the information and the author of a document ancestral to the papyrus he had just translated. He hypothesized that this individual could well have been the priest-physician Imhotep, but he knew this was only a speculation.

The second contributor lived several hundred years later, probably during the closing years of the Old Kingdom. Because some of the language in the original medical treatise had become archaic, this unknown individual took on the job of commentator and produced an annotated copy of the earlier text. He attempted to explain many of its archaic words and phrases to his contemporaries.

The third person made his contribution around 1650 B.C., just prior to the territorial expansion of the New Kingdom. He used red and black ink to make a copy of what must have been a later copy of the commentator's manuscript, the original

FIGURE 2.5.

A section of the Edwin Smith Surgical Papyrus showing the hieratic script. Light and dark text are the result of the use of two different types of ink. (Courtesy of the University of Chicago Press.)

almost certainly having been destroyed. This contributor did not write using hieroglyphs (Fig. 2.5). By choosing scriptlike hieratic writing, as opposed to pictures, he
was able to work more rapidly. Most likely he was a professional scribe hired to re-
produce an interesting and important ancient document.

For some unknown reason, the third contributor stopped writing in the middle
of a word, leaving a good part of the lengthy scroll empty. Even stranger, when he returned to the papyrus, he began to copy some unrelated material about magic and
folk cures on its reverse side.

The Papyrus and the Brain

Working with patience and great skill, James Breasted found that the Edwin Smith
Surgical Papyrus contained descriptions of forty-eight individuals with physical injuries. The document begins with descriptions of patients with head injuries. It then
moves progressively lower in the body, eventually to injuries of the thorax and spine,
where the writing abruptly ends.

All told, there are twenty-seven cases of trauma to the head. There is no real evidence of brain damage in fourteen of the cases. The remaining thirteen patients,
however, suffered fractures and "smashes" of the skull, as well as a host of neurological abnormalities. Some of these individuals appeared to have contracted meningitis,
a finding that suggests they were examined some days or weeks after sustaining their
brain injuries.

A description of the brain (the "marrow of the skull") appears near the beginning
of the papyrus. The convolutions of the human brain are compared to corrugated
metal slag, the residue left from smelting copper ore, which was then being used to
make works of art, religious objects, and useful tools. The writer also mentions the
"sack" covering the brain, an obvious allusion to the meninges. There is even a statement about the fluid underneath the sack, a reference to the watery cerebrospinal
fluid.

The descriptions of the head-injury cases show that the early Egyptians recognized that damage to the central nervous system can have effects far from the site
of injury. For instance, there are examples of head injuries causing problems in eye-
hand coordination and instances of paralyzed limbs on the opposite side of the body.

One individual even seems to have suffered a "contrecoup" injury, one in which a
blow to one side of the head causes the brain to compress against the opposite side of
the skull. Such an injury produces some symptoms on the same side of the body as
the external blow. In contrast, most blows to the head are associated with contralateral effects, meaning sensory and motor difficulties on the opposite side of the body.

Breasted also found that the Egyptians classified their cases as (1) an ailment to be
treated, (2) an ailment which may or may not be treated, or (3) an ailment not to be
treated. These categories most likely corresponded to situations in which (1) recovery was expected, (2) recovery was possible but not certain, and (3) there was little
hope for a successful outcome. Especially under battlefield conditions, where there
could be a massive influx of wounded soldiers, it was important for the physicians to
have the right to choose. In this respect, the Egyptian classification system has much
in common with a modern triage system, in which the wounded are sorted according to severity of injury and expectations for survival.

For severe head injuries that caused a loss of speech, the prognosis was poor and the classification was usually "an ailment not to be treated." One such case was described in typical telegraphic style as follows:

> If thou examinest a man having a wound in his temple penetrating to the bone [and] perforating his temple bone . . . if thou puttest thy fingers on the mouth of that wound and he shudder exceedingly; if thou ask of him concerning his malady *and he speak not to thee* while copious tears fall from both his eyes . . . [this is] an ailment not to be treated.

In contrast, Case 8 in the papyrus involved a head injury that could be treated. Here the writer simply stated: "His treatment is sitting, until he gains color, [and] until thou knowest he has reached the decisive point." Could the Egyptian physicians have realized that simply sitting up would slow or reduce the chances of intracranial hemorrhaging?

Although the Edwin Smith Surgical Papyrus has been called a "surgical" papyrus, there is no evidence of major operative treatment anywhere in the document. The only "surgeries" mentioned are the simple removal of a fragment of bone from the external ear canal and a blood clot from the nose. In the land of the Nile, we find little evidence of skulls being sawed or drilled open, although many such trepanned skulls, some thousands of years older, have been found in other locations.[4]

The prescriptions recommended for head wounds in the papyrus are very specific. For example, one requires the fats from a lion, a hippopotamus, a crocodile, a snake, and an ibex. The instructions call for mixing these fats into one mass and then anointing the head with the mixture. In another recipe, powdered ostrich egg is specified to fill in a crack on a man's forehead.

The ostrich-egg fracture case is especially interesting because it is the only one in the papyrus that is accompanied by a prayer. After the head wound was treated with the egg poultice, the following prayer was said:

> Repelled is the enemy that is in the wound! Cast out the [evil] that is in the blood, The adversary of Horus, [on every] side of the mouth of Isis. This temple does not fall down; There is no enemy of the vessel therein. I am under the protection of Isis; My rescue is the son of Osiris.[5]

This prayer is like those found in other medical papyruses. It shows that the Egyptians occasionally appealed to the supernatural for help in cases of injury. But why are there no other prayers in the Edwin Smith Surgical Papyrus? And were illnesses treated as practically as war or construction injuries, or was there greater reliance upon the gods when it came to diseases from unrecognizable sources?

Looking back, we can see that by the Third Dynasty there were already two orientations in Egyptian medicine, although they intermingled and overlapped. One is what some modern writers have referred to as the "rational" approach. This orientation was most likely to be used for treating acute injuries rapidly and efficiently. It dominates in the Edwin Smith Surgical Papyrus, where the wounds were easily seen and their causes known. The contrasting approach is more religious and mystical in nature. It is more likely to be found in cases of disease, where the problem was not always visible and the cause was mysterious.

But before turning to the role of the supernatural in Egyptian medicine, we must first ask how much these early healers really knew about the brain. Were they relatively advanced in their understanding, or was everything tied to some grand metaphysical scheme?

The Heart and its Channels

The Egyptians looked upon the heart, not the brain, as the central organ of the body and the seat of the soul. They further believed in a system of channels to connect the heart with the other bodily organs. They called these channels from the heart *metu*, and they included not only the arteries and veins, but also the nerves and tendons.

The *metu* from the controlling heart went to the nose, bladder, rectum, ears, and other organs. They carried air, liquids, and solid matter. Although each channel and organ was important, as can be witnessed with head injuries, all basically served the ruling organ, the pulsating heart in the center of the body.

There is no evidence to show that the ancient Egyptians understood the functions of the nerves or saw the brain as the real organ of mind. They believed only that the nerves must form one part of an intricate system of channels. In effect, the heart was the organ responsible for feeling, thinking, and all other functions now associated with the central nervous system; in short, the heart was firmly associated with the soul, both in life and after death.

The Egyptians viewed the human body as analogous to the fertile land of the Nile, with its vital basins, dikes, and irrigation canals. They knew that lack of moisture or its opposite, uncontrollable flooding, could be disastrous to their crops and the survival of their communities. Legend has it that Imhotep won great praise when he showed his pharaoh how to make the sacrifices demanded by the gods to end a famine caused by the failure of the Nile to flood at the usual time. Reasoning by analogy, he also maintained that it was important to keep the channels of the body open, but not too much, to ensure optimal health.

Seen in this context, bloodletting was rational therapy to the Egyptians. A wall painting in the tomb of Userhat, a scribe who lived about 1500 B.C., depicts a leech being taken from a bowl for application to a patient's head. We also know that venesection, the cutting of veins to draw blood from the body, was performed. For the same reason, the bowels were regularly cleansed and "quieted."

Disease was attributed to an abnormal state of the channels and the organs served by them. Feelings of sickness could occur if a channel were either blocked or opened too wide. As for the causes of disease, it is here, more than in cases of external injuries, that the Egyptians pointed to the roles played by sinister supernatural forces.

Of Demons and Disease

Having little understanding of the body's physiology, the Egyptians attributed diseases to evil spirits, sorcery, and witchcraft. In the land of the Nile, it was believed that menacing demons could enter the body through any opening, although evil was usually associated with the left nostril or left ear in Egyptian mythology. The demons would then cause pain and suffering by attacking the beating heart or the canal-like *metu*, which project to the different bodily organs. Working from this orientation, the Egyptian priest-physicians treated their sick patients by calling upon the powers of omnipotent deities, by employing "magic," and by writing seemingly strange prescriptions.[6]

The healers used rituals, fasting, incantations, potions, and ointments to help contact their gods for help. They drew magical pictures, put messages into the mouths of dead black cats, and even had their patients go to special "incubation sites," such as the temple in Memphis dedicated to the healing god Imhotep.

The medical papyruses contain many spells and incantations for the priest-physicians to use. These spells urge the malicious demons to leave the body, and appeal to benevolent gods, such as Isis and Thoth, for help. In mythology, Isis was the wife of Osiris, the Egyptian ruler who was assassinated by Set (Seth), his evil brother. Set dismembered the body of Osiris and scattered its parts throughout the land so he never could be made whole again. But Isis, with the help of her sister Nepthys, found and collected the parts of her husband's body. Then, with the assistance of the jackal-god Anubis, she prepared it for proper burial. She uttered prayers and chants so powerful that Osiris, with his body now intact, was allowed to pass into the sacred world of the gods.

In a later form of the story, which has many variations, Isis performed her miracle while giving birth to a son named Horus (Fig. 2.6). Once Horus attained manhood, he set forth to battle his evil uncle, who now ruled Egypt. Horus won the fight that followed, but was blinded by Set. This time Thoth, the ibis-god who recorded all knowledge of healing in his secret books, stepped forth to restore Horus' sight, allowing him to secure his rightful place on the throne.

Small wonder that sick Egyptians appealed to Isis and Thoth for help. These powerful gods could restore health to those who are worthy. These legends also explain why Osiris and Horus are often mentioned in prayers and incantations. Not only

FIGURE 2.6.

Bronze statuette showing Isis with young Horus on her lap. Both Isis and Horus were associated with healing in Egyptian mythology and were frequently mentioned in prayers for the sick. (Courtesy of the British Museum, London.)

were they important gods, they were also figures with whom the sick and injured could identify. In ancient Egypt, Horus was often represented as a falcon-headed god and, more than any other symbol, his eye became the symbol of well-being. The eye of Horus can be found on many amulets as well as on Egyptian coffins, where it might have been painted to make sure that the deceased had full use of their senses in the next world.

The Ebers Papyrus

Many prescriptions appear in the Ebers Papyrus, another treasure acquired by Edwin Smith when he was in Egypt.[7] This papyrus received its name when Smith sold it to Georg Ebers, a German Egyptologist, in 1873. Possibly discovered along with the Edwin Smith Surgical Papyrus, the Ebers Papyrus was written around 1555 B.C. and, like the Edwin Smith Surgical Papyrus, it is based on records from the Old Kingdom. In fact, it is thought that some of the material in this papyrus may even be traceable to the First Dynasty (c. 3100–2890 B.C.).

The Ebers Papyrus measures an incredible twenty meters (sixty-five feet) long. It starts with the statement that it will deal with the preparation of medicines for all parts of the body. Indeed, the massive work contains almost nine hundred very specific prescriptions for a wide range of ailments, including migraine headaches and possibly even Parkinsons disease. Many of the prescriptions were probably passed on by oral tradition well before the Egyptians turned to written records.

The prescriptions are extremely specific and call for ingredients such as milk from a woman who has given birth to a boy, certain parts from animals of different ages, and minerals from certain localities. In all, some five hundred different animal, mineral, and vegetable substances are mentioned. The papyrus also specifies the exact dosages and modes of administration for each of the various medicines.

Some of the prescriptions in this papyrus have been found to contain medicinal agents considered helpful today. Nevertheless, the unsuspecting reader of the Ebers Papyrus will be surprised to discover that fifty-five prescriptions contain urine or feces, substances we would ordinarily associate with the spread of disease, not cures. In some cases, the healers rubbed excrement—most often crocodile dung—over the body of the afflicted. The logic behind "coprotherapy" (the use of feces and urine for healing) was to make the body uninhabitable for those demons thought to be responsible for illness. The Egyptians hoped the demons would have at least a modicum of self-respect, so that they would be anxious to leave the body under such abhorrent, disgusting, and intolerable conditions.

The healers prescribing these medicines and therapies knew that it was important to intervene rapidly, before the demons could cause irreparable damage to the vulnerable *metu* and the vital parts they served. As far as they were concerned, the malevolent spirits could leave through feces, urine, sweat, gas, or breath. The important thing was to get them to leave as quickly as possible, before irreversible harm could be done. Rapid intervention, of course, is also a basic feature of modern medicine.

Looking back, there must have been numerous cases where the chants, amulets, and prescribed medicines failed to restore health to the afflicted. We know that the healers had alternatives to turn to if their preferred methods failed. In addition, a priest-physician could always claim that the deities were not willing to listen to pleas for help or requests for forgiveness in some cases because the patient's sins were too serious. This tactic, of course, was an all-important trump card that could be played by a physician to protect himself from upset family members and friends should things go wrong. Such claims would also serve to protect the healer's reputation in the wider community.

The Brain After Death

The ancient Egyptians believed that the heart will faithfully record each person's good and evil deeds. Their mythology held that, at the time of death, the heart would be weighed against a feather to see if it were "heavy" from guilt, or "light" because it was unburdened by sin (Fig. 2.7). This special weighing ceremony would determine if the spirit would go to heaven or whether it would be fed to a dreaded mythological creature called the Devourer. This fearsome monster is depicted in Egyptian art as part crocodile, part lion, and part hippopotamus.

At the time of judgment, Anubis, the god of mummification (usually shown with a jackal's head), would manage a scale with the heart on one side and the feather on the other. Thoth would record the answers given by the heart to forty questions. An anonymous Great God, or Osiris, who became the supreme deity of the dead during the Middle Kingdom, would then pass sentence. It was essential to keep the heart with the body for this all-important ceremony. For this reason, the heart was never removed when the corpse was prepared for burial.

Although the early Egyptians wrote little about their embalming procedures, some depictions of embalming can be found in the Book of the Dead. In addition, the Greek historian-traveler Herodotus, who lived in the fifth century B.C., described embalming procedures used in the past.[8] He wrote that embalming was an elaborate process in Egypt, one that took seventy days to complete. Some historians

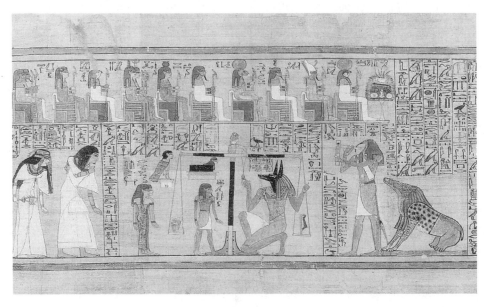

FIGURE 2.7.

Vignette from the Book of the Dead of the scribe Ani (c. 1250 B.C.), showing the weighing of the heart. (Courtesy of the British Museum, London.)

think, however, that seventy days is more likely to have been the time between death and the final burial ceremony.[9]

The heart was so important to the Egyptians that the *paracentetes*, the men who prepared bodies for mummification, sometimes fled in fear of their lives if they caught a glimpse of it.[10] In contrast, the Egyptians treated the brain with almost complete indifference. During the time of Imhotep and throughout the Old Kingdom, the brain was simply left in place when the body was prepared for burial. During the New Kingdom, however, Herodotus reported that a common procedure among the wealthy and noble classes was to have most of the brain scooped out with an iron hook inserted through the nostrils or the hole (foramen magnum) at the base of the skull. It was then simply discarded. Writers after Herodotus suggest that a small chisel could have been used to pierce the bones of the nose, making the removal process easier.[11] Brain tissue that could not be scooped out was simply rinsed away with chemicals or just left in place.[12]

Even the lungs, liver, intestines, and stomach received better treatment than the lowly brain. For the wealthy, these organs were either packed into four parcels to be placed back into the body or stored in four canopic jars, sometimes sculpted to represent the four sons of Horus (Fig. 2.8). The poor, who could not afford the fees for surgical removal and preservation of these organs, simply had the internal organs of deceased family members flushed away with cedar oil and their bodies preserved with natron, a type of salt.

The brain will eventually be thought of as the seat of the soul and the organ of mind, but these ideas would not originate from Egyptian soil. They are ancient Greek achievements. Hence, let us now cross the world's largest inland sea, the Mediterranean, to see how perceptions slowly began change about the functions of the heart and those of the brain—changes usually associated with a name well known to all medical students, that of Hippocrates.

FIGURE 2.8.

Canopic jars from the Twenty-First Dynasty (c. 1000 B.C.). The jars represent the four sons of Horus and were used to house the liver, the lungs, the stomach, and the intestines. The brain was discarded, and the heart was considered too important to remove from the body. (Courtesy of the British Museum, London.)

3

Hippocrates
The Brain as the Organ of Mind

> Wherefore the heart and the diaphragm are particularly sensitive, they have nothing to do, however, with the operations of the understanding, but of all these the brain is the cause.
>
> —*Hippocrates, c. 400 B.C.*

Three Faces of Greece

In order to understand the Hippocratic revolution in medicine and how it led to the perception of the brain as the ruling member of the body, we must think of ancient Greece at three different but unequal points in time. These epochs can be designated Early Greece, the Golden Age of Greece (or the Classical Period), and the Hellenistic Era.

Hippocrates lived from approximately 460 to 377 B.C., during the middle period or Golden Age of Greece. This period is usually defined as starting in 480 B.C. and ending in 336 B.C., the year in which Alexander first occupied the throne in the northern kingdom of Macedonia. Early Greece will be defined as the period preceding the Golden Age. The Hellenistic Era, in contrast, followed the Golden Age and ended in 150 B.C., although some authorities prefer 60 B.C., the year of Cleopatra's death.

As we shall see, these three time periods are very different culturally. In each one, not only are there differences in government, religion, and the arts, but they also provide us with distinctly different views of the mind and the brain.

Early Greece and Homer's Gods

Before Greece entered its famed Golden Age, people in the Greek world, like their counterparts across the Mediterranean in Egypt, viewed the heart as the seat of the soul and the command center for the body. Religion was the driving force in many aspects of their lives, as was also true for the people living in the land of the Nile. The early Greeks believed that humans were created by the gods, controlled by the gods, and subject to the caprices of the gods. Angry gods could even cause diseases, including the deadly plagues that could sweep across the land and annihilate large groups of terrified people.

In 776 B.C., the Olympic Games originated as a way to worship and placate Zeus, the most powerful of the gods. The *Odyssey* and *Iliad,* traditionally associated with the name Homer, date from this same period. It is by reading these two epic poems that we know something about the Greek mind and medicine in the centuries before Hippocrates.[1]

In the *Odyssey,* Phemius says, "No one has taught me but myself, and the god has put into my heart all kinds of song." His statement is one of many that show the perceived importance of the reigning heart at the time. It also illustrates just how indebted the early Greeks were to the mighty Olympian deities. In a world controlled by Zeus and a myriad of lesser gods, all of whom were capable of good deeds as well as jealousy and revenge, there was little free will, little need for a psyche.

In the *Iliad,* which chronicles the Trojan Wars fought approximately four hundred years earlier, there are passages that deal with the wounds of war. We read that serious head wounds always proved fatal, although other injuries were sometimes treated successfully by other soldiers or men skilled in the craft. Nevertheless, the phrase "swiftly Apollo cured the wound of Ares" shows that the god Apollo (Fig. 3.1) was perceived to be the real healer in the eyes of Homer's men. Faithful obedience to his laws meant protection against disease and possibly his assistance when injured.

Asklepios

In Greek mythology, Apollo, the son of Zeus and the god of medicine, was the teacher of the centaur Chiron, a creature who was half human (head, arms, and trunk) and half horse. This friendly centaur discovered many medicinal herbs and

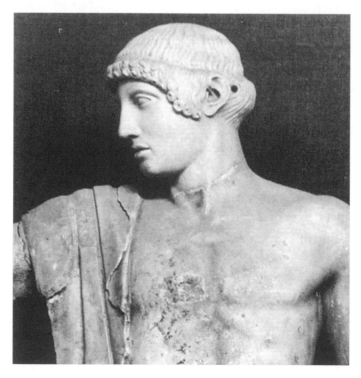

FIGURE 3.1.

Apollo, Greek god of medicine. (Detail from the temple of Zeus at Olympia.)

drugs and learned magical songs, all of which he conveyed to his pupils. In Hesiod's slightly later extension of the Homeric epics, Chiron taught the art of healing to Asklepios, who was the son of Apollo and the nymph Coronis.

Asklepios learned rapidly and became a gifted physician. He cared for the sick and injured with the help of a bevy of children, offspring who would make even the most demanding father proud. His daughter Panacea possessed knowledge of the earth's remedies for curing and rehabilitating. Hygeia, another daughter, excelled in cleanliness and disease prevention, whereas Telesphoros, a son, was in charge of recovery. Even his wife, Epione, had an important job: She was responsible for soothing hurt and pain.

Asklepios was so skilled in the healing arts that he was able to bring a man back from the dead. This transgression into Zeus' domain so angered the powerful god that he slayed the skillful healer with a bolt of lightning, ending his days on earth and sending him back to the heavens.

Was there really an Asklepios about whom some aspects of these myths were based? The answer appears to be yes. A very real Asklepios lived at the time of the Trojan Wars. He was a Thessalian chief highly skilled in medicine and a man praised by Homer in his *Iliad*.

With time, as the myths and legends about Asklepios spread, sick and injured Greeks began to pray to him, much as they would pray to Apollo. At first they built simple altars in the open or in caves to request his help. As beliefs about his healing powers continued to grow, he became looked upon as a god, and special temples, called *asklepieia*, began to be erected in his honor.[2] The sick and injured were brought to these temples, where those admitted underwent special purification rites, prayed, and made sacrifices in his name. They listened to soothing hymns, ate prescribed foods, and had an opportunity to relax.

The afflicted sometimes took part in sacred ceremonies performed by the *asklepiadae*, the priest-healers whose name means "sons of the god of medicine." Most notable among the ceremonies was the incubation ritual that took place in the evening. The *asklepiadae* had their patients dress in white linen vestments and then led them to a sacred hall in the complex called the *ábaton*, where they went to sleep.

The belief was that Asklepios himself, a member of his immediate family, or even his sacred snake or dog would "visit" them while they were dreaming. All could heal by touch and possibly just by being seen. It was not beyond the priests in the incubation sites to dress like Asklepios or to have an assistant dress like Hygeia. Nor was it unusual for them to release a trained dog or a harmless snake into one of the resting chambers.[3] We can only imagine how inspiring the nightly drama with its mythical cast would have been to the true believers in the ancient Greek world.

The spiritual treatments and directives given by the *asklepiadae* clearly helped some patients. Whether their successes on a case-by-case basis were due to suggestion, to changes in diet, to the quiet setting, or to the mobilization of the body's own defenses will never be known. All we can say with certainty is that short descriptions of the most successful cases, including individuals who were cured of such things as blindness, loss of speech, and paralysis, were occasionally carved onto tablets at the *asklepieia* for onlookers to witness.[4]

Today, the staff of Asklepios with a large but harmless constrictor wrapped around it remains the symbol of the medical profession (Fig. 3.2). Because the snake

emerged from the ground, the early Greeks looked upon it as a holy animal graced
with the secret healing powers of the earth. In addition, because it sheds its skin and
then appears rejuvenated, it was revered as a symbol of eternal life.

The Awakening

Although the religious cult of Asklepios continued to grow in popularity during the
Golden Age, the way many Greeks thought about themselves, their gods, and the
universe slowly began to chang between 600 and 480 B.C. One dimension of this
change can be seen by comparing the lyric poetry that emerged in this period with
the earlier Homeric epics. In lyric poetry, which was intended to be sung to the
music from a single instrument, the characters begin to express emotion and show a
nascent sense of individuality, personality, and self.[5] In the poems of Sappho, for instance, the people are no longer as puppetlike as Homer's men. Now they mourn,
grieve, and desire, although they still remain very dependent on their deities.

The demand to know more about nature and the place of humanity in the natural
world also began to grow as the once-tight grip of the gods slowly began to loosen,
allowing the human spirit more freedom. It began in Ionia, then a part of ancient
Greece but now a western coastal region of Turkey. Although Ionia had been settled
by Greeks from the mainland, the Ionians were exposed to other cultures and were
unusually open-minded and interested in new ideas. Indeed, it was from this part of
the Greek world that lyric poetry originated.

The revolution in thinking was led by the philosopher Thales (Fig. 3.3). He contended that the gods were not responsible for earthquakes or flooding, and argued
that one needed to study nature to understand such events. He looked for universal

FIGURE 3.3.

Thales (c. 652–548 B.C.), the Ionian philosopher who tried to identify the basis of all matter without recourse to the supernatural. (Courtesy of the Museo Capitolino, Rome.)

laws of nature and concluded that everything in nature had its basis in water. But being a stimulating teacher, he encouraged his students to develop their own philosophies about physical matter without recourse to the supernatural.

The men associated with Thales and his school did precisely this, and brought earth, fire, and air into their discussions. The early, freethinking Ionian philosophers, however, did not "package" these four elements together. This was left for the Pythagoreans a few decades later.[6]

Pythagoras (Fig. 3.4) himself was born in Ionia in 531 B.C. but emigrated to Croton, a Greek city in southern Italy. He and his students were intrigued by numbers and the notion of balance. In particular, four was a particularly important number to the Pythagoreans. It represented a perfect balance or harmony between two pairs of opposites.

For Empedocles, an early-fifth-century Pythagorean, the universe was composed of two pairs of elements: earth and water plus air and fire. Although these paired elements were initially associated with dry and wet, and cold and hot, respectively, they slowly began to be associated with two qualities each and bodily fluids called humors. It would be within this Pythagorean framework, and without recourse to the supernatural, that the Hippocratic physicians of Greece's Golden Age would try to maintain health and tend to their sick or injured patients.

The Golden Age of Greece

By the fifth century B.C., the philosophers were turning away from the cosmos to focus on the individual. In addition, democratic reforms took hold in many parts of the Greek world. Rule by elected citizens, as opposed to kings or powerful nobles (then called "tyrants"), found its ultimate champion in Pericles, who fought to democratize Athens, the rapidly growing capital of Attica. In Athens under Pericles, free men could choose and direct their government. But with their new power, they also had to shoulder the responsibility for what may happen.

Some of these change were mirrored in a new art form, the tragedy.[7] In the plays of Aeschylus, Sophocles, and Euripides, the characters appear as thinking men and women, not puppets controlled by the gods. Traditional beliefs are scrutinized and, for the first time, the central figure must choose between moral (usually a sworn oath) and rational courses of action—always with their own destinies and the lives of many other people hanging in the balance.

By appreciating parallel events in philosophy, politics, and the arts during the Golden Age of Greece, the changes characterizing Hippocratic medicine become easier to understand. As we shall see, the Hippocratic physicians recognized that Greek citizens had the right to choose any type of treatment and any healer. Like the citizens in the democracies, these physicians were also willing to take responsibility for their actions; they had no intention of being like the temple healers, who were quick to duck behind "the will of the gods" when it came to explaining their failures. But most important, the Hippocratic physicians were students of nature offering a rational alternative to religious healing—one deeply rooted in Pythagorean natural science and philosophy.

It has been claimed that the break between medicine and religion was not as abrupt as some writers have indicated—the point being that the Greeks were already on the way to more naturalistic explanations of disease prior to the Hippocratic epoch.[8] Although the exorcism of the demons from medicine might well have started earlier, the fact remains that demonology, purification, prayer, dream incubation, and magic are notably absent from Hippocratic medicine. At no time in the past was such a shift in orientation so clear, so apparent. This will become most evident when we turn to the subject of epilepsy, a disorder long thought to be caused by the forces of darkness.

But Who Was Hippocrates?

Hippocrates is one of the most mysterious characters in all of medicine. In this regard, he is a lot like Homer, the Greek epic poet whose name is a household word but about whom we really know very little.

It is generally believed that Hippocrates was born in 460 B.C. on Cos, a small island off the Doric coast of modern Turkey, just south of Ionia.[9] He opened his ever-observant eyes to the world not long after the Athenians and their allies repelled the occupying Persian forces from the coast of Asia Minor in the Battle of Marathon, an event that helped to launch Greece into her Golden Age.

Hippocrates began his study of medicine under the tutelage of his father, continuing a family tradition. After his father died, he crossed the Aegean to the Greek mainland. He settled in Athens, then the cultural center of the Greek city-states, where magnificent new buildings such as the Parthenon were being erected.

In 430 B.C., a great plague broke out in Athens. (Within two years one-third of the population, including Pericles himself, would perish.) In addition, the disastrous Peloponnesian War between Athens and the militant Greek state of Sparta had begun, leading to civil unrest and increasing fear. After evaluating the deteriorating situation, Hippocrates concluded that it would be wise to leave the city. He took what he needed and set forth to study, teach, and practice medicine in safer localities.

One of the most important works ascribed to Hippocrates is a treatise called *Epidemics,* consisting of seven books. In ancient times, the word *epidemios* meant nothing more than "characteristics of a certain region," "indigenous," or "confined to a certain people." Thus, *Epidemics* is not really a collection of books about epidemics in the sense in which the word is used today. Rather, these are commentaries about the sick—where they lived and how they were treated—collected by itinerant physicians such as Hippocrates, who often journeyed far from home.[10]

Hippocrates, who amazed people with his many successful treatments, fathered two children and supposedly lived to be over 90. Some sources even say he lived to be 104. He probably died near Larissa, a city in Thessaly, and was supposedly laid to rest under a stone depicting a hive for bees that made a special honey given to sick children. More than five hundred years later, Galen, a physician who found praising others exceedingly difficult, honored Hippocrates by calling him "the ideal physician."

Other than these scattered bits of information, very little else is known about the real life and accomplishments of Hippocrates. This is partly because the name Hippocrates appears only a few times in the literature from the period in which he lived. With so little factual information to draw from, it is often difficult to distinguish the truth from the stories fabricated by admirers to fill in the gaps.

Until recently, scholars remained uncertain about what Hippocrates even looked like. Sculptors and painters working well after he died depicted him as tall, handsome, and noble with a high forehead. This impressive "made-for-Hollywood" image was finally debunked by the discovery of some coins from Cos bearing his name and the portrait of a rather unattractive man (Fig. 3.5). They depict the revered healer as homely and far from godlike in appearance.

The *Corpus Hippocraticum*

Around 300 B.C., Ptolemy Soter, the first of the Greek rulers in Egypt, ordered scholars at the new library in Alexandria to collect documents containing all of human knowledge.[11] The works of Hippocrates, who was now acknowledged as the "father of medicine," were among those most sought. Over time, the Alexandrian scholars accumulated everything they could even remotely associate with Hippocrates. They then tied a figurative knot around the package by attributing all of the writings to him.

The *Corpus Hippocraticum,* or Hippocratic Collection, is the name given to these writings.[12] This heterogeneous collection of medical manuals, speeches, notes, fragments, and books, including *Epidemics,* date from approximately 450 to 350 B.C. Most of the writings were probably intended for other physicians, although some might have been written for well-educated Greek citizens.

The uncertainties surrounding Hippocrates' life and the contributors to the *Corpus Hippocraticum* have not detracted from the importance of the collection. This is because these treatises expound a very different philosophy for the *iatroi* (physicians) to abide by than the one followed in the Asklepian temples.

In terms of ethics and morality, the Hippocratic practitioners adhered to a strict codes of conduct based largely on Pythagorean ideals.[13] Without question, the best-known part of the *Corpus Hippocraticum* is the Hippocratic oath. Although almost certainly not written by Hippocrates himself, this was a sacred oath that at least some physicians swore they would follow.[14] After declaring that they would respect their teachers and practice medicine to the best of their abilities, they vowed that they would not aid in a suicide, perform abortions, or make personal information public.

If the *Corpus Hippocraticum* contained only this sacred oath, with its ethical code and covenant, its overall importance would be diminished. This is because some healers swore adherence to other ethical codes at and before this time. Further, the oath did not represent the law of the land, and because many physicians did not agree with all parts of the strict oath, it was not even an oath for the entire medical profession.

The true glory of the *Corpus Hippocraticum,* scholars agree, rests on its many careful observations and thoughtful comments about how disease can be prevented, diagnosed, and treated by those physicians willing to be guided by nature, not religion. Its greatness lies in the idea that a physician should be an astute student of nature and an expert craftsman, rather than a god-inspired priest. It also lies in the fact that the brain is now presented as the organ of the mind, not the heart.

The Brain Assumes New Importance

Hippocrates and his followers were convinced that the brain is the major controlling center for the body. This belief marked a major change in thinking from Egyptian, biblical, and earlier Greek views, which were based on the ruling heart. The newly elevated role of the brain is best illustrated in the following passage:

> Men ought to know that from nothing else but the brain come joys, delights, laughter and sports, and sorrows, griefs, despondency, and lamentations. And by this, in an especial manner, we acquire wisdom and knowledge, and see and hear and know what are foul and what are fair, what are bad and what are good, what are sweet, and what are unsavory. . . . And by the same organ we become mad and delirious, and fears and terrors assail us. . . . All these things we endure from the brain. . . . In these ways I am of the opinion that the brain exercises the greatest power in the man.[15]

This passage is significant, but we should remember that the Hippocratic physicians still knew very little about how the brain actually works. In fact, although they wrote such things as "it is the brain which interprets the understanding," physiological explanations from Hippocratic times will seem bizarre to the modern reader.

For this reason, it is important not to take a few lines out of context. A good example of reading too much into a few isolated lines relates to the now well-accepted idea that the right and left sides of the brain are unequal contributors to higher mental functions. Today this concept is called "cerebral dominance," and it is based partly on the fact that damage to the left hemisphere is more likely to disrupt speech, whereas certain spatial functions (such as assembling a jigsaw puzzle) are more likely to be compromised if the right hemisphere is damaged.

Our specific example comes from a medieval manuscript (c. A.D. 1100) thought to reflect ideas from the third century B.C.[16] The following lines suggest that cerebral dominance was recognized at this early time:

> Accordingly, there are two brains in the head, one which gives understanding, and another which provides sense perception. That is to say, the one which is lying on the right side is the one that perceives; with the left one, however, we understand.[17]

Although these lines sound wonderfully modern, we would have second thoughts about this insight if we read on. Among other things, we would discover that the unidentified author believed the noble heart also participates in higher cognitive functions and also exhibits this sort of functional asymmetry. To quote:

> Because of this, this is also being done by the heart, which is lying under this [i.e., the left] side [of the head], and which is also continually vigilant, listening and understanding, because it also has ears to hear.[18]

The idea that the heart can play a major role in perception suggests that this author was really not a follower of Hippocrates after all and had a poor understanding of brain and heart functions. Rather than being a lustrous gem based on fresh clinical observations, his imaginative theory seems to be little more than archaic philosophical conjecture.

Brain Damage

Some of the Hippocratic writers concerned themselves with brain damage, which was probably what led them to view the brain as the organ of mind from the start. In the *Corpus Hippocraticum,* there are numerous references to disturbances of movement, including various types of paralyses and seizures. Some can be found in the treatise *On Injuries of the Head.*[19] In it, the author correctly associates wounds of one side of the head with convulsions on the opposite side of the body. To quote:

> And, for the most part, convulsions seize the other side of the body; for, if the wound be situated on the left side, the convulsions will seize the right side of the body; or if the wound be on the right side of the head, the convulsion attacks the left side of the body.[20]

Knowing that convulsions and paralyses can follow brain injuries, the unnamed writer warns against tampering with the brain.

There are also occasional references to disturbances of speech in this treatise. One patient had loss of speech accompanied by paralysis of the right side of the body. Today, this is precisely what neurologists would expect to find in most patients with strokes affecting the front of the left hemisphere.

The *Corpus Hippocraticum* also deals with diseases of the brain, including epilepsy. To the earlier Greeks, this frightening seizure disorder was thought to be a sign of demonic possession caused by a fall from divine favor. Hercules, for all his strength, supposedly suffered from epilepsy. In fact, epilepsy was even called *morbus Herculeus* before it became known as *morbus sacer,* the dreaded "sacred disease."

The Hippocratic physicians looked upon epilepsy as a naturally occurring brain disease. From the beginning of the treatise *On the Sacred Disease* to its last line, the reader can sense this radically different emphasis. To quote:

> It is thus with regard to the disease called Sacred: it appears to me to be nowise more divine nor more sacred than other diseases, but has a natural cause from which it originates like other affections. Men regard its nature and cause as divine from ignorance and wonder.[21]

Later in the same work the author tells us that every disease has its own nature and arises from external causes. *External* was broadly defined at the time to include heredity. These physicians believed epilepsy ran in families and had something to do with blockages in the vessels to the brain.

Interestingly, the Hippocratic writers did not think of epilepsy as a distinctly human condition. They wrote that it can be seen in pets and domestic animals and is commonly encountered in goats. Most likely some of their goats were affected by other diseases that could have made them fall, such as myotonia congenita, a genetic disorder affecting the muscles.[22]

The Four Humors and Treatment

The author of *On the Sacred Disease* not only presents new ideas about epilepsy, but editorializes about the state of religious medicine. He writes that charlatans and priest-healers have for too long invoked the twin concepts of sin and divine punishment as convenient ways to mask their inadequacies in treatment.

But with religion out of the picture, how did the Hippocratic physicians treat brain diseases, strokes, and head injuries, including fractures of the skull? The answer to this question is that most turned to humoral theory to explain the inner workings of the body. This theory emanated from the premise that all things are composed of just a few essential materials. It is a theme that began with Thales and his students in Ionia and was adopted by the Pythagoreans. It still characterizes chemistry today.

During the Golden Age of Greece, each of the original four elements became associated with two qualities: Earth was perceived as cold and dry, air as hot and moist, fire as hot and dry, and water as cold and moist (Fig. 3.6). In addition, each element was linked to a specific bodily fluid called a humor. These pairings also took place gradually, but within the later Hippocratic writings we find fire paired with yellow bile, air with blood, water with phlegm, and earth with black bile. Today, black bile is the only humor not recognized as a physiological fluid. Most likely, the Greeks observed the darkening of certain bodily fluids in states of disease, and concluded that there must be a dark, turgid humor.

Each of the four humors would next be tied to a specific internal organ. Blood would be associated with the heart, black bile with the spleen, yellow bile with the liver, and phlegm with the brain. The elements, the humors, and the organs would also be linked to the seasons in an almost unlimited expansion of this system, which was based on two pairs of opposites.

Seen through Pythagorean eyes, health, or *eucrasia,* would signify a harmonious balance among the humors, whereas illness, or *dyscrasia,* would reflect a fault with the mixture and a loss of needed balance. In some cases the source of the *dyscrasia*

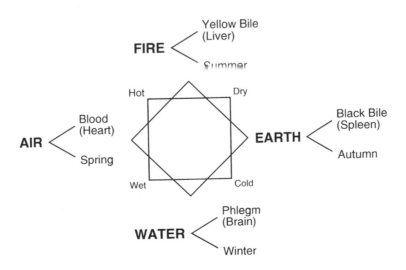

FIGURE 3.6.

The theory of four elements as expanded to include dual qualities and single associated humors, body organs, and even seasons of the year.

would be immediately apparent. The yellowing of jaundice, for example, would reflect too much yellow bile, whereas looking flushed or having bloodshot eyes would signify too much blood. The basic idea was to restore the balance by treating with opposites, such as bloodletting in cases showing flushing.

Today we define a stroke as a break in one of the blood vessels of the brain. In the *Corpus Hippocraticum,* however, strokes were looked upon as acute events due to blockages (not ruptures) of the blood vessels to the head. Because the causal agent was usually thought to be an overabundance of cool black bile, warming was seen as the most effective treatment. In this context, the Hippocratic physicians hoped that a fever would develop or could be induced to warm and loosen the sluggish bile.[23] As put in *On Diseases II,* still another treatise from the *Corpus Hippocraticum:* "If fever does not supervene, he succumbs in seven days, but if it does he usually recovers."[24]

The philosophy behind dealing with head injuries is just as interesting, in part because the Hippocratic physicians treated injuries associated with broken skulls very differently from injuries without breaks in the skull. Specifically, the *iatroi* tended to leave large skull fractures (open-head injuries) alone, but frequently recommended drilling holes in the skull for serious injuries that did not break the cranium (closed-head injuries). From all indications, the cutting and drilling were performed by an assistant. Upstanding Greek citizens typically relied on slaves and foreigners for manual labor, and those who took the Hippocratic oath swore to keep their own hands bloodless.

But why should surgery be done for closed-head injuries? The answer to this question again lies in humoral theory. The Greeks believed that a blow to the skull could cause blood and other humors to accumulate. These excessive humors could form a harmful dark "pus," which could affect the brain and disrupt its functions. With a break in the skull, the unwanted humors would have a way out. But if the skull did not break (read closed-head injuries), a craniotomy would have to be performed to provide the dark humors with an opening for proper drainage.

When the physician was unsure whether the skull had been broken, he turned to a black ointment (like our shoe polish) to clarify the situation. The ointment was rubbed on the exposed skull and then wiped off to expose new cracks.[25]

The author of the Hippocratic treatise *On Injuries of the Head* calls for performing craniotomies as soon as possible after closed-head injuries—before pus could form and drain the life out of the patient. By this time, the surgeons knew enough to stay away from areas associated with major blood vessels. They also learned how to drill slowly in order to prevent burning. The Greeks even designed special tools to reduce the chances of breaking the overlying meninges and penetrating the brain substance—slips likely to result in meningitis and other deadly infections.[26]

By draining pools of blood on the brain surface and reducing pressure, these operations, performed for what we may think is an absurd reason, probably saved more lives than were lost. In turn, increased survival assured that craniotomies would be performed again and again, even if the underlying theory of humoral imbalance makes little sense to us today.

Although the Hippocratic Greeks also had a multitude of pharmacological agents and other interventions at their disposal, the guiding axiom of Hippocratic medicine was always to let Mother Nature heal the sick and wounded if possible. In fact, the Greek word for nature was *physis,* from which our word *physician* is derived.

Because unnatural treatments were looked upon with skepticism, the Hippocratic physicians stressed the benefits of a sound diet, a healthy environment, and exercise (physical trainers often worked with physicians). Drugs and physical interventions, in that order, were regarded as backup remedies—the reserves ready for summoning from the armories by the conservative generals should Mother Nature need a helping hand. The primary responsibility of the physician was to know when to help and to judge how much of an unnatural treatment each patient could tolerate. Above all, the whole patient was to be treated, and he or she was not to be harmed physically or mentally by the physician or any of his assistants.

Although dedicated to their craft and philosophy, the Hippocratic physicians were not offended if patients wanted to pray; it was, after all, their right to do so. In fact, cases viewed as hopeless were occasionally encouraged to turn to religion. In the Hippocratic treatise *On Regimen,* we are told that prayer is good, but while calling on their gods for help and comfort, humans should also treat each other.

In retrospect, Hippocratic medicine was well suited for its primary clientele—the wealthy and educated citizens of ancient Greece. These individuals could afford special foods, changes of scenery, time for exercise, and considerable fees. They were also ready to endure more dramatic treatments when necessary. As a result, many Hippocratic physicians developed lucrative practices helping sick patients return to a balanced, wholesome way of life. These practitioners, however, also treated the less fortunate—they saw it as morally wrong to deny care to any person in need of services.

Hippocrates on Dissection

One of the most interesting questions that must now be asked is whether Hippocrates and his followers learned their craft by conducting human autopsies. In the *Corpus Hippocraticum,* there is no evidence of human autopsies being conducted. In fact, Hippocrates lived during a time when autopsies were impossible. The Greeks of this period believed that the human soul would not find peace until the body was put to rest. They also thought that destruction awaited those who tampered with a body. Thus, it was important to bury the dead quickly and with proper respect.

If not by human dissection, how did these physicians learn about anatomy, much less physiology? To some extent, they must have learned by observing gladiators, soldiers, and commoners who sustained wounds that exposed their bones and internal organs. Additionally, they might have watched men butchering domestic animals in the marketplace and hunters preparing wild animals they had killed.

As for performing their own animal dissections, there is no evidence for this at all in the Hippocratic writings. We must remember that Hippocratic medicine was born from Pythagorean philosophy and that the Pythagoreans were vegetarians who, like some of their neighbors to the east, believed in reincarnation. For these reasons, it seems highly doubtful that the Hippocratic physicians killed or dissected animals as a part of their medical training.

The Hellenistic Period and Dissection

Systematic human dissections began after Hippocrates died in 377 B.C. They emerged during the Hellenistic Era, the period associated with Alexander the Great and his conquest of Greece and then much of the known Mediterranean world. These dissections took place in Alexandria, the model city that Alexander himself helped lay out after he conquered Egypt in 332 B.C.

Not only did Alexandria house the magnificent library containing the *Corpus Hippocraticum*, it also had botanical and zoological gardens and a celebrated museum.[27] Almost overnight it had risen to become one of the world's greatest centers of learning. Studies in Alexandria fell into four groups: literature, mathematics, astronomy, and medicine. It was here that Euclid and Archimedes worked on mathematics and mechanics, while their counterparts studied philosophy and history, or perhaps the life sciences.

Alexandria was the ideal place to learn about anatomy and physiology during the Hellenistic period. The study of medicine was not tied down by the civil laws, taboos, or moralism that prevailed on the Greek mainland. Although the practice of human dissection would have been decried by the priesthood throughout the rest of Egypt and would have enraged the vocal citizens of Athens, the situation was very different in this well-insulated, distant center of learning.

Paving the way for human dissection was the fact that the Greek view of the soul had undergone significant changes over the centuries.[28] Most Greeks no longer believed that the future of the immortal soul was tied to the condition of the body. The body was now perceived as only a temporary abode or shelter for the soul. As a result, inquisitive scholars wanting to dissect human bodies no longer feared doing so—there was no longer a compelling reason to treat a corpse with the care and reverence granted to it in the past.

The leading Alexandrian dissector was an anatomist named Herophilus of Chalcedon, who practiced around 300 B.C. He reportedly dissected hundreds of cadavers.[29] Erasistratus, a slightly younger scientist, also participated in human autopsies.[30] By dissecting dead humans, both men brought forth a new approach to medicine, one that went well beyond what the Hippocratic physicians were willing to do during the Golden Age of Greece. And by doing so, they opened new doors to understanding the soft organs of the body, including the brain.

Nevertheless, it has been alleged that Herophilus and Erasistratus did more than simply dissect the bodies of dead people. The first two Ptolemies, we are told, sent condemned criminals to them for vivisection, hoping their physicians could learn new facts about the body. The Greek rulers of Egypt, it was said, did everything possible to help make Alexandria's physicians the envy of the world.

Celsus, the Roman encyclopedist who lived in the first century A.D., wrote about anatomy in Alexandria four centuries earlier.[31] He said the following about Herophilus and Erasistratus:

> They hold that Herophilus and Erasistratus did this in the best way by far, when they laid open men whilst alive—criminals received out of prison from the kings—and whilst these were still breathing, observed parts which beforehand nature had concealed.[32]

Celsus presented arguments both for and against vivisection, but in the end concluded that human vivisection was cruel and unnecessary. To Celsus, even the practice of dissecting the dead was repulsive. Tertullian, another Roman, did not mince words when it came to vivisection, calling Herophilus a butcher.

None of the original writings of Herophilus and Erasistratus has survived into the present. Consequently, the claims made by Celsus and Tertullian cannot be verified. In addition, some scholars have pointed out that Galen said nothing about human vivisection during the Hellenistic period, although he repeatedly referred to the discoveries of Herophilus and Erasistratus (see Chapter 3). In part because Galen was not one to remain silent on any issue, the naysayers contend that the two Alexandrians were only given the bodies of recently executed criminals to dissect and were being "framed" by jealous adversaries in other locations.[33]

The Alexandrians on the Nervous System

Still, by reading Galen, we have a reasonable appreciation of what Herophilus and Erasistratus learned in their early explorations of the nervous system. Looking at Herophilus first, he separated the nerves from the tendons and blood vessels, a considerable source of confusion prior to this time. He also studied many of the cranial nerves and attempted to distinguish between motor and sensory nerves. We also know that he described the anatomy of two most obvious parts of the brain, the cerebrum and the cerebellum.

Herophilus also examined the blood vessels at the base of the brain and the cavities within the brain. After studying the hollow ventricles, he raised the possibility that the posterior (fourth) ventricle, located close to the cerebellum, might house "the dominant principle of the soul." One factor behind this strange and seemingly novel speculation (which some others would later find attractive), is that he saw many motor nerves leaving from this region, which is close to the spinal cord.

For his part, Erasistratus attempted to deduce the functions of some parts of the brain by comparing these parts across species. He found that deer, hares, and other fast-running animals have more intricately folded cerebellums than less active animals. His association between the cerebellum and movement might have been the first attempt to link this organ with a specific behavioral function. But as his critics pointed out, a few animals not noted for being fleet of foot, such as sloths and oxen, possessed cerebellums just as convoluted as the speedy deer.

Erasistratus also had ideas about the superior intellect of humans. He theorized that we humans are smarter than other animals because our cerebral hemispheres have more convolutions than those of the brutes. Whether this was his own idea or one that really originated with Herophilus is unclear. We do know, however, that this theory was also challenged. Four centuries after the idea was presented, Galen entered the fray pointing with fingers on both hands to donkeys. Although donkeys possess highly convoluted brains, they had to be, in Galen's not-so-humble opinion, among the stupidest creatures on the face of the earth.

It is important to realize that "proofs" during the Hellenistic period were still really little more than hypotheses. They originated from thinking by analogy, and claims were made by citing other often dubious proofs. The guiding principles of modern science—formulating testable hypotheses, experimenting under controlled conditions, and being objective—were not in place at this time. That is why there were still some learned men in the Hellenistic world who were not ready to side with the Hippocratic physicians and those others who were declaring the brain the true organ of mind.

The Debate Goes On

The idea that the brain is the command center of the body gained more adherants after Hippocrates died. But although it became the dominant idea in most circles, not everyone was swayed by the logic of the Hippocratic physicians or the philosophers who went along with them. One doubter was Diocles, a fourth-century B.C. philosopher who adhered to the older "cardiocentric" position. But the most notable exception of all was Aristotle, the greatest natural philosopher of the ancient world.

Born in 384 B.C., Aristotle was a student of Plato (who believed in brain supremacy) and served as a tutor for Alexander the Great. He also was an accomplished animal dissector, having worked on some forty-nine species, ranging from elephants to sea urchins. His cardiocentric view was based largely on what he had observed with his eyes and felt with his hands, but it was also rooted in the nuances of Greek metaphysics and in his knowledge of different cultures.[34]

Observing that the heart moved and contained blood, whereas the brain was insensitive and bloodless, had a profound influence on Aristotle's thinking. He believed that sensation and movement were key features of animal life, and looked upon the blood as essential for these functions. Moreover, he reasoned, because certain primitive animals are capable of movement and sensation but do not have brains, the brain can not be responsible for these functions.

In addition, he pointed out, the heart is centrally located and feels warm, whereas the brain is far from the center of the body and feels cool. Warmth, he emphasized, is significant because it distinguishes living organisms from dead ones. Further, the beating heart can be seen well before the brain in embryos.

Additionally, the heart was recognized as "the acropolis of the body" by the Egyptians, the Mesopotamians, the Hebrews, the Hindus, and the early Greeks. No culture he knew of, past or present, had ever looked upon the brain as the organ of mind. Why should he be any different?

Yet Aristotle did not completely ignore the brain. In fact, he viewed the brain as the second most important organ in the body. Its role, he believed, was to temper "the heat and seething" of the heart, much like a radiator. Because the cool brain is surrounded by hollow blood vessels, it is perfectly designed for this task.

But what about the large size of the brain, especially the cerebrum, relative to body size in humans, the most intelligent of all animals? Humans have huge brains, reasoned Aristotle, because they produce more heat than other animals. The large, cool human brain with its many surface blood vessels provides the right amount of cooling needed by the spirits coming from the warm human heart, ensuring us the most rational of souls.

Aristotle's cardiocentric view faced formidable opposition, but it did not die a quick death due to lack of supporters.[35] A few medical sects continued to side with Aristotle during the Roman era. In the ninth century, there was even a movement in the Middle East to revive his authority. Although the heart was written off in many quarters, some scholars still confused heart and brain functions well into the European Renaissance.

William Shakespeare, for one, remained thoroughly confused in 1596. He did not question the organ responsible for reason, but he had serious questions about the organ governing "fancy," meaning emotion. At one point in his *Merchant of Venice,* he has Portia ask: "Tell me where is fancy bred, Or in the heart, or in the head?"

Even today, shades of the centuries-old debate over the functions of the heart and the brain remain with us more than we may think. For example, despondent lovers still suffer "broken hearts," close friends and loving parents continue to speak to us "from the heart," and we continue to convey our most "heartfelt" thanks when those who care deeply do nice things for us.

In fact, even the role that the heart was believed to play in memory has remained a part of our language. *Record* has its roots in *cor,* Latin for "heart." But even more telling, think how often students complain about having to memorize material—you guessed it—"by heart"!

4

Galen

The Birth of Experimentation

How the phenomena revealed in the brain and cord can best be observed in the dead
and the living respectively will be made clear in this book. The anatomy of the dead
teaches the position, number, proper substance, size, and construction of the parts.
That of the living may reveal the functions at a glance or provide premises for
deducing them.

—*Galen, second century* A.D.

Galen of Pergamon

Pergamon was one of the colonies that the Greeks established along the northwest-ern coast of Ionia. Then located fifteen miles inland, it was graced by steep moun-tains, fertile plains, and three rivers. Renowned far and wide for its beauty, for its famous temple complex, and as a commercial hub, Pergamon, now a part of Turkey, became a territory of Rome in 133 B.C. Approximately two hundred and sixty years later, in either A.D. 129 or 130, Galen was born in this scenic city.[1]

Galen's father was an architect by the name of Nicon, and he had considerable influence on his son's education. One night, as legend has it, Nicon dreamed that he was visited by Aesculapius (the Roman name for Asklepios). The medical god in-structed him to persuade his son to become a physician. Nicon took the directive to heart and convinced Galen, then seventeen, to attend the city's *aesculapion*. This beautiful and inspiring complex, with its religious sanctuary and healing center, was considered one of the man-made wonders of the world.

At Pergamon's *aesculapion*, Galen met physicians representing many different medical sects and orientations.[2] For example, some, but not all, stressed the impor-tance of anatomy in medicine. The sects also differed over how much practical experience was needed to practice medicine. Some felt that all medical wisdom was already packaged in the *Corpus Hippocraticum*, whereas others called for more. There were Dogmatists, who based their system on reason, Methodists, who be-lieved in reductionism, and Pneumatists, who stressed the importance of air and gases entering and leaving the body. There were also the Empiricists, who argued that medical knowledge must stem from actual observation.

Galen listened intently. He knew that he wanted to become far more than a spiri-tual healer or priest-physician who would offer sacrifices, sing hymns, and prescribe music to his patients. He also rejected the idea of becoming a medical compiler like the two great Roman encyclopedists of the first century, Celsus and Pliny the Elder. Of all the possibilities, he found himself most attracted to a medicine based on anatomy and physiology.

Although Galen concluded that these two sciences were the legs on which medicine must stand, he did not close his eyes to the best qualities of the other schools. His idea was to pick and borrow from each school, hoping to unify medicine under a single banner. He became so dedicated to his goal that he never gave serious thought to marriage, social activities, or other "distractions" that could interfere with his career.

After Nicon died, Galen traveled to Smyrna, Corinth, and other Greek centers of learning in an effort to increase his knowledge of medicine, the life sciences, and philosophy. His longest stop was a five-year stay in Alexandria.[3] Although in decline from the splendor it had enjoyed centuries earlier, it was still the best place to study anatomy and physiology.

At the age of twenty-eight Galen returned to the city of his birth. With twelve years of education, he was finally prepared to practice medicine. From A.D. 159 to 168 he served the gladiators in Pergamon's amphitheater. Far from modest, he attributed his seven successive appointments as physician to the gladiators to his remarkable skills and unprecedented record when called upon to treat the wounded.

While not attending the wounded gladiators from the arena, Galen treated patients in his private practice and diligently dissected animals. He compared his ambition to be a knowledgeable physician to how his father had functioned as an architect. As far as Galen was concerned, a good physician must have a thorough knowledge of anatomy and physiology, just as a good architect must know enough about physics and mechanics to make a set of plans for his builders to follow. But he did not just want to be good—he was intent on being the best.

Galen's Rome

When Galen was thirty-two, the Parthian War broke out between the Pergamenians and the Galatians to the east. Given the uncertainty of the situation, Galen decided to leave Pergamon for Rome, now the hub of government and a thriving commercial center for the Mediterranean world. He was appalled by what he found. The medical profession seemed to be in disarray, and the city was overrun with sects of every sort and many charlatans.[4] Just about anyone was able to call himself a physician; there seemed to be no principles guiding these so-called professionals or their various cults.

In particular, Galen's broad and extensive training stood in striking contrast to the education garnered by those physicians who trained with Thessalus of Tralles, a man who was first a weaver by trade. Called the "conqueror of physicians," Thessalus became one of Nero's "quack" doctors. He stated that for a substantial fee he would teach his medical students all that they had to know in just six months! In a frightening way, Thessalus exemplified what had happened to medicine in Rome. His very name made Galen's blood boil. He called Thessalus "impudent, insolent, stupid, barbarous and asinine," and his name-calling did not stop with these invectives.

Galen was so disgusted with the situation in Rome that he considered immediately returning to Pergamon. But after thinking things through he decided to try to make the best of the bad situation. Little did he realize that he would spend most of his professional life in Rome. Nor did he know that he would serve as *medicus* to four successive emperors, beginning with Marcus Aurelius.

Although Galen became one of the most imposing medical figures of all time, his popularity in Rome was limited. Some of the sniping he endured was probably due to professional jealousy, but much was the result of his gigantic ego and abrasive disposition. Although his name derived from the Greek *galenos*, which meant "peaceful" or "gentle," Galen was an elitist, and he was extremely castigating and abusive to people with opposing views.

It is estimated that Galen wrote between five and six hundred treatises on science, medicine, and philosophy while in Rome. Totaling some four million words, these works were written in Attic Greek, then the preferred language for scholars in these fields. Unfortunately, Galen's library burned in 191, destroying a large number of his books and treatises. From the works that have survived, however, it is clear that medicine was no longer the craft it had been during the time of Hippocrates. Medicine in Galen's hands had risen to become applied science, a trade rooted in dissection, observation, and experimentation.

Animal Dissection

Galen agreed with Aristotle that nothing should be recognized except that which can be experienced through the senses. In his treatise *On Medical Experience*, which he composed when he was only twenty years old, Galen wrote:

> I am a man who attends only to what can be perceived by the senses, recognizing nothing except that which can be ascertained by the senses alone with the help of observation and retention in the memory, and not going beyond this to any other theoretical construction.[5]

With such a philosophy, it is easy to understand why Galen ardently practiced dissection. His dissections, however, did not include human bodies. Religious sentiment against conducting human autopsies (a word he coined) was strong in the Roman Empire. In addition, the Roman legal system also protected the corpse from the dissector's knife.

As a result, Galen's study of the human body was limited to "chance encounters" and what he was able to observe when gladiators or soldiers were wounded. His serendipitous discoveries included a decomposed human corpse found after a flood, which he said was well prepared for study. He was also able to inspect the remains of a robber who had been murdered in the woods, a man whose flesh had already been picked bare by birds and animals (Fig. 4.1).

Recognizing the rarity of such discoveries, Galen advised his pupils to travel to Alexandria, where they could still see an articulated human skeleton, although not the soft organs. From all indications, human autopsies had stopped in Alexandria in about 30 B.C.

With his insatiable thirst for more knowledge—especially about the soft inner organs—Galen turned to animals.[6] Ideally, he wrote, one should study those animals most closely resembling humans. Not knowing about the great apes, Galen thought

FIGURE 4.1.

There are no pictures of Galen (130–c. 200) from his lifetime, but he has been depicted in many ways at later times. In this fanciful and relatively modern illustration, Galen finds a human skeleton while taking a walk.

the tailless Barbary ape—able to "walk properly and run swiftly"—came closest to appearing human. For this reason, the Barbary ape, which is really a tailless macaque monkey, became his favorite subject for dissection. When they were not available, he turned to tailed monkeys, which were easier to find and not as hard to keep or handle.

Still, Galen did not restrict his studies to primates. He dissected barnyard animals, cats, dogs, weasels, camels, lions, wolves, stags, bears, mice, and even one elephant. Heading down the evolutionary ladder from mammals, he also dissected fish, birds, and reptiles. He drew the line at gnats, flies, and other bothersome insects.

One of Galen's most fascinating lectures is entitled *On the Brain.* In it he tells medical students how to conduct a systematic dissection of an ox brain. He issued the following directives in the year 177:

> Ox brains, ready prepared and stripped of most of the cranial parts are generally on sale in the large cities. If you think more bone than necessary adheres to them, order its removal by the butcher who sells them. . . . When the part is suitably prepared, you will see the dura mater. . . . Slice straight cuts on both sides of the mid-line down to the ventricles. . . . Try immediately to examine the membrane that divides right from left ventricle [septum]. It has a nature like that of the brain as a whole and is thus easily broken if stretched too vigorously. . . . When you have exposed properly all the parts under discussion, you will observe a third ventricle between the two anterior ventricles with a fourth behind it. You will see the duct on which the pineal gland is mounted passing to the ventricle in the middle.[7]

In this short but notable fragment, Galen not only instructs his pupils how to cut a brain, but identifies many parts that deserve their attention. We see that the cavernous ventricles of the brain clearly intrigued him. Galen also mentions the pineal gland, a tiny structure named for its resemblance to a pine nut.

Galen was an astute observer and noticed some of the changes that accompany old age. Among other things, he recognized that the brain becomes smaller in old animals. Although he did not have the material to confirm this change in humans, he deserves credit for providing an early description of brain atrophy in old age.

Vivisection and Experimentation

In addition to dissecting animals that had been killed, Galen conducted many experiments on living animals. He performed his vivisections on oxen, horses, sheep, and swine, as well as on monkeys. It is especially in the domain of experimental physiology that Galen stood far above his predecessors. His experimental work was deemed so important that royalty, including the emperor himself, sometimes attended his demonstrations.

In one set of experiments, he set out to find and cut the nerves to the lungs.[8] He noticed that the struggling pig he was operating on stopped squealing but kept breathing immediately after he severed one pair of nerves in the throat (Fig. 4.2). Galen confirmed this surprising finding before incredulous onlookers using bleating goats, barking dogs, and even roaring lions from Rome's Coliseum. He knew he had found the "nerves of voice." These nerves, now known as the recurrent laryngeal nerves, are sometimes called Galen's nerves in his honor.

This discovery allowed Galen to explain what had happened to two patients operated upon for goiter. With characteristic vanity, he wrote:

A surgeon while operating for deep glandular swellings of the neck . . . without realizing what he was doing tore through the recurrent nerves. As a result, he rendered the patient voiceless, though the boy was cured of his glandular trouble. Another person operating on another boy in a similar way, left his patient

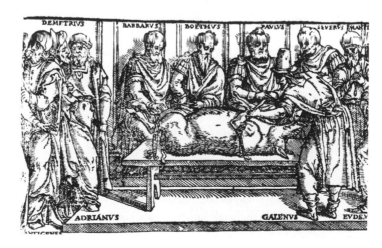

FIGURE 4.2.

Galen shown vivisecting a pig, possibly to demonstrate the nerves controlling the voice. This illustration comes from a 1609 edition of Galen's works.

with half a voice as the result of injury to one of the recurrent nerves. These results seemed perplexing to everybody, because it was seen that although neither larynx nor the trachea was injured, the voice was profoundly affected. But when I demonstrated to them the vocal nerves, they ceased to be bewildered.[9]

Galen cut the spinal cord at different levels in another set of experiments. One of his objectives was to determine which body parts below the cut would become paralyzed. He also cut halfway through the spinal cord in various animals. Here he noted that paralyses appeared on the same side of the body as the cut. After doing this on a monkey, he wrote:

> You have seen further in our dissection that transverse hemisections, which do not cut deeper than the center of the cord, do not paralyze all the inferior parts of the body, but only those directly beneath the incision, the right when the right side has been cut and vice versa.[10]

In a related experiment, he showed that severing the spinal cord close to the skull leads to an immediate cessation of breathing. In contrast, the diaphragm continues to function if the cut is made lower. These findings shed light on why one gladiator died immediately after a neck injury, whereas another was able to continue breathing after enduring a slightly lower sword wound.

Galen's inability to dissect human brains, combined with a willingness to generalize from the brains of barnyard animals to humans, led him to draw some incorrect conclusions about human neuroanatomy and neurophysiology. Nevertheless, his dissections and experiments were instrumental in leading many learned men away from the older but still competing idea, once supported by Aristotle, that intellect, reason, and perception are functions of the noble heart.

The Brain, Not the Heart

Galen idolized Hippocrates, whose words, he declared, were "the voice of a god."[11] He said that Hippocrates had never written anything untrue and bemoaned the fact that physicians in his own time could not hold a candle to the ideal physician of days past. He added that he hoped to meet Hippocrates after his own death, feeling confident that his own soul would ascend directly to heaven.

He also venerated Aristotle, the most outstanding natural philosopher of Greece's Golden Age. But despite his appreciation of Aristotle's accomplishments, his empiricism, and his search for the truth, he could not accept Aristotle's assertion that the brain's job is to cool the passions of the heart. He also found fault with Aristotle's belief that the nerves originate from the heart. Both notions, he concluded, are utter nonsense.

Galen believed that Nature would have put the brain closer to the heart if Aristotle's "refrigeration" or "radiator" theory had merit. He made this point in one of his most important treatises, *De usu partium,* translated as *On the Usefulness of the Parts of the Body.*[12] This work, presented as a "hymn of glory to the Creator of Man," was written when he was forty-four years old and at the peak of his intellectual vigor. In the text, he maintained that a crude, inert, and formless sponge, rather than a complex brain, would suffice if a large cooling organ were really necessary.

He also argued that another of Aristotle's basic premises is false. He put his own hands on the brain of a living animal and it did not feel cool in the least.

Another reason Galen gave for rejecting Aristotle's theory is that he had traced the nerves from the various sense organs to the brain, not the heart. In addition, his related discovery that nerves from the brain control voice served to debunk the idea that thought, which is closely related to speech, is a function of the central heart. But Galen was not ready to stop at this point. He also showed that stupor can result from pressure on the brain. Moreover, he pointed to the changes in sensation, perception, and cognition that can follow human head injuries.

Thus, Galen had every reason to believe that the brain is not simply a cooling machine, and further, that it is the true organ of mind. As he saw it, Hippocrates and Plato had been correct. Aristotle, who was observant and brilliant in so many ways, was wrong when it came to the brain.

The Nerves

Galen wrote many treatises about the cranial nerves, the paired nerves from the face and upper body that go to and from the brain.[13] He was not, however, the first to approach these nerves systematically. Extensive work on the cranial nerves had been conducted late in the first century A.D. by another anatomist from Pergamon, a man named Marianus. Galen frequently cited Marianus and spoke favorably about his earlier work, calling him the "restorer of anatomy."

One of Galen's favorite subjects was the eye and vision.[14] In *On the Usefulness of the Parts of the Body,* he wrote a section entitled "On the Eyes and Their Accessory Organs."[15] In it he provided descriptions of the aqueous and vitreous humors and many parts of the eye, including the conjunctiva, cornea, iris, lens, choroid, and sclera. He also described the retina, an anatomical term that has as its root the Latin word *rete,* meaning "net." Galen likened the retina to a net because it fell around the lens during an eye dissection in the same way that a limp net might collapse around a large fish.

Galen distinguished between motor and sensory nerves and contended that the eye is supplied with both types. His dissections allowed him to see the optic nerves, which he thought were hollow. He also saw the optic chiasm, a structure on the underside of the brain that was given its name because of its resemblance to the Greek letter chi (the letter *X*). Galen did not think, however, that any of the optic nerve fibers actually crossed in the chiasm (we now know that in humans about half of the axons cross). The chiasm, he reasoned, is only a place for the spirits from the two eyes to communicate with each other before heading up into the brain on the same side—it allows us see a single image, not two competing images.

Why, then, do we need two eyes? Galen explained that two eyes permit us to see a wider area than would be possible with just one "cyclopean" eye located in the center of the head. Moreover, he pointed out, if one eye were injured, we would still have the other for sight, our most important and reliable sense.

Galen was obviously not perfect when it came to the visual system. Indeed, he was also mistaken about the receptive part of the eye. Like everyone else at the time, he pointed to the lens. The receptive role of the retina, now known to contain rods and cones, would not be understood before the Renaissance.

Galen wrote in similar ways about the cranial nerves involved in hearing, taste, and touch.[16] He thought that all of these nerves were softer than the motor nerves. This quality, he reasoned, makes them better suited for taking on sensory impressions. The stronger and harder nerves, in contrast, must be designed for movement.

Although the olfactory nerve will be found listed as the first cranial nerve in contemporary books, Galen did not classify it as a cranial nerve. Instead, he thought that odorants inhaled through the nose go straight to the olfactory bulbs under the front of the brain, and then on to the hollow ventricles. He reasoned that the brain expands to bring fresh air and odorants into the system and contracts to expel the stale air. He based this idea on brains actually pulsating. It was consistent with the teachings of a sect called the Pneumatists.

Galen also studied the nerves of the sympathetic nervous system.[17] These nerves are now associated with control of the internal organs, where they play an especially important role in "fight-or-flight" situations. He described both the sympathetic chain running alongside the spinal cord and the major sympathetic ganglia. He correctly guessed that the sympathetic nervous system works as a whole, but was less accurate when it came to its function. Instead of associating the sympathetic nervous system with efferent nerves to the smooth muscles, he thought it was designed to foster communication among the internal organs of the body. He reasoned that extensive interconnections among the nerves in the sympathetic chain allow the spirits to convey messages from one internal organ to another, so that all could experience the change. The sympathetic division of the autonomic nervous system was given its strange name because of this early belief in sharing or "sympathy."

A Theory of Brain Function

Let us now return to the brain itself. From his lecture on how to dissect an ox brain, it is clear that Galen recognized many different parts of the brain. Further, while dismantling Aristotle's heart theory, he made it clear that he regarded the brain as the definitive organ of mind. But did Galen have a clue about how the brain actually functions, or did he propose just another fanciful theory?

In truth, Galen's brain anatomy was well ahead of his physiology. His physiology was based on bodily spirits, called *pneumata*, and the teleological (*telos* = "end," *logos* = "discourse") idea that a supreme being designed every organ with a specific end or purpose in mind. Particularly when it came to brain physiology, the empiricist in Galen was pushed aside by another aspect of him, that of the speculative philosopher who was governed by ideas about designs, purposes, and final causes. The latter Galen was so eager to put forth a complete physiology of the body that he violated his own oath not to recognize anything he could not experience through his senses.

Galen probably started with Plato's assertion that there are three fundamental organs—the liver, the heart, and the brain—each associated with a different spirit or soul. He then blended Plato's theory with the pneumatism of Erasistratus, the Alexandrian who had suggested that tiny spirits are able to travel through hollow nerves to and from the brain and its large hollow cavities, the ventricles.[18]

Galen described three kinds of spirits. He said the least about the natural spirits, associated with the liver and its veins. These spirits were thought to regulate nutrition, vegetative functions, and other fundamental needs. Some natural spirits, however, reach the heart, where they are converted into "vital" spirits. These higher-level spirits generate the internal heat of the body and are at least partly responsible for our base emotions. In turn, some of the transformed spirits from the heart go the brain, where they are converted into *pneumata psychikon,* or "animal spirits." These are the spirits of mind.

Galen had the most to say about the animal spirits. He maintained that the final conversion occurs in the *rete mirabile* (miraculous net), a network of fine arteries surrounding the pituitary gland at the base of the brain. The term *rete mirabile* has the same Latin root as the word *retina.* Galen wrote: "It is not a simple network but [looks] as if you had taken several fisherman's nets and superimposed them."[19] Galen probably saw the net-like *rete mirabile* while dissecting oxen, sheep, and pigs. What he did not realize was that humans, or monkeys for that matter, do not possess a comparable meshlike network of blood vessels at the base of the brain.[20]

Galen also suggested that animal spirits may be manufactured in the vascular linings of the anterior ventricles of the brain (called the choroid plexuses). Whether they originate in the *rete mirabile* or the ventricle linings, they are stored in the ventricles until called into play by the brain.[21] When needed, they enter the nerves and move the muscles. Conversely, the highest spirits can also carry sensations from the eyes, ears, tongue, and skin back up to the brain via the nerves and tracts.

In his *Commentaries on Hippocrates and Plato,* Galen made it clear that he associated the brain substance itself, and not the ventricles, with our highest functions. As we have seen, by experimenting on animals and by studying people with head injuries, he recognized that brain damage can affect the rational mind. Hence, he maintained that although the spirits in the ventricles act as the physical "instrument of the soul," the true seat of the rational soul must be the brain.

Galen considered perception, cognition, and memory to be the functions of the rational soul. He acknowledged that brain diseases and injuries can affect these functions independently. Nevertheless, he stopped short of localizing these three distinctive functions in different parts of the brain, at least in his surviving writings. In addition, there is no evidence to show that he associated these functions with the anterior, middle, and posterior ventricles.

In many respects, Galen would have had reason to agree with Erasistratus, who stated that the size and complexity of the cerebral convolutions can be correlated with intelligence. But, as noted in the previous chapter, once Galen examined the donkey's large and highly convoluted brain, such an association was out of the question. "Donkey intelligence" was an oxymoron to him, a true contradiction in terms. This being the case, he was forced to conclude that the "temperament" of the thinking body is a more important contributor to intelligence than the size or complexity of the brain.

The Practicing Physician

By associating internal diseases with specific organs, Galen's medicine differed from that practiced by the Hippocratic physicians, who were inclined to treat the body as a whole. In this regard, Galen was more modern in orientation. But like many other physicians in Rome, Galen still accepted the Hippocratic doctrine of the four humors and their qualities. He even diagnosed and treated his own painful colic in a manner consistent with humoral theory.

One of the medical subjects of considerable interest to him was apoplexy, or stroke, which in the ancient literature referred to one of two things: either a sudden loss of consciousness with a paralysis of the whole body, or a partial paralysis of rapid onset.[22] Galen believed that strokes resulted either from an accumulation of a thick, cold humor (most likely sticky phlegm or dense black bile) in the ventricles, which in turn obstructs the flow of the animal spirits, or from a rush of cool blood to the head. When he practiced medicine, no one was correlating apoplexy with broken blood vessels in the brain; the association of apoplexy with cerebral hemorrhage would not emerge for several hundred years.

As for treatment of apoplexy, this had to do with reducing the noxious dark humor by diet, purging, vomiting, and bleeding. Since the "rule of opposites" was followed and apoplexy was seen as a "cold" disease, agents that would heat the body were also given. Such remedies included thyme, mustard, and oregano.

In agreement with Aristotle, Galen felt that dreams provide clues about the status of the four humors and hence about diseases. In his short treatise, *Diagnosis from Dreams,* he writes:

> To dream indicates an abnormal condition of the body. To see fire in sleep means this condition is due to yellow bile, to see smoke or fog or profound darkness, black bile. . . . We conclude that the wrestler who dreams that he is standing in a cistern of blood which he barely overtops is affected with an overabundance of blood and requires blood letting.[23]

As can be surmised from the last sentence, Galen continued to cup, bleed, and leech to restore balance to the damaged nervous system. Being a frugal man, he even suggested clipping the tails of leeches to allow more blood to be drawn from each. This creative recommendation did not last long; when tail clipping markedly increased the leech mortality rate and the prices of leeches rose, the cutting was abandoned.

In addition to these traditional therapies and dietary and hygienic recommenda-

FIGURE 4.3.

An ancient Greek ceramic plate (Campanian ware) showing three fish. The odd-looking fish on the right is the electric ray, Torpedo ocellata.

tions, Galen turned to the use of electricity generated by fish.[24] We know that the Egyptians and the Greeks already knew about the shocking powers of the Nile catfish and the electric ray or "torpedo" (Fig. 4.3). There is no concrete evidence, however, to show that they used the discharges from these fish as therapy.

In contrast, Scribonius Largus, who lived just before Galen, placed a large electric ray across the brow of a patient suffering from severe headaches.[25] He allowed the ray to discharge its electricity "until the patient's senses were benumbed." Successful in reducing headache pain, he also benumbed patients who complained of other types of pain.

Galen followed Scribonius Largus when it came to using the discharges of the ray, recommending it for headache and other pains. He also promoted the electric ray as a good remedy for epilepsy, the best-documented and most feared of all brain diseases in ancient times. But it was not always possible to obtain these creatures, and for this reason Galen worked hard to develop more effective pharmacological cures.

Mithradates and Theriac

Galen oversaw his own pharmacy, mainly because he felt he could not trust the pharmacists in Rome. He had many medicinal agents at his disposal but was especially fond of a medicine called theriac. The story behind this compound drug makes for fascinating reading. One part of its history can be traced to Mithradates VI, the indomitable ruler of Pontus, now a region of northern Turkey.[26]

Mithradates Eupator (Fig. 4.4) was born in 132 B.C. and blended royal Persian blood and Eastern passion with classical Greek values. Historians remember him

FIGURE 4.4.

A coin showing the head of Mithradates VI of Pontus (132–63 B.C.), the king who was obsessed with the study of poisons. The coin was minted in 74 B.C.

best for skillfully fighting three long wars against Rome. In fact, he was the last recognized ruler in the eastern region capable of standing his ground against the mighty onslaughts of the Roman legions.

Even before assuming the throne, Mithradates feared that some of his enemies might try to poison him. Gradually killing a person by poisoning food was a relatively common practice at this time, so Mithradates' paranoia was probably justified. Poisons from spiders, scorpions, sea slugs, and the plant aconite (also called monkshood or wolfsbane) were among the agents that most worried him.

In order to protect himself from these poisons, he studied animals that seemed to be immune to these substances. Soon he accumulated a number of seemingly protective agents. He tested them separately on other animals and on condemned criminals. If the agent effectively blocked the poison, he took it himself. He occasionally followed it by a tiny dose of the poison to make sure it was doing its job.

Given the inconvenience of taking large numbers of drugs each day, Mithradates VI began to mix his substances into a compound. His concoction eventually contained forty-four ingredients and became known as a *mithradaticum*. The list of ingredients would probably strike the modern reader as particularly unappetizing. For example, he put duck's blood in the mixture because ducks can eat certain plants that are poisonous to us and yet feel no ill effects.

The *mithradaticum* seemed to be so effective in counteracting the effects of deadly substances that the king did not hesitate to drink poisons in public to show his invincibility and enhance his stature. The hapless criminals and captured soldiers who were forced to join him in these demonstrations were not so fortunate and provided the ghoulish contrast he desired.

In 63 B.C. Mithradates' forces were finally cornered by the legions of Pompey. Having no intention of being brought to Rome to be crucified, he turned to a powerful poison hidden in his sword hilt. Although his two daughters who took the poison died immediately, he remained unaffected. In the end, he had to order one of his own chieftains to slay him.

The Romans thus failed to take the king alive, but they did find his coveted secret formula and brought it back to Rome. They then proceeded to modify it, as they did to other compound drugs of this type. Andromachus, Nero's chief physician, added ingredients to the formula, including a large dose of viper. He also multiplied its opium content by five. His sixty-four-component concoction was called *galene*, because it really seemed to have tranquilizing effects.

Galen wrote a book about the compound wonder drug, and in it he described his own revision of the formula. His universal cureall contained one hundred ingredients, each differing in quantity. He was so convinced of the broad powers of his theriac as a preventative and as a treatment for bites and many diseases that he personally made and administered it to three of the emperors whom he served.

Marcus Aurelius, one of his patients, preferred the theriac fresh rather than aged.[27] He probably became a bona fide opium addict by taking it daily. In a treatise entitled *Antidotes,* composed just a few years before he died, Galen wrote:

> Marcus Aurelius, for whom we first prepared it, showed by daily employment his good common sense, and his knowledge of the dangers to which he was exposed. He took an amount the size of an Egyptian bean, either in water or

wine, or mixed with a little honey. When he found that he became drowsy over his daily duties he omitted its opium constituent, but when . . . he suffered from insomnia, he added it again.[28]

During the Dark Ages and through the Renaissance theriac and closely related compounds continued to be recommended for epilepsy and other brain disorders, as well as for just about everything else. The drugs also underwent further changes. For example, Jean Fernel, the sixteenth-century Paris physician whose medical biography would later be written by Sir Charles Sherrington (see Chapter 14), used only fifty ingredients in his compound.[29]

In some towns at even later dates, festivals were held when a new batch of theriac was prepared. Its popularity was so great that it was still present in many French and German pharmacies in the nineteenth century. As put by medical historian Guido Majno:

> Theriac died ever so slowly. The German official pharmacopoeia still included it in 1872, vipers and all; the French in 1884. By then you could order your *Theriaca Andromachi* by telephone.[30]

Legacy

The traditional dates of Galen's death are either A.D. 199 or 200, although it has also been contended that he died as late as 216.[31] Yet the duration of his grip on anatomy, physiology, and medicine persisted well after his death. Remarkably, Galen's teachings served as the guiding force in science and medicine for over thirteen hundred years.

The incredible durability of what has been called Galenism can be attributed to Galen's many accomplishments as a scientist and to his bold attempt to build a single system of medicine based on science. His belief in a single god who created everything, including each bodily organ, for a purpose—a deity who could not be praised enough—added to his appeal, especially as the pagan religions gave way to monotheism.

Galen met the need for absolute authority that emerged as Europe entered the Dark Ages. For Westerners in particular, many of his ideas about medical science became dogma. In this regard, we can say that his drive to unify the field of medicine was, in fact, finally realized. Indeed, late in life Galen himself wrote:

> I have continued my practice on until old age, and never as yet have I gone far astray whether in treatment or in prognosis. . . . If anyone wishes to gain fame through these, and not through clever talk, all that he needs is, without more ado, to accept what I have been able to establish by zealous research.[32]

In the next chapter we shall discover just how powerful an influence Galen had on the physicians, clerics, and scholars who followed him. We shall also see that relatively little more of lasting value will be learned about the brain before scientists began to return to the dissecting table in the turbulent period known as the Renaissance.

Andreas Vesalius
The New "Human" Neuroanatomy

His central achievement is the epoch-making *De humani corporis fabrica*, "On the workings of the human body." . . . This immense work is a milestone in the history of thought; it is the first great work of science in the modern manner; . . . it introduced a sound and positive basis into medical education; . . . it established exact graphic treatment as an essential adjunct of biological research.

—*Charles Singer, 1952*

The Intellectual Climate

After his death, Galen's thoughts about the brain and behavior were treated as dogma. This situation contributed to an incredibly long period of relative stagnation, one that extended from the third to at least the thirteenth century. There was little in the way of experimentation or new science during this protracted period of darkness and superstition. The greater interest was now in spiritual matters, not in discovering the secrets of anatomy or physiology. Like a bear approaching the cold of winter, the spirit of scientific inquiry slipped into hibernation.

Although some men wrote medical treatises during this long interval, most simply paraphrased Galen or cited the authority of Hippocrates. In the ninth century, there was also a notable attempt to revive Aristotle.[1] Unlike Aesculapius, the healing god who performed miracles, Hippocrates, Galen, and Aristotle were not perceived as pagan deities or rivals of Jesus. All three were earthly men who venerated nature, a wonderful creation of God. Thus, the science of Galen and Aristotle would continue to be taught, and the Hippocratic art of healing would be seen as an acceptable supplement to prayer.[2]

The Church Fathers and the Ventricles

The most significant conceptual change about brain function to take place during the Dark Ages seems to have occurred in the fourth or fifth century. At this time, some scholarly Christians began to present the hollow ventricles of the brain as the abode of the rational soul. The brain substance forthwith assumed a secondary role to that of its cavities. Significantly, each cavity was now associated with a different function.

One of the earliest known advocates of localization in the ventricles was Nemesius, Bishop of Emesa. He blended Galenic medical science with Christian theology around the year 400. He looked upon the two symmetrical anterior (or lateral) ventricles as the first cavity. Here, he maintained, is where the faculty of perception resides. He assigned cognition to the middle ventricle and then delegated memory to the posterior ventricle, near the cerebellum. Nemesius might have envisioned a logical front-to-back progression, one in which information is received, processed (interpreted), and finally stored.

Nemesius made it sound as if these structure-function associations were based on hard clinical facts:

> If the front ventricles have suffered any kind of lesion, the senses are impaired but the faculty of intellect continues as before. It is when the middle of the brain is affected that the mind is deranged. . . . If it is the cerebellum that is damaged, only loss of memory follows, while sensation and thought take no harm.[3]

He did not mention anything more about the evidence for ventricular localization. But if Nemesius knew of clinical cases, they probably were not his own. He most likely relied on various statements made by Galen, who associated the front of the brain (the cerebral hemispheres) with sensory functions because it was softer than the cerebellum. Galen also contended that imagination, intellect, and memory could be dissociated. But, at least in his surviving writings, he did not go any further. The "prince of physicians" maintained only that the cavernous ventricles store the spirits and thus serve the rational soul, which he firmly associated with the brain substance itself.

The theory of localization in caverns was also promoted by Poseidonius of Byzantium, a slightly younger contemporary of Nemesius, as well as by St. Augustine of Hippo, sometimes considered the greatest thinker of Christian antiquity. Although opinions differed as to which specific function belonged in each of the three cavities, the basic idea of ventricular localization grew and was accepted for well over a thousand years (Fig. 5.1).[4]

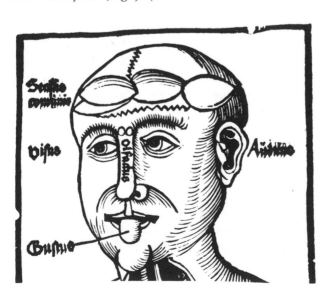

FIGURE 5.1.

The ventricles of the brain as envisioned from the Dark Ages into the Renaissance. (From J. Peyligk, Compendlosa capitis physici declaration. *Leipzig: W. Monacensis, 1516.)*

The Detour Through the Middle East

During the fifth century, the Nestorian Christians founded a school in Jundi-Shapur, in what is now Iran. For the better part of two centuries, the men of this school translated ancient medical texts from Greek into Syriac and Arabic. Stimulated by the translations of the works of classical scholars into languages they could understand, the translators and their clients became the leading trustees of the great ancient heritage by the eighth century.

The medieval physicians from the Middle East did not do dissections, nor did they develop radically new theories of brain function. They accepted the Greco-Roman theory of four humors and the early Christian idea that the individual ventricles are critical for perceiving, thinking, and remembering. In summing up the situation, one well-known medical historian wrote: "The anatomical knowledge of Islam was merely that of Galen in Moslem dress."[5]

The most important medical figure in the Middle East at the end of the first millennium was Abu 'Ali al-Husain ibn 'Abdullah ibn Sina, known simply as Avicenna in the Christian world. He produced approximately a hundred books. His five-volume *Canon* served as a standard reference for medicine for more than five centuries. Rhazes, Albucasis, and Averroës, to use their Latinized names, were among the other important Moslem writers on the nervous system during this period.

The Language Problem

While the Eastern physicians and philosophers were reading and translating classical medicine, the Europeans managed to lose their ability to read ancient Greek, long the language of science. In terms of scientific inquiry and learning in general, the Dark Ages in Europe were well deserving of the name. To awaken Europe from its long intellectual slumber, not only would the ancient material have to be rediscovered, but it would also have to be translated from Syriac and Arabic into yet another language, Latin.

When the Europeans drove south in Spain to conquer Toledo in 1085, Cordoba in 1236, and Seville in 1248, they came upon some of the Arabic translations of the ancient writings, as well as some of the newer works these documents had spawned. In particular, the conquerors discovered the books of Avicenna and Rhazes, the prolific medical writers from what are now Iran and Iraq, respectively. They also came across the works of Albucasis and Averroes, two scholars from Moorish Spain itself.

Constantinus Africanus and Gerard of Cremona were important translators of the newly discovered material into Latin. The former was a Tunisian who converted to Christianity and, in about 1063, affiliated with the Italian school at Salerno. Gerard of Cremona, who appeared a century later, worked in Toledo, the scenic Spanish walled city. Their new Latin translations of Greco-Roman science with Byzantine and Arabic overtones stimulated Westerners to begin to write their own books. Although the first such books did not represent radical departures from the past, they served as important stepping stones for the increasingly original works that slowly began to appear.

The Return to the Dissecting Table

Dissections and demonstrations were sorely needed if scientists were to acquire new knowledge about the internal organs of the body. In Italy, Frederick II recognized this need early in the thirteenth century. He controlled the medical school at Salerno and decreed that no student could practice his craft before studying anatomy for at least one year. Supposedly he also broke with the past by ordering a public dissection of a human cadaver every five years, although this claim is not as well documented.[6]

Mondino de' Luzzi, a professor at Bologna, conducted human dissections early in the 1300s and wrote a popular guide to anatomy in this new, more encouraging, early Renaissance environment.[7] He made several discoveries and along with them many mistakes, one of which was a detailed description of the human *rete mirabile*, Galen's fine network of blood vessels at the base of some animal brains. His stated purpose was to examine what Galen and his faithful disciple Avicenna had to say. Even if he came across some things to question, he was not ready to challenge the revered authorities of the past.

Most of the human bodies that he and other anatomists used for dissections came from executed criminals and were provided by sympathetic civil authorities. Tight controls still limited the number of cadavers that could be obtained each year. Slowly, however, the autopsies began to increase, mainly because the one or two dissections each year drew such large crowds that viewing was obstructed. Interestingly, more students were permitted to watch when a woman was dissected than when the body was that of a man. The reason for this was that more men were accused of serious crimes and were given death sentences—cadavers of executed women were rare.

Besides performing dissections for medical training, the Italians also began to conduct autopsies to determine how best to stop the frightful epidemics that were sweeping across the country.[8] Without question, the most devastating epidemic was the "black death," which spread from Asia into Europe in 1347. Estimates are that between one-quarter and one-third of the population of Europe succumbed to the plague within a ten-year period, leading to yet more autopsies.

A third reason for the increase in autopsies was the need to settle legal disputes. Sometimes inquisitive authorities ordered an autopsy to determine whether a sickly person might have been murdered by a subtle poison, or to see whether a wound was really the cause of death.

The Church and the Surgeon

An interesting issue is the extent to which the clergy accepted the religious dictate *ecclesia abhorret a sanguine* ('the Church abhors the shedding of blood'). General edicts to this effect appeared in several European cities in the twelfth and thirteenth centuries (for example, the Council of Tours in 1163). But exactly what was demanded and to whom the dictates were really directed were not always clear. Did they really mean that no members of the Church could perform surgery?

That the Church authorities kept repeating their bans on the drawing of blood would suggest that such edicts were not followed too rigidly. Many local Church officials probably viewed them as applicable to higher religious orders but not to

FIGURE 5.2.

An illustration of a human dissection. The physician is shown sitting on his cathedra, or high chair, while the cutting is done by an individual of lower rank. This woodcut is from a 1495 edition of the anatomy book written by Mondino de' Luzzi around 1316. (From Mondino de' Luzzi, Anatomia corporis humani. Venice: Bernhandis Venetus de Vitalibus.)

lower levels of clergy.[9] For the most part, however, the craft of surgery was in the hands of barbers, executioners, and bathkeepers during the Renaissance. These surgeons "of the short robe" were of a lower social class than the more erudite physicians, men who distinguished themselves by reading Latin and wearing fine, long robes.

What, then, took place during a dissection? From what we can gather, the physician usually directed the dissection without even touching the cadaver. Most often he sat on a high chair called a *cathedra*, usually behind and to the side of the body (Fig. 5.2). From his perch, he probably read from an authority on Galenic anatomy or from a Latin version of one of Galen's books. The cutting itself was usually performed by an "unlearned" barber. An assistant with a pointer might also be present. He directed the attention of the students to the different parts of the dismembered body as they were described by the physician doing the reading, while simultaneously being exposed by the barber-surgeon.

Leonardo as a Transitional Figure

Medical texts began to display illustrations during the Renaissance. Prior to the sixteenth century, however, the artwork lacked realism and, like the texts, was heavily influenced by the words of Galen and by the modifications made by his Christian and Moslem followers. Thus, although human dissections were now being performed, much of the accompanying art remained simplistic and cartoonlike, especially when it came to the brain and the ventricles.

Leonardo da Vinci was a pioneer in the movement toward realism. This remarkable painter, sculptor, architect, and engineer dissected many cadavers and made over fifteen hundred high-quality drawings in his lifetime. He displayed an accuracy regarding the human body and its parts that had not been seen before.[10] The only problem—and in terms of his impact it was a big one—was that he did not publish his anatomical drawings for others to appreciate.

Leonardo da Vinci was able to draw body parts more accurately because he did dissections and also because he was an ingenious technician. Among other things, he conducted a series of experiments to reveal the true structure of the ventricles. Knowing how statues could be made by pouring molten metal into a mold, he reasoned that a similar procedure might be used to produce a faithful cast of the ventricles. Thus, he injected hot wax through a tube thrust into the ventricular cavities of an ox that had just been killed, inserting a second tube to allow existing substances in the ventricles to be flushed out. He gave his injected wax time to harden, then he cut, peeled, and scraped away the overlying brain to obtain a faithful cast of the ventricles.

Leonardo da Vinci performed these experiments between 1504 and 1507. His resulting drawings of the ventricles differed dramatically from the simplistic depictions of other anatomists. In fact, with an eye on man, he even sketched the ox ventricles he had cast onto images of the human brain. Nevertheless, he was not prepared to reject well over a thousand years of Church-sanctioned doctrine about the ventricles. Instead, he chose to compromise between what he was observing and what he had been taught. Although the artist and scientist in him produced anatomically correct drawings of the ventricles, he proceeded to assign imagination, cognition, and memory to the different brain cavities (Fig. 5.3).

Slowly, however, the groundwork was being prepared for a revolution in thinking about the mind and the body. At about the same time as Leonardo was making casts of the ventricles, Jacopo Berengario da Carpi, a member of the highly regarded faculty at Bologna, wrote that he had performed many human dissections and was un-

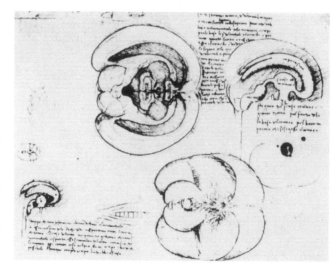

FIGURE 5.3.

Drawings by Leonardo da Vinci of the dissected brain of an ox. Leonardo worked very early in the 1500s. Although his realistic drawings broke new ground, his physiology was that of Galen and his followers.

able to find the *rete mirabile*.[11] Nor could he find certain other structures mentioned by Galen and his Middle Eastern followers. Something was very wrong and he knew it. Similar confusions would soon be experienced by another free-thinking medical man of the Renaissance, an inquisitive anatomist named Andreas Vesalius.

The Student Years of Andreas Vesalius

Andreas Vesalius (Fig. 5.4) was probably born on December 31, 1514, in Brussels.[12] Although the original family name had been Witing, during the fifteenth century it was changed to Van Wesele, Vésale, and in Latin to Vesalius. The intent was to reflect the family's place of origin, the town of Wesel on the German-Dutch border.

During the 1520s Vesalius attended schools in Brussels. When he was about fifteen years of age he left for the nearby town of Louvain, where a university had opened its doors about a century earlier. Seven years later he found himself in Paris studying medicine with an insatiable curiosity. His goal was to become the fifth physician in his family line.

The Paris medical school was modest and extremely conservative at this time.[13] The teachings of Galen were stressed, and his was considered the final word. In this environment, human dissections were performed much less frequently than in progressive Italy. Vesalius would later recall:

> My study of anatomy would never have succeeded had I, when working at medicine in Paris, been willing that the viscera should be merely shown to me and to my fellow students at one or another public dissection by wholly unskilled barbers, and that in the most superficial way. I had to put my own hand into the business.[14]

FIGURE 5.4.

Andreas Vesalius (1514–1564), professor of anatomy at Padua, whose dissections and observations overthrew ideas based on Galen and rekindled the study of the human body. This portrait is from his Fabrica *of 1543.*

The Paris teacher most frequently mentioned by Vesalius was the anatomist Jacobus Sylvius. He was a devout student of Galen and an established anatomy teacher.[15] Sylvius taught by dissecting animals, and he stood apart from other teachers by doing this with his own hands.

Dedicated medical students such as Vesalius, however, did not confine their studies to the classroom. Foraging in the cemeteries for human bones was common, and we know that Vesalius was quite familiar with the rich bounty that could be obtained in this way. The biggest problem facing these adventurous students came from the savage dogs that roamed the cemeteries. These dogs were always ready to defend their territories and finds against intruders, be they of the four-legged variety or ambitious medical students.

Vesalius did not complete his studies in Paris, choosing instead to return to Louvain three years later.[16] He left because war had broken out and rumors were spreading that Paris would soon be sacked. Soon after his return to Louvain, he stealthily climbed a gallows to obtain the remains of a hanged criminal. He then avoided arrest by convincing everybody that the skeleton found in his room was imported rather than stolen. This incident, like his trips to the cemeteries, reveals much about Vesalius. Even as a young medical student, he was willing to take chances, and perhaps even put his own life in danger, for firsthand knowledge about the body.

The Awakening in Padua

In 1537 Vesalius headed south to Padua, located close to the Italian city of Venice. The University of Padua had been founded in 1222 by disgruntled students and teachers from the University of Bologna. In 1250 it added a medical faculty. After it came under the control of the Republic of Venice in 1440, the school continued to improve and rose to such prominence that it succeeded in attracting some of the best students and teachers in all of Europe.[17]

After passing an examination, Vesalius accepted the chair of senior lecturer in surgery. This appointment was exceptional both because he was a foreigner and because he was only twenty-three years of age. His duties were to give surgical lectures and anatomical demonstrations to students, other physicians, and honored guests.

Not one to let unskilled barbers do the cutting, Vesalius acted as lecturer, demonstrator, and dissector when human bodies were available. The Venetian authorities not only provided their anatomists with bodies from the gallows, but even set the times and conditions for some of the executions to benefit the dissectors, who did not have the means to preserve bodies during the hot summers.

Shortly after settling into his position, Vesalius published three brief works. The most important was his *Tabulae anatomicae* of 1538.[18] Also known as *Tabulae sex* because it contained six large illustrations (woodcuts), this work contained the most highly detailed drawings of the human body yet published. The first three figures, showing the internal organs, were drawn by Vesalius himself. The views of the human skeleton, however, were drawn by Jan Stephan van Calcar, a Dutch pupil of Titian, the great Italian painter of the high Renaissance.

The opportunity to dissect human bodies and the preparation of *Tabulae anatomicae* slowly led Vesalius to realize that Galenic anatomy was far from perfect. Although there were only hints of this in the *Tabulae* itself, with each passing day he became more and more perplexed by how an observer as skillful as Galen could have made so many errors in the field of human anatomy.

This matter continued to bother him, especially as he became more confident about what he was seeing with his own eyes. Early in 1540 he conducted some dissections in Bologna at the invitation of the faculty. A medical student by the name of Baldasar Heseler, who had studied theology with Martin Luther at Wittenberg, was present at the first demonstration. It lasted about two weeks and involved three human bodies, six dogs, and several other animals.[19]

From Heseler's copious notes we know that Vesalius now openly questioned some of Galen's statements. Nevertheless, Vesalius still accepted much of what Galen had to say, including the reality of the *rete mirabile* in humans (although he was forced to use a sheep's head to show this vascular structure). Heseler was impressed by the fact that Vesalius attempted to open the eyes of his students to things they had not seen or even thought about before. He contrasted the spirited and careful demonstrations given by Vesalius with the anatomy lectures given by an authoritative Galenist named Matthaeus Curtius, a teacher he called ignorant on many matters.

While in Bologna, Vesalius assembled a human skeleton as a present for his hosts. He also put together the bones of an ape as a measure of his appreciation. Side by side, the two articulated skeletons could be examined and compared. The differences led Vesalius to the greatest insight of his life.

He suddenly realized that Galen had not been a poor anatomist prone to mistakes. Instead, he was a dedicated, astute scientist who, quite simply, was not describing humans! Essentially limited to the study of animals by the laws of Imperial Rome, Galen's anatomy was restricted to oxen, pigs, dogs, and other animals. We can imagine Vesalius wanting to kick himself in anger for not recognizing this earlier. Finally, many of the Galenic "mistakes" that been plaguing him were understandable. It was now essential to spread the word and sound the call for a new human anatomy.

The *Fabrica*

In political history there are important dates that stand out as if written in large, bold type. The same is true in science. One such date is 1543, when two dedicated scientists broke in very different ways with the past. As a result, how people thought about man, nature, and the universe changed forever.

In that year Nicolaus Copernicus, the Polish physicist and astronomer, published *De revolutionibus orbium coelestium* (Concerning the Revolutions of Heavenly Bodies). In this book, he challenged the theory that the solar system revolves around the earth. In the same year, Andreas Vesalius published his landmark *De humani corporis fabrica* (On the Workings of the Human Body), one the most important medical science books ever written.[20] The *Fabrica*, with its 663 folio pages, was actually completed in 1542, when Vesalius was twenty-eight years of age.[21]

As he did in his earlier *Tabulae anatomicae*, Vesalius had high-quality illustrations made for the *Fabrica*. This time, however, there were many more drawings and a marked improvement in their fine details. Among other things, twenty-five magnificent illustrations present the human brain and its parts as they were revealed with a skillful top-to-bottom dissection (Fig. 5.5). But this time Vesalius did not identify the talented artist or artists who worked for him. Nor are there identifying marks on the plates. As a result, considerable speculation has revolved around who did the artwork; the only certainty is that they were individuals so masterful as to be able to make even piles of dry bones spring to life.[22]

The *Fabrica*, with its wonderful cover plate showing a rare dissection of a woman (Fig. 5.6), consists of seven small "books" (really chapters) on human anatomy. Of these, Book IV deals with the nerves and Book VII with the brain. Along with the larger text, Vesalius also had his publisher in Basel, Switzerland, release an abridged text in 1543. It was entitled *Epitome* and was primarily intended for students.[23]

Both books were based on the deep-seated belief that because all earlier descriptions of the human body might be faulty, it was necessary to verify everything previously written by carefully examining human cadavers. Vesalius identified some new structures by conducting autopsies. Just as importantly, by comparing human findings with those from dogs and other animals, he was able to show his readers that Galen and his followers had never adequately described human beings.

Vesalius found some two hundred errors in Galenic "human" anatomy. Yet he never tried to humiliate his predecessor and never corrected him or his disciples without hard facts to back up his statements. To Vesalius, Galen was neither entirely right nor completely wrong. As for his own philosophy, it was one we have encountered before: Do not accept things that you cannot verify with your own eyes.

In addition to writing about anatomy, Vesalius also discussed brain disorders. For example, in his *Fabrica* he described some individuals with very large heads and provided what may be the first good descriptions of hydrocephalus. One was a beggar

FIGURE 5.5.

Cover plate from the 1543 edition of Vesalius' Fabrica.

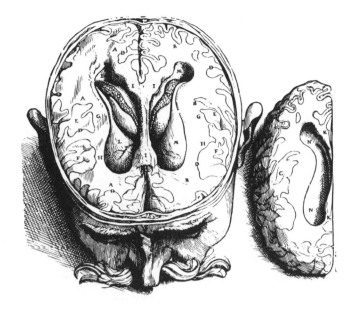

FIGURE 5.6.

A dissected brain from Vesalius' Fabrica *showing the ventricles.*

who worked the streets of Bologna, and another was a mentally deficient boy who lived in Genoa, Italy. Vesalius commented that the "small boy is carried door to door by a beggar woman and was displayed by actors in Brabant; his head is without exaggeration larger than the heads of two men and bulges on either side."[24]

The Fate of the Ventricles and the *Rete Mirabile*

The ventricles and the *rete mirabile,* two structures basic to Galenic physiology, merited special attention in the *Fabrica*. Vesalius wrote that the human ventricles are not different in shape from those of other mammals, although other animals do not compare to humans when it comes to reasoning powers or other attributes of the rational soul. In his words:

> All our contemporaries, so far as I can understand them, deny to apes, dogs, horses, sheep, cattle, and other animals, the main powers of the Reigning Soul not to speak of other [powers] and attribute to man alone the faculty of reasoning; and ascribe this faculty in equal degree to all men. And yet we clearly see in dissecting that men do not excel those animals by [possessing] any special cavity [in the brain].[25]

After questioning the roles assigned to the ventricles, Vesalius turned to the *rete mirabile,* the network of blood vessels at the base of the brain supposedly responsible for the production of animal spirits. In his earlier *Tabulae anatomicae,* he depicted this network of blood vessels in a way that corresponded to Galen's descriptions; so strong had been his earlier belief in the reality of Galen's *rete mirabile* in humans that he never dissected a human head in public without an animal at hand to demonstrate what he could not find in the human cadaver. But in the *Fabrica,* he denied its existence in humans. To quote:

How much has been attributed to Galen . . . by those physicians and anatomists who have followed him, and often against reason! In confirmation there is that blessed and wonderful *plexus reticularis* which that man everywhere inculcates in his books. There is nothing of which physicians speak more. They have never seen it yet they describe it from Galen's teaching.[26]

Finally, the enduring myth of the human *rete mirabile*, as well as many other errors of the past, could be exposed.

The Less Knowable World of Function

Although he had great faith in what he was able to see, Vesalius was uncertain when it came to how the brain functions. Hence, he did not immediately reject the idea of animal spirits residing in the ventricles, even though he had doubts about the source of the spirits and Galenic physiology in general.

For example, he was clearly uncomfortable with the theory that different functions of mind could be associated with spirits in the anterior, middle, and posterior ventricles. The theory made little intuitive sense to him. But because his anatomy could not shed light on the problem, he only stated that he could not form an opinion about how the brain might regulate the functions of the rational soul. In his own words:

> I can in some degree follow the brain's functions in dissections of living animals, with sufficient probability and truth, but I am unable to understand how the brain can perform its office of imagining, meditating, thinking, and remembering, or, following various doctrines, however you may wish to divide or enumerate the powers of the Reigning Soul.[27]

Nevertheless, Vesalius felt he could associate intelligence with brain size, something Galen was unwilling to do after seeing how large donkey brains proved to be. In fact, Vesalius sounds wonderfully modern when he mentions how well the human brain stacks up to that of animals, especially when the size of the body is taken into account:

> There is hardly any difference that we have detected except in bulk, [although] the brains do vary according to the intelligence with which the animals are endowed. For to man has been given the largest brain; next to him the ape, the dog, and so on, according to the order that we have learned of the power to reason in animals. And to man's lot falls a brain not only bigger in proportion to the bulk of his body, but actually bigger than the brain of any other animal.[28]

Another questionable tenet of Galen's physiology was that the nerves are hollow. The basic idea was that spirits could shoot through the great network of tubes to bring sensory information to the brain as well as to control the muscles. Galen wrote that he had actually seen the hollow central canal while studying the optic nerve. More than thirteen hundred years later, Vesalius stated that he could not verify this observation, even though he had carefully examined the optic nerves of humans, dogs, and various large animals. But again he was unable to come forth with an alter-

native theory to explain how the optic nerves may work. He could say only that existing theories seemed faulty.

In short, Vesalius was an empiricist who showed complete faith only in the observable. Issues of mind or soul, endlessly debated by philosophers and theologians,
and ideas pertaining to physiology, were beyond his immediate grasp. He seemed to
know that the old physiological theories were wrong, but could not formulate new
ones of his own with so few observable facts to go on.

The Reaction to the *Fabrica*

The *Fabrica* represented a dramatic break with Galenic anatomy, which had been
treated as dogma for over thirteen hundred years. It also raised serious questions
about the corresponding physiology. As Vesalius saw it, the time had come for the
ancient authorities to abdicate the throne of science to a new leadership—unbiased,
learned men who would make their own observations and take nothing for granted.

The success of the *Fabrica* and the strong message it conveyed can be measured in
many different ways. Records show that the book sold well in many countries. It and
the *Epitome* were also plagiarized and translated from Latin into modern languages,
such as German and French. In addition, Fallopio, Eustachio, Varolio, and many of
the other great anatomists of the Renaissance drew heavily from the *Fabrica* and
used it as a starting point for their own important explorations of the human body.

Still, these two books were not embraced in all quarters. In fact, the receptions
given to the *Fabrica* and *Epitome* became increasingly less favorable as one headed
north from Padua into cooler Central Europe. Many hard-line Galenists were not
particularly receptive to new concepts or ideas, even facts they could easily verify for
themselves.

At the head of the ranks of the conservative opposition was Jacobus Sylvius (Fig.
5.7), the Galenist who had taught Vesalius anatomy in Paris. In the preface to the
first edition of the *Fabrica*, Vesalius paid homage to his mentor, calling him "the
never-to-be-sufficiently-praised Jacobus Sylvius." Nevertheless, the underlying thesis of the *Fabrica* enraged Sylvius. He was not willing to concede that many things
Galen had to say about the human body were wrong; he was not about to permit
anyone, not even his most talented pupil, to get away with such heresy.[29]

In his revised translation of Galen's *On the Bones,* Sylvius rose to Galen's defense.
The year was 1549, and although he did not mention Vesalius by name, his intended
target was no secret when he remarked: "I honest reader, I urge you to pay no attention to a certain ridiculous madman, one utterly lacking in talent who curses and in-
veighs impiously against his teachers."[30]

His wrath continued unabated in another work, one translated as *A Refutation of
the Slanders of a Madman Against the Anatomy of Hippocrates and Galen.*[31] This
book, with its overly descriptive title, was published in Latin in 1551. In it, Sylvius
wrote: "Let no one give heed to that very ignorant and arrogant man who, through
his ignorance, ingratitude, impudence, and impiety denies everything his deranged
or feeble vision cannot locate."[32] He even appealed to the authorities to "punish severely, as he deserves, this monster born and bred in his own house, this worst example of ignorance, ingratitude, arrogance, and impiety, to suppress him so that he may
not poison the rest of Europe with his pestilent breath."[33] Although Sylvius' reaction

JACOBVS SYLVIVS .
D. der Arzneykunde geb 1478 gest 1555 .

FIGURE 5.7.

*Jacobus Sylvius,
the Paris teacher
who turned
against Vesalius
for contending
that Galenic
human anatomy
was faulty.*

may be viewed as resistance to change in the extreme, it was not all that unusual for the tumultuous Renaissance.

But on what grounds did Sylvius actually base his defense of Galen? He gave many reasons. First, he claimed that Vesalius made a mess of things because he misunderstood what Galen was saying. Second, he pointed to faulty editions and translations of Galen's original books. His third point was conjecture at its worst: He postulated that human anatomy had undergone significant changes since the fall of the Roman Empire. The message was clear: Galen's anatomy had been perfect for the second century, even if some of his facts may not hold in the present!

In his "Letter on the China Root," written in 1546, Vesalius responded to some of the disparaging remarks made by his fanatical teacher. Without the hatred and venomous language displayed by his mentor, his strategy was simply to distance himself from him.

When a revised edition of the *Fabrica* was published in 1555, he no longer mentioned Sylvius. Perhaps he felt that Sylvius had already been defeated. In addition, Vesalius was now a member of the imperial court, a man expected to act in a dignified way. Another consideration is that it simply might not have been in his nature to continue to argue with such an irrational person.

Although Vesalius did not receive accolades from his mentor in Paris, applause did come from other quarters. For example, in 1551 Leonhart Fuchs, a professor at Tübingen, published two volumes in which he praised Vesalius before borrowing extensively from his *Fabrica*.[34] To quote from the preface to his book:

Although Galen provided no little help to anatomy, yet it is clear that he composed his account from apes and dogs rather than from men, as anyone will understand if he uses his eyes as witnesses and is willing to use his hands in dissection. Among those who in our age wrote about the structure of man, Vesalius is the only one to write with exquisite care and to describe in a suitable manner, and had his commentaries not been published we should have been deprived of the true account of many parts of the human body. . . . He ought to be praised for his zeal in searching out the truth.[35]

After suffering the brutal assaults from the enraged Sylvius, these were words that must have comforted Vesalius' wounded soul.

Later Years

One year after the publication of the *Fabrica*, Vesalius relinquished his professorship and left Padua. He then traveled to Bologna to visit friends and assist in dissections. After turning down a faculty position at the University of Pisa, he joined the imperial medical service of Charles V, the Holy Roman Emperor whose European kingdom included the Netherlands and parts of Italy and Spain. By so doing, he continued a long family tradition.

Unfortunately, being a military surgeon and a physician to a sickly ruler effectively ended his career as a creative scientist. Vesalius must have known this was coming. After previously stating that a dedicated scientist should not wed, he now married Anne van Hamme and fathered a daughter. At the same time, he burned several unfinished manuscripts and vowed not to conduct new research.[36]

Vesalius traveled extensively during his years of service to the crown. Yet somehow, and in contradiction to his earlier statements about not writing anymore, he completed a revision of the *Fabrica*.[37] It was published in 1555, the year in which Jacobus Sylvius, the most vocal opponent of Vesalian anatomy, was laid to rest in the Cemetery of the Poor Scholars in Paris.

In 1556 Charles V abdicated his throne and gave his son, Philip II, the kingdom of Spain. This took Vesalius to Madrid, where he served the Netherlanders in Philip's court and occasionally the king or his family.[38] It was in Spain that Don Carlos, the eldest son of Philip II, suffered a head injury while falling down a flight of stairs in pursuit of a young servant girl. Philip II sent for Vesalius to join the physicians who were already at his unconscious son's bedside.

When they learned that the prince was still comatose, the townspeople got into the act. They brought the century-old remains of Diego d'Alcalá from his resting place at a Franciscan monastery to the royal palace and laid them next to the bed of Don Carlos. Since the friar had performed miraculous cures during his lifetime, they hoped that he might do the same after death.

When the indiscreet prince finally awakened, the commoners as well as the royal family attributed his recovery to the powers of the dead friar, not to anything the many physicians had done. Soon afterward Diego d'Alcalá was even canonized; he became San Diego.

Vesalius left Spain soon after this incredible encounter with superstition. He was also bothered by the jealous behavior shown toward him by many of the Spanish physicians, the overly conservative nature of the medical establishment, and the power of the Inquisition to bring intellectual pursuits to a terrifying halt. He might even have feigned illness to obtain permission to leave. All we know with certainty is that he next showed up in Venice in 1564. He was headed on a pilgrimage to Jerusalem, while the rest of his family headed back to Brussels.

After visiting the Holy Land, Vesalius boarded a ship back to Italy. A terrible storm ravaged the ill-fated vessel, killing many people on board. The ship finally reached a small island off the coast of what is now Turkey. Vesalius survived the harrowing ordeal at sea, but crawled ashore on the island of Zante in extremely poor health. He died shortly afterward, probably in April 1564. His remains were buried in the cemetery of the Church of Santa Maria delle Grazie, on Zante, far from home.

The Faces of Change

The sixteenth century was clearly a time of great upheaval. Above all, it was a period in which people began to question time-honored beliefs. In anatomy, Vesalius led the movement to describe the human brain more realistically than ever before, both in words and with pictures. His advancement was paralleled by that of Paracelsus, an argumentative man who fought to replace humoral theory with a new approach, *iatrochemistry* (meaning "medical chemistry"). The new orientation was based on folk medicine and on the "principles" of sulfur (combustibility), mercury (vapor), and salt (solidity).[39]

Another dimension of the revolution was a greater willingness to operate upon patients with severe head injuries.[40] In this domain, Jacopo Berengario da Carpi is one of many names that stand out. In a book published in 1518, he described six patients who survived severe brain damage.[41] The most interesting was Sir Paul, who had been hit on the top of his head by a halberd, a weapon with a long shaft topped by an ax and a spike. The wound broke his skull and sent pieces of bone deep into his brain. Berengario used a forceps to remove the bone fragments. He then repeatedly drained the wound before finally packing and closing it. Sir Paul slowly regained his health and lived a long and productive life, showing that it is in fact possible to survive a deep brain wound; his recovery, in turn, stimulated more surgeries of this kind.

Yet not all facets of the brain sciences advanced at an equal rate. Although neuroanatomy sailed ahead rapidly with Vesalius at the helm, and the willingness to intervene neurosurgically quickly changed for the better, the Galenic pharmacy was not about to yield so quickly, especially to the likes of Paracelsus. Furthermore, on a physiological level, even the most intellectually gifted scientists freely admitted that they had no real understanding about how the brain could mediate perception, cognition, or memory.

With this in mind, we shall now turn to two men who attempted to shed new light on the relationship between brain and behavior. The first is René Descartes, a French philosopher, and the second is Thomas Willis, an English physician. Although the methods and ideas of these two seventeenth-century figures contrast sharply with each other, both drew new attention to the working brain, the concept of mind, and the ever perplexing issue of function.

6

René Descartes
The Mind-Body Problem

What a piece of work is man! How noble in reason! how infinite in faculty! in form, in moving, how express and admirable! in action how like an angel! in apprehension how like a god! the beauty of the world! the paragon of animals!

—*William Shakespeare,* Hamlet, *c. 1601*

There is a little gland in the brain in which the soul exercises its functions in a more particular way than in the other parts.

—*René Descartes, 1649*

Seventeenth-Century Ideas

Scientists in the seventeenth century were the beneficiaries of the cultural changes ushered in by the Renaissance, which began in Italy early in the fourteenth century. They were now increasingly guided by the belief that natural laws can explain the workings of the material world. In particular, the new breed of scientists turned to the burgeoning fields of physics and mathematics to explain how things worked.

Western European science was entering an exciting new era. Many investigators were clearly influenced by William Harvey's 1628 publication describing the perpetual circulation of the blood. For these individuals, the blood would no longer be considered one of the four humors, constantly produced and then quickly dispersed and used. Also generating tremendous interest was Galileo's use of the telescope to study the rotation of the earth and the movements of the planets within the solar system.

These landmark events and the changing atmosphere stimulated others to take a fresh look at brain function. The new challenge facing the scientific and medical community was to understand how the nervous system permits us to perceive, move voluntarily, and remember. These functions took investigators into the domain of physiology, a considerably more mysterious realm than that of anatomy, the study of structure.

One seventeenth-century explorer who developed a system to explain how the nervous system may work was not a man of medicine, but a philosopher who turned to geometry, physics, and mechanical models for his ideas. His name was René Descartes, and he was a man of great intellect and unbridled spirit (Fig. 6.1). He strove to understand not only the body, but the physical world in all its glory. Driven by reason, he was firm in the belief that he could unlock even the most hidden secrets of Nature by thinking in terms of mechanics and motion, and by applying rock-solid logic.

Descartes asked how brain activities can account for behaviors as diverse as scratching an insect bite or memorizing a book. His labors resulted in an all-encompassing theory of brain physiology—one that goes far beyond the observable. But to emphasize the speculative nature of his theory of how the mind interacts with the body would be to lose sight of the even more important fact that Descartes focused new attention on the brain and forced people to think about how it may control both automatic and voluntary acts.

The Unconventional René Descartes

René Descartes was born in the small town of La Haye, taking his first breath of French country air in 1596.[1] His mother suffered from "a dry cough and pallid complexion" and died when the boy was just thirteen months old.[2] Fortunately, his father, Joachim, recognized the importance of a good education and had the finances to provide his son with one.

In 1606 René was sent off to school at La Flèche, about sixty miles from home. Here, talented Jesuit scholars educated the boy, who seemed to have a natural inclination for learning. In addition to studying literature, languages, and philosophy, he took classes in physics and mathematics. The analytic methods of these two related disciplines were of particular interest to him and remained so throughout his life.

Legend has it that the young student had already developed the habit of staying in bed through the morning hours. Supposedly this behavioral trait began early in life, when his health was fragile and he was compelled to rest for fear that he had inherited his mother's chest weakness. It is often said that the faculty at La Flèche admired

FIGURE 6.1.
*René Descartes
(1596–1650).*

his incredible intellect so much that they allowed him to continue this habit. This idea, however, may be an exaggeration, since the Jesuits prided themselves on producing students tough enough to be soldiers. What we do know with certainty is that Descartes considered himself to be in excellent health by the time he was twenty, and that he preferred to spend his mornings thinking and writing in bed, at least when not forced to do other things.

After graduation from La Flèche, Descartes studied law for two years and received a degree in civil and canon law. He then packed his bags and made his way to Paris. He lived for a while in or near the suburb of St. Germain, located on the banks of the River Seine. Here he visited the royal gardens, where he saw one of the man-made wonders of the day, the spectacular animated fountains (Fig. 6.2). These remarkable moving figures of Neptune, Diana, and other mythological entities would affect his thinking about the machinery of the body.

During this phase of his life, however, Descartes was not interested in physiology or even in anatomy. He was in crisis. He questioned the value of his erudite education, because it seemed to have no utility in the real world, and he had no idea what to do with his life. As a result, he became increasingly reclusive and depressed.

At age twenty-two he began to emerge from his anguish with the realization that he had to examine "the great book of nature" with his own two eyes.[3] Hoping that seeing the world around him would prove to be more satisfying than reading academic treatises, he decided to join a foreign army. Hence, in 1618 he enlisted in the service of Prince Maurice of Nassau, commander in chief of the Dutch armies fighting for independence from the Spaniards. The main attraction of the prince's unit was his school of military architecture and engineering, located in the Dutch city of Breda.

While in Holland, Descartes met Isaac Beekman, a gifted thinker who engaged him in physical and mathematical studies. Later he would praise Beekman for waking him from his intellectual slumber. Nevertheless, remaining with the Protestant forces in Breda for a prolonged length of time was not for him. A year after arriving, he took a slow and meandering overland journey through Poland and northern Germany, hoping to join the forces of his own religion, the Catholics, under Maximilian I in Bavaria.

FIGURE 6.2.

An engraving from 1615 showing the automated mythological figures in the royal gardens of St. Germain. (From S. de Caus, Les Raisons de forces mouvantes. *Frankfurt: J. Norton, 1615.)*

The traveling afforded Descartes ample time for reflection, especially during the cold of winter when the military campaigns associated with the Thirty Years' War ground to a halt. One of his first discoveries, and without question his most enduring, came in 1619. It was analytic geometry, a way of solving algebraic problems geometrically and geometric problems algebraically. He developed a system of reference lines, including the familiar ordinate and abscissa, that are now known as Cartesian coordinates. His other mathematical innovations included the use of numerical exponents to indicate the powers of numbers, and letters, such as x, for unknown quantities.

During this period, Descartes became increasingly obsessed with two beliefs. One was that the natural sciences must have the certainty of mathematics. The second was that knowledge must be based on simple ideas that cannot be doubted. Possibly inspired by a dream he had while confined to a small room far from home on a wintry night, he now believed his calling was to develop a universal science, one based on the clear and precise methods of geometry and physics—a science about which there could be no doubt.[4]

After visiting Italy, Switzerland, and several other European countries, Descartes meandered back to France as a civilian in 1625. In Paris, he continued to develop his analytic methods and questioned everything around him, taking nothing for granted. He concluded that spatial extension and motion are two properties basic to physical matter; these cannot be doubted. Even the human body, he thought, can be explained by these two properties.

The further development of these ideas, however, did not take place in Paris. The philosopher's growing reputation as a thinker led to too many intrusions on his privacy. In 1628 he packed up once again and slipped back to the quieter confines of Holland. He also hoped to find more tolerance for his new ideas in Holland, the more permissive Protestant country where his thoughts about science first began to take shape.

The Power of the Church

In 1633 Descartes penned a letter to Father Marin Mersenne in Paris to tell his scholarly friend, also a graduate of La Flèche, that he had just finished writing *Le Monde* (The World). In this work, he presented his mechanistic analysis of inanimate and living bodies. The first book of *Le Monde* dealt with physics and light, and the second book was his *Traité de l'homme* (Treatise of Man), which examined the human body. There was probably even more to *Le Monde* than these two books, but only they have survived.[5]

In 1634, just as Descartes was ready to head to the printer, a return letter came from Father Mersenne (Fig. 6.3). It informed him that Galileo had been arrested and was being brought before the Inquisition for presenting his progressive theories in public as facts. Looking privately through a telescope was not a problem. But Galileo's book, *Dialogue of the Two Chief World Systems,* which publicly supported Copernican theory in writing, had pushed the conservative guardians of the Church too far.

Because the first part of his own *Le Monde* endorsed the theories of the solar system held by Copernicus and Galileo, Descartes now chose to withhold publication

FIGURE 6.3.

Marin Mersenne (1588–1648), the priest who acted as a sounding board for new ideas and a disseminator of new information in the first half of the seventeenth century.

of his multivolume work. Even though he was living in a country where the Pope's authority was not officially recognized, he wished to avoid controversy and especially any charge of heresy from distant Rome. As a result, he made a decision to publish only his "safer" writings. These included his famous *Discours de la méthode* (Discourse on Method) and his essays on geometry, meteorology, and optics.

In his heart, Descartes probably hoped that these works would pave the way for the publication of *Le Monde*. Unfortunately, his strategy did not work; even these seemingly nonconfrontational writings irritated conservative forces inside and outside of Holland. Some of his enemies shouted that he was a promoter of atheism, while others contended that he was daring to question the authority of Aristotle.[6] Descartes, who only wished to be left alone, now feared being thrown out of Holland or, even worse, being murdered by some self-appointed inquisitors.

The threatening atmosphere created by the conservatives affected Descartes so much that the public had to wait for years before being able to read his best writings on the mind and body. Descartes sent *Les Passions de l'âme* (The Passions of the Soul) to press in 1649, just months before he died.[7] Even more disconcerting, *Traité de l'homme*, which was written in 1633 as a part of *Le Monde*, did not make its debut until he had been dead for twelve years.[8]

The fears that Descartes had about formally publishing some of his ideas, however, did not mean that all leading intellectuals were unaware of his thinking when he lived. The lines of communication were kept open through Father Mersenne. He was one of a special group of learned men who served the intellectual elite as critics, sounding boards, and disseminators of scientific information before the advent of scientific journals. Although a cleric, Father Mersenne kept an open mind and did not see Descartes' theories as particularly threatening to the Church.

Descartes wound up spending almost twenty years in Holland, but never at the same location very long. To keep a step ahead of his tormentors and to protect his privacy, he resided at some twenty-four different addresses in at least thirteen different towns. Only a few close and respectful friends knew where he could be found from one month to the next.

Although he was not the sort of man to settle down in a small house with a wife, children, and a garden, Descartes did father an illegitimate daughter named Francine. Little is known about his relationship with Francine's mother, a Protestant servant who lived with him for a few years. From his letters, we know only that he felt the most profound loss of his life when Francine contracted scarlet fever and died at the age of five.

Queen Christina and Sweden

The final chapter in Descartes' life opened when he allowed himself to be lured to Stockholm in 1649.[9] He was drawn north by an offer to teach philosophy to the promising young Swedish queen, Christina, to whom he had sent his newest work, *The Passions of the Soul*. He was impressed by the queen's desire to understand philosophy and her ambition to turn Stockholm into a major center of learning. He probably also thought the move would put some distance between him and his enemies in Holland.

The winter of 1649–50 was one of the most frigid in Scandinavia's recorded history. The climate, coupled with the queen's desire to discuss philosophy at five in the morning, proved too much for the frail philosopher. Descartes contracted pneumonia and died at age fifty-four, before the ice could even melt.

The remains of the Catholic philosopher were first placed in a Swedish cemetery for unbaptized children. A few years later they were disinterred for delivery to French soil, but not without incident. Someone severed a finger as a souvenir, and the metal casket for shipping was so small that the head had to be detached from the torso. The skull was then lost for 172 years. When it was finally recovered, it was presented to the Musée de l'Homme (Museum of Man) in Paris, where it was put in a cabinet for public display. The torso, after being moved around, was eventually buried in the churchyard of St. Germain des Prés, a fair distance away. In an ironic way, as we shall now see, the disconnection of Descartes' head and torso was reflective of, and consistent with, his own belief that mind and body must be treated as separate entities.

The Nervous System as a Machine

René Descartes was greatly impressed by hydraulic machines that animated the wonderful mythical figures in the royal gardens of St. Germain, as previously described. He was also stimulated by the mechanical dolls in the windows of fancy shops and the complicated clocks with moving figures in the bigger cities. It seemed perfectly natural for one witnessing the incredible mechanization of the physical world to think that natural bodies may also possess clocklike mechanisms to propel them into motion.

To Descartes, Galileo understood the mechanics of motion when he described the workings of the universe, as well as when he compared the bones and joints of the body to a system of levers. William Harvey also understood the mechanics of motion when he explained the pumping action of the heart and the flowing of blood through the arteries, veins, and capillaries. Truly inspired by the mechanical revolution, Descartes now strove to present a theory of brain function based on similar principles.

The French philosopher was not naive when it came to biology and medicine. He had an interest in anatomy and physiology and, soon after arriving in Holland, made trips to slaughterhouses to obtain animal heads and organs for dissection. He might even have performed experiments on live animals.[10] His studies usually began late in the afternoon and so engrossed his attention that he often worked well into the night. Once, when a visitor asked to see his library, he supposedly pointed to the sheep parts he had dissected and remarked, "These are my books."[11]

In the domain of physiology, Descartes, like many others of this era, still believed in animal spirits. For the French philosopher, however, these spirits were not manufactured in the *rete mirabile,* as Galen once believed. Instead, a tiny gland associated with the brain was singled out for their production. The pineal gland received its name because it looks like a Mediterranean pine nut.[12] In ancient Greece, it was called the *soma konoeides.* During the seventeenth century, however, learned men preferred the term *conarium,* the Latin form of the word.

Today, we know that the pineal gland is innervated by the sympathetic nervous system and that it can function like a photoreceptor, especially in season-breeding reptiles, amphibians, and birds. In reduced light—namely, during the winter months, when the weather is poor and food is scarce—it releases more of a hormone called melatonin, which indirectly affects the gonads, inhibiting reproduction.[13]

Of course, photoreceptors, melatonin, and biological clocks were not even remote thoughts in the minds of scientists in the seventeenth century. Instead, these men were concerned with the nature of life itself and with a related issue—how the human soul, or mind, could activate the physical machinery of the body. Descartes was no exception.

He knew that the pineal gland could be found close to the ventricles. In fact, he erroneously drew it inside the ventricles rather than above them, where it really belongs (Fig. 6.4). He hypothesized that fine capillaries tethering the pineal gland permit only the tiniest particles to filter into it from the blood. These particles, he believed, would be converted into pure animal spirits for release into the ventricles.

Descartes further assumed that the nerves from the sensory organs to the brain contain extremely delicate filaments inside their lengthy canals. He thought that,

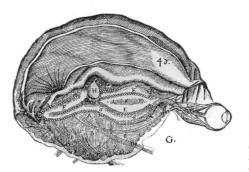

FIGURE 6.4.

Descartes' erroneous location of the pineal gland (H) in the hollow ventricles (E) of the brain. (From R. Descartes, De homine. . . . *Leyden: F. Moyardeim and P. Lefton, 1662.)*

when pulled at the far end, these threads would open small valves in the walls of the ventricles, allowing animal spirits to enter the nerves and then flow to the muscles.

Another important aspect of his theory is the notion that the pineal gland can rock or tilt to direct the flow of animal spirits toward specific openings made in the ventricle walls (Fig. 6.5). This tilting would result partly from the vortex or draft created by the spirits flowing rapidly into the nerves. Pineal movements could also be caused by spirits leaving from different locations on the gland's surface. He believed it would take very little to make the pineal move, because it is suspended only by fine capillaries.

Although Descartes did not use the term *reflex,* he was clearly intrigued by the idea that automatic actions—such things as pulling a limb away from a flame or unconsciously swatting an insect—could easily be explained by these purely mechanical actions. But Descartes did not look upon this sort of automatic, stimulus-response activity as just one part of animal behavior. For the brutes, he maintained, these mindless actions are everything. Animals may be warm and cuddly or menacing and fearsome, but underneath it all, each is little more than a stimulus-response machine.

Witness what Descartes had to say about the brutes resembling the automated fountains in the royal gardens:

> Similarly, you may have observed in the grottoes and fountains in the gardens of our kings that the force that makes the water leap from its source is able of itself to move diverse machines and even to make them play certain instruments or pronounce certain words according to the various arrangements of the tubes through which the water is conducted.
>
> And truly one may very well compare the nerves of the machine which I am describing to the tubes of the mechanisms of these fountains, its muscles and tendons to diverse other engines and springs which serve to move these mechanisms, its animal spirits to the water which drives them, of which the heart is the source and the brain's cavities the water main.[14]

With such machinery firmly in hand, it then became fairly easy for him to explain sleep and wakefulness. He postulated that sleep occurs when the brain becomes relatively devoid of spirits. This makes the brain somewhat limp, slackening the at-

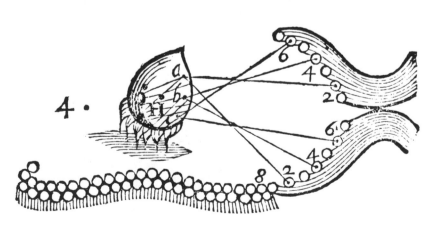

tached nerve fibers and reducing their ability to respond to external stimuli, such as a tug on the arm. In contrast, when an abundance of animal spirits entering the brain expands it and makes the nerves tighter, wakefulness occurs and there is greater sensitivity to external stimulation. Indeed, the basic features of all animal life—eating, breathing, walking, reproducing, responding to stimuli, and the like—can now be reduced to simple mechanical actions based on established laws of physics.

We Are More than Animals

Descartes presented the aforementioned theory to account for all animal behavior, but only one aspect of human behavior—involuntary or automatic activity. Examples of the latter include quickly dropping a hot object, squinting in bright light, and digesting food. In contrast, he asserted that voluntary, thoughtful, and conscious acts are distinctly human traits because they demand a mind.

As possessors of minds, humans are truly unique. Only humans can think and, by thinking, only humans can know with certainty that they must exist. This logic is summarized in what may be the most famous sentence in Western philosophy. As stated in French in Descartes' *Discourse on Method:* "*Je pense, donc je suis*" (in the Latin version, "*Cogito, ergo sum*"), "I think, therefore I am."[15]

The idea that beasts can reason, Descartes repeatedly emphasized, is ridiculous. In a letter to Henry More, Descartes explained: "The greatest of all the prejudices we have retained from infancy is that of believing that brutes think."[16] If they could think, he argued, they would have language. This, of course, is something the beasts simply do not have.

Thus, animals can react in a mechanical fashion, like a pump organ or a mechanical fountain, but unlike us, they lack abstract thought, do not experience real emotional states, such as love and remorse, and are unable to think or respond voluntarily. The beasts are machines, complex beyond our imagination, but stimulus-response machines nonetheless.

In contrast, Descartes saw humans as more than just flesh and blood. Each one of us is blessed with a rational soul that can override the reflexes that drive the mindless brutes. We alone can think and make rational choices after considering the alternatives.

A Solution for the Mind-Body Problem

Descartes now faced the very thorny issue of explaining how the body and soul could affect one another. The issue of how something material (the body) could interact with something immaterial (the soul or mind) was an enigma to philosophers, theologians, and naturalists. Although the so-called mind-body problem seemed to defy scientific solution, he had an explanation.

Descartes lived during the reigns of the French kings Louis XIII and Louis XIV, absolute rulers who exercised complete control over their empires. He believed in the divine rights of kings and the need for centralized authority. Reasoning by analogy, he also believed that the human soul must have a well-defined operational center in the brain—one that could control the movements of the animal spirits through the nervous system. (Actually, he did not really think that the human soul can be confined to any one bodily location, but it was simpler to envision it acting through a centralized structure.)

He concluded that the pineal gland was the most likely candidate to take on this important controlling function in humans. It is likely through this gland, he reasoned, that the rational soul, with its innate ideas of infinity, unity, and perfection, influences its temporary abode, the machinery of the body. Moreover, it could be through the pineal gland that the soul is made aware of the flow patterns of the animal spirits, which in turn allow it to perceive, imagine, and generate new ideas.

Why the Pineal Gland?

The pineal gland was selected by Descartes to be the *siège de l'âme* (seat of the soul) for many reasons.[17] Although he did not come out and say so directly, one was probably parsimony or simplicity. He already had the pineal gland distilling pure animal spirits from blood. This being the case, he had a central mechanism in place; there was no need for additional parts.

In his writings, however, Descartes placed more emphasis on the fact that the pineal gland is a singular structure. There is only one pineal gland, and a single brain structure is certainly best suited for melding the impressions from the two eyes or from the two ears into a single, unified conscious experience.

To modern brain scientists, the first thought that may now come to mind is this: Why did Descartes not seize on the pituitary gland rather than the pineal gland as this "seat of the soul"? After all, the pituitary is singular, and it was certainly better known at the time than the pineal gland. In fact, Descartes did consider the pituitary, now recognized as the true master gland of the body. But in the end he rejected it in favor of the pineal gland for several reasons. One was that he felt the seat of the soul must be in the brain; physicians observing brain-damaged patients and scientists studying animals left little doubt about the brain being responsible for sensation and movement. The pituitary, buried under the brain, failed to meet this basic criterion; it was not considered a part of the brain in the seventeenth century.

A second reason was that the pituitary already had an assigned function about which there seemed to be little disagreement. From ancient times, people thought it acted as a drain or pump for removing waste (phlegm) from the brain. As Thomas Willis put it later in the same century, it was the brain's "sink."[18] In contrast, the tiny pineal gland was a veritable orphan at the time, a structure not firmly tied to any function, one awaiting an assignment.

The pineal also had the advantage on two other fronts: location and mobility. Being located near the center of the brain was ideal. It not only made ruling easier, but also gave it, like the centralized royal enclaves, an added degree of protection. Moreover, it was positioned near the ventricles, poised like a fountaineer well situated to control the pipes and valves of the complex machinery.

As for mobility, Descartes believed the pineal to be suspended so perfectly that little force would be necessary to make it move. This is extremely important because the rational soul, being immaterial, would need all the help it could get to move even the smallest physical object a little bit. Moving the pituitary would be much harder to envision than moving this tiny tethered gland.

With such reasoning, Descartes concluded that the rational soul—by sensing movement and causing the pineal gland to release and direct animal spirits—could in fact affect the machinery of the body. This interaction of mind and body would allow an archer to shoot an arrow at a target or a butcher to point to some meat on a hook. In contrast, all automatic and reflexive actions could be accounted for without bringing the rational soul into the picture. As Descartes envisioned it, a body without a soul is merely an automaton, whereas a soul without a body should still have innate ideas, if not the richness of sensory experience.

A Frequent Misinterpretation

Before we turn to how Descartes' solution to the mind-body problem was received, one point about his choice of the pineal gland as the seat of the soul should be clarified. Some authors have stated that Descartes also selected the pineal gland because he believed this tiny gland is unique to humans, the only organism with a soul. This assertion, however, is a misinterpretation of the philosopher's thoughts.[19]

Descartes must have known that animals have pineal glands. The gland was described as a basic part of animal anatomy in books written by respected authorities familiar to the philosopher. One such person was Galen. In the year 177 as we have seen, Galen described how to find the pineal gland when systematically dissecting the brain of an ox (Chapter 4). Similarly, the pineal gland was described in sheep by Renaissance anatomist Andreas Vesalius (see Chapter 5).

There exists, however, even more convincing evidence to show that Descartes must have seen with his own two eyes the pineal in freshly killed animals. Specifically, he complained that this gland is much easier to find in animals than in humans, a statement that still holds true today. This is because the pineal tends to calcify (early anatomists preferred "turns to sand") relatively early in life in humans. In a letter penned to Father Mersenne on April 1, 1640, Descartes lamented:

> Three years ago at Leyden, when I wanted to see it [the pineal] in a woman who was being autopsied, I found it impossible to recognize it, even though I looked very thoroughly, and knew well where it should be, being accustomed to find it without difficulty in freshly killed animals.[20]

Earlier the same year Descartes sent a letter to the physician Lazare Meyssonnier. He presented his thoughts on memory and the pineal gland, and postulated that the pineal may be less mobile in people whose minds are "sluggish." He then went on to discuss very bright minds and here made another direct reference to the gland in animals:

> As for very good and subtle minds, I think their glands must be free from outside influence and easy to move, just as we observe that the gland is smaller in man than it is in animals, unlike the other parts of the brain.[21]

Unfortunately, Descartes was less than clear about animals having pineal glands in his major treatises. But as his letters show, he never meant to imply that humans are unique because they alone possess a pineal gland. His point was simply that only the human pineal gland can be associated with a special inhabitant—an inhabitant called the rational soul, a spiritual entity not found in any other living creature.

Reaction to the Theory

A number of open-minded physicians embraced the concept of the body as a machine. One early supporter was Louis de la Forge,[22] and another, at least on many points, was Henricus Regius, a professor of medicine at the University of Utrecht.[23] Regius was especially intrigued by the idea that all animal behavior and some human behavior could be reduced to simple cause-and-effect actions and treated scientifically.

The more humanistically oriented philosophers were somewhat slower than Regius to accept the idea that the body might work like a complex machine, yet the idea gradually found its share of supporters. The French took the lead, but national boundaries meant little as this new, mechanistic orientation toward the physical body took hold and spread across Europe.[24]

Nevertheless, Descartes' solution to the mind-body problem disturbed more than a few thinkers.[25] The interaction of a nonmaterial substance with a physical machine was seen as hopelessly contradictory by many of his scientific and medical critics. Even some of his admirers had trouble warming up to his interactionism.

Books have been written about the criticisms of the clergy, the battles fought at universities, and even the trials that took place in municipalities between the old guard and the revolutionary Cartesians, as they came to be known.[26] To appreciate more fully how some members of the medical and scientific communities felt about the new model, let us look in greater detail at two central issues. The first will be the reaction to Descartes' selection of the pineal gland as the seat of the soul, and the second will be the separation of humans from the brutes.

The Logical Seat of the Soul?

To most anatomists and physiologists, the elevation of the pineal gland to its lofty position as seat of the soul made little sense. Among the first men of medicine to criticize this selection were Christophe Villiers and Lazare Meysonnier. Through Father Mersenne, they explained to Descartes that his choice of the pineal gland cannot possibly be correct.[27] Postmortems had been performed on individuals who exhibited "stones," "chalk," and other deformities of this structure. Yet the minds of these men and women often seemed perfectly fine before they died.

These French physicians also thought the pineal gland was too small for the noble function assigned to it. In fact, Villiers went on to suggest that the cerebellum was the more logical candidate for the job. Others interjected that the true seat of the rational soul must be far bigger in humans than in the brutes. Indeed, pineal size differences across species, with humans on the short end, had been noted by Andreas Vesalius. In his celebrated *Fabrica* of 1543, he decried: "I wish that a sheep's brain be at hand, since it shows the gland . . . more distinctly than does the human head."[28] Descartes could hardly disagree after witnessing human autopsies in Leiden, but reasoned that a smaller pineal made more sense for humans, since it would be easier for the soul to move a small object than a large one. This logic, however, was not particularly convincing to many of the intellectuals who were pondering the issue.

Thomas Bartholin, a Danish anatomist, not only brought up the issue of size but also mentioned a host of other reasons why the pineal theory should be discarded.[29]

The fanciful idea that the pineal is a mobile structure capable of dancing like a balloon above a flame was one of several issues he singled out for severe criticism. This assertion, he said, was complete nonsense, as was the notion of ventricular pores and valves for the dissemination of the spirits. These structures had never been seen; even skilled anatomists well equipped for the search could not find them. Another problem for the theory, remarked Bartholin, was the discovery of cerebrospinal fluid in the ventricles. Any fluid, he noted, even a thin, watery one, would impair the movements demanded of the gland.

To some of his more scholarly critics, Descartes' pineal theory also had an all-too-familiar and archaic ring. The notion of pineal gland movements regulating the flow of animal spirits probably originated with Herophilus, the Alexandrian anatomist who lived around 300 B.C. (Chapter 3). Several hundred years later, Galen would write in his own disparaging way that the idea that the pineal could rise and fall to regulate the flow of spirits within the ventricles is absurd:

> The notion that the pineal body is what regulates the passage of the pneuma is the opinion of those who are ignorant. . . . Since this gland . . . is by no means part of the encephalon and is attached not to the inside but to the outside of the ventricle, how could it, having no motion of its own, have so great an effect on the canal? . . . Why need I mention how ignorant and stupid these opinions are.[30]

Nevertheless, some later scientists—Renaissance physiologist Jean Fernel among them—continued to keep the ancient idea from falling into complete oblivion. True, Descartes modified and further developed the model, but to some scholars his efforts were no more than just another variation on a theme that should have been discarded centuries earlier.

Hence, the pineal theory proposed by Descartes was seen as hopelessly naive, unreal, and even unoriginal by many intellectuals in the seventeenth century, including Nicolaus Steno. In 1665 he gave a lecture in Paris in the house of Monsieur Thévenot in which he explained that Descartes' theory was physiologically too speculative and made no sense.[31] Steno, a Danish anatomist, had more to say about Descartes in later years. In 1680 he wrote: "Descartes' method is praiseworthy, but blameworthy is a philosophy where the author forgets his own method and takes that for granted which he has not yet proven by reason."[32]

Our Place in Nature

A more overlooked facet of Descartes' legacy is that his theory forced people to wonder whether the differences between man and beast are as great or absolute as the philosopher made them out to be.[33] Many individuals who owned dogs had a difficult time with the idea that man's best friend is simply an unconscious automaton. After all, is a faithful dog just a reflexive creature? Is a bothersome, irritating neighbor really more worthy of charity than a cuddly, caring family pet?

Given his mechanistic views about animals, it may seem hard to believe that Descartes could warm up to an animal, but he did have a pet dog of his own.[34] The dog's name was Monsieur Grat (Mister Scratch), and Descartes took him on walks and treated him with affection. He even sent Monsieur Grat along with a little female dog to his friend the Abbé Picot, hoping to present the cleric with some offspring of the breed.

Still, Descartes was not willing to maintain that Monsieur Grat could think, reason abstractly, or have true conscious awareness. As he put it in a letter to the Marquess of Newcastle in 1646: "If they [animals] thought as we do, they would have an immortal soul like us," and if one type of animal has a rational soul, all must. He then added: "Many such as oysters and sponges are too imperfect for this to be credible."[35]

If pet owners had trouble with the mechanistic philosophy presented by Descartes, they must have had a considerably more difficult time with the views adopted by the extremists who followed in his footsteps. The most disturbing contention came from a man of the cloth, Father Malebranche. This "unfeeling" cleric wrote that animals "eat without pleasure, cry without pain, grow without knowing it; they desire nothing, fear nothing, know nothing"[36]

One day, while walking down a Paris street with some of his friends, including the famed writer of animal fables La Fontaine, Father Malebranche came upon a pregnant dog. The priest, who could ooze sweetness at times, drew attention to the poor animal and then kicked her squarely in the stomach. The men walking with him were horrified by such an act of cruelty, but Father Malebranche felt no qualms or remorse. Instead, he turned to his companions and chastised them for wasting their sentiments on a mere "machine" when there were human souls to be saved. To some, Father Malebranche blessed cruel acts and unmentionable experiments on cats and dogs with his behavior and the words that came from his mouth.

The sharp separation between "them" and "us" did not go unopposed. Formidable opposition came from two very different camps, both of which brought man and beast much closer together: the extreme materialists and the cognitivists.

The extreme materialists played down or even eliminated the role played by the rational soul in humans. They found their ultimate champion in Jules Offray de La Mettrie, a physician who lived a century after Descartes.[37] In 1748 La Mettrie went so far as to write that human beings, like their animal cousins, are nothing more than soulless automata. According to La Mettrie, humans are superior to animals because of their better-developed brains, not because of some spiritual substance called the rational soul. The so-called father of materialism gave his best-known book on the subject an audacious title. He called it *L'Homme-machine,* meaning "The Man-Machine."

In striking contrast to the extreme materialists, the cognitivists argued that higher animals really can perceive, think, and remember. To these individuals, the materialists had it backward. Humans are not more like the beasts; rather, the beasts are more like us. Because animals have sensory organs, God must have wanted them to perceive and experience the world to the fullest extent possible, just as we do.

Pierre Gassendi, the influential seventeenth-century French philosopher, was a leading cognitivist.[38] He believed that higher animals can reason, just not as well as humans. As he saw it, the difference between our intelligence and theirs is simply one of degree. For the common people—namely, those men and women who never came close to accepting the idea of the beast machine—the idea that a pet can have something like a mind was a welcome, attractive idea.

When poets and men of letters entered the fray, it was usually to glorify the virtues of animals.[39] For La Fontaine, beasts possess both feelings and intelligence and are worthy of champions such as himself to defend their cause. His approach, however, was not a new theme in French literature. Michel de Montaigne was glorifying animals even before René Descartes was born.

The "them"-versus-"us" controversy climaxed long after the Cartesian pineal gland theory had faded into oblivion. During the middle of the nineteenth century, humans and animals were tied to each other as never before by the evolutionary theories of the English naturalist Charles Darwin. Most scientists followed Darwin in acknowledging a gradual steplike progression from animals to humans, much as Gassendi had envisioned. The majority also agreed that issues about the immortal human soul and any afterlife really should be left for the theologians.

But the seventeenth century was not yet over, and the immediate need was for someone to provide a more realistic model of brain function than one tied to the tiny pineal gland and spirits flowing from the ventricles into the hollow nerves. The call was for a man of science to step forth to associate higher mental functions with newer brain structures, and to tie more basic functions, such as breathing and involuntary acts, to circuitry lower in the system.

In the next chapter we shall meet Thomas Willis, a practicing English physician and enthusiastic scientist, who will do just this. Descartes knew that observation and experimentation were important, but he was so deeply committed to the separation of mind and body that he let his philosophy dominate his science. With Willis, theory will still play a big role, but his ideas about brain function will be based more firmly on human case studies, animal research, and a newer, more detailed neuroanatomy.

7

Thomas Willis
The Functional Organization of the Brain

The Cerebel [cerebellum] is a peculiar Fountain of animal Spirits designed for some works, wholly distinct from the Brain. Within the Brain, Imagination, Memory, Discourse, and other more superior Acts of the animal function are performed. . . . But the office of the Cerebel . . . seems to be for the animal Spirits to supply some nerves; by which involuntary actions . . . are performed.

—*Thomas Willis, 1664*

The Medical Physiologist

Thomas Willis was born in England only twenty-five years after the birth of René Descartes. Although both men were products of the seventeenth century and loved to speculate, their approaches to the study of the brain could not have been more different. Descartes can be thought of as the last of the old guard whose brain physiology was still strongly guided by humoral theory and ancient philosophical notions. In contrast, Willis can be considered the first of the new breed of medical physiologists, men more influenced by clinical and laboratory insights than by Greco-Roman theories held sacred by their forefathers.

Willis did not set forth to figure out how the spiritual mind could interact with the physical body, as did Descartes. Nor did he hope to understand or "dissect" the immortal human soul. Nevertheless, he agreed with "the great Cartes" that considerably more had to be learned about the brain. To quote from the opening of one of his books:

> Among the various parts of the animated Body, which are subject to Anatomical disquisition, none is presumed to be easier or better known than the Brain, yet in the meantime, there is none less or more imperfectly understood.[1]

Like Descartes, Willis accepted the idea that there are specialized controlling structures in the brain. But unlike Descartes, who singled out the pineal gland in this context, he emphasized that the brain can best be understood by envisioning different levels of function. He thought that those structures located higher in the brain must do those things that are unique to advanced organisms, whereas lower structures, such as the cerebellum, must be responsible for more elementary functions, those that vary little across vertebrates.

Exactly who was Thomas Willis, the physician who developed these ideas? And on what sort of evidence was his thinking based? For answers to these and related questions, we must begin in England during the first half of the seventeenth century.

The "Orthodox, Pious, and Charitable Physician"

Thomas Willis (Fig. 7.1) was born in 1621 in the English town of Great Bedwyn, not far from Oxford.[2] His biographers tell us that he was a charitable, church-going lad who grew into a charitable, church-going man. As a schoolboy, he acquired the habit of giving his own food to the poor. His father worried that the saintly boy would go hungry; consequently, he had to make sure that his son ate amply at home before heading off to school. Later in life, Willis gave free medical treatment to the sick who could not pay for his services. He also donated fees earned on Sundays to the poor, supported his less prosperous siblings, and gave generously to the Church of England.

Although Willis aspired to become a physician, he had surprisingly little formal medical schooling.[3] In fact, his medical education at Oxford probably lasted less than six months. The medical curriculum that he began in 1643 was among the many casualties of the English Civil War (1642–1651), which had broken out a year earlier and pitted Oliver Cromwell and his Parliamentarians against Charles I and his loyalist forces. Among the educational changes demanded by the Parliamentarians were an overhaul of the previously stodgy university curriculum and greater acceptance of new ideas.

Oxford was turned into a garrison town when Charles I moved there, and Willis, like many other students, joined the Royalist volunteers. These students trained on the university grounds (often in archery) and spent long nights on guard duty as academic endeavors effectively ground to a halt. In 1646, when Willis received his bachelor of medicine degree, his formal "medical" education included little more than a

FIGURE 7.1.
Thomas Willis
(1621–1675).

master of arts degree, a good working knowledge of Latin, and an above-average mastery of Aristotle.

After obtaining his medical license, Willis worked on building his private practice in Oxford. He had an unusually difficult start, partly because he was not a man of means. He was unable to dress as nicely as most other physicians and even had to share a horse with another man. His physical appearance did not compensate for his shabby clothes. He was of average height, had dark red hair, and was described by those who knew him as being less than handsome. His tendency to stammer was a further drawback. In time, he would overcome these impediments and become one of the most sought-after physicians in all of England.

Following in the footsteps of Paracelsus, the iatrochemist who lived over a century earlier, Willis based his medicine on the active principles of mercury, sulfur, and salt, to which he added the inert substances water and earth. Like his Swiss predecessor, he hoped that a new medical chemistry would replace the Greco-Roman humoral pharmacy, with its heavy reliance on plants and its philosophy of reestablishing a balance by providing opposites (bleeding for a flushed appearance, and so forth). He expressed this desire very clearly in 1659, when he published his first books, which dealt with fermentation, fevers, and urine.[4] Because he embraced new chemical principles and had little use for the four humors of times past, some historians have dubbed Willis the leading English "Paracelsian."[5]

Yet the fame associated with Thomas Willis today does not stem from his acumen at diagnosis, his bedside manner, or the fact that he made more money than any other English physician. Nor is he well remembered for his chemical theories of medicine. Without question, it was his new way of looking at the brain and behavior that is his most significant contribution.

The Allure of the Brain

The nerves and the brain were not foremost on Willis' mind when he received his bachelor of medicine degree. The nervous system attracted his interest only after he began to associate with a circle of natural philosophers at Oxford, some of whom had strong interests in neuroanatomy and neuropathology. These remarkably talented men called themselves the *Virtuosi*.[6] They strove to surpass all others in their search for new knowledge and hoped to establish a new science, even if this meant overthrowing the time-honored Aristotelian system.

A second factor that drew Willis to the brain was a series of epidemics that swept through Oxford during the 1650s.[7] Some outbreaks involved meningitis and sleeping sickness, two disorders that affect the nervous system. When autopsies were performed on the brains of some of the dead, Willis was quick to recognize that existing descriptions of the human brain were still hopelessly inadequate.

Willis received the title Doctor of Medicine in 1660, the year of the restoration of the English monarchy. By this time he had been married for three years to Mary Fell, the daughter of the former dean of Christ Church College. The recommendation for the higher degree came from Gilbert Sheldon, the Archbishop of Canterbury. Along with his doctorate came a prestigious elected appointment as Sedleian Professor of Natural Philosophy at Christ Church, Oxford.

The Sedleian Professorship was a high honor, but it was not a true medical appointment. It required the holder to lecture at least twice each week in the "Aristotelian tradition." Willis, however, interpreted his job description with considerably more latitude than did his predecessors. He used his Oxford professorship to study the senses, the nerves, and "affections of the soul" in his own special way.

His research in preparation for lecturing required extensive human autopsy material, mainly from the gallows, and a wide variety of animals for dissection. Following in the footsteps of Aristotle and Galen, Willis managed to dissect pigs, horses, goats, sheep, foxes, dogs, cats, hares, and monkeys, as well as fish and fowl. Lobsters, oysters, lowly earthworms, and even silkworms also fell victim to his scalpel. Few other anatomists at the time even cared about invertebrates, unless they were the succulent shellfish that occasionally graced their dinner plates.

Indeed, Oxford University had changed dramatically between the time Willis had begun as a student and the start of his Sedleian appointment. The shackles that had bound it to its medieval ways had been broken, and thanks to men like him, the once stodgy school was now on its way toward becoming a leading center for new ideas. The precipitating event was the fall of the royalist garrison in 1646 and the subsequent firing of many of the school's ultraconservative faculty members—men who had accepted the authority of Aristotle as supreme. In their places, the Parliamentarians appointed scholars interested in fresh new ideas. Even after the restoration of the English monarchy under Charles II in 1660, most of the long-overdue changes at Oxford remained in place.

On the world scene, the knots that bound neurology to many ancient notions continued to loosen. Inquisitive scientists now questioned older ideas, such as whether brain size is affected by phases of the moon and whether phlegm is really a waste product of the organ of mind. In addition, the ancient theory that associated perception, cognition, and memory with the different ventricles continued to lose support.

The Books

Willis' marvelous descriptions and insights about the nervous system can be found in two books, the best known of which is *Cerebri anatome* (Cerebral Anatomy).[8] Written in Latin while he was living in Oxford, it made its debut in 1664 and focused on the brain, the spinal cord, and the nerves. From the moment it came off the press, it was recognized as a work on the nervous system unlike any that had come before it. It ran through nine separate printings or editions in its first twenty years. Although its title would suggest that it was solely about anatomy, Willis also concerned himself with the physiology of the nerves and brain throughout his text.

Willis was clearly influenced by William Harvey, who had been trained in Italy and achieved great fame by showing that blood, one of the original four humors, continuously and rapidly circulates through the veins and arteries of the body. Harvey's book, *De motu cordis,* was instrumental in bringing about the downfall of humoral theory. Previously blood was thought to move sluggishly through a system of canals, being used up by the organs and being replaced by new blood. Now the heart was looked upon as nothing more than a mechanical pump.

Harvey resided in Oxford between 1642 and 1646 and served as a physician to King Charles I. While in residence, he took the time to demonstrate the circulation of the blood through the arteries and veins and to explain the importance of the heart's valves to the faculty. Whether Willis went to some of these demonstrations and lectures, or even knew Harvey personally, is uncertain. What is important is that Willis heard the famous physician's clarion call for a fresh new physiology, and in *Cerebri anatome* he responded to it.

In his epilogue to *Cerebri anatome*, Willis promised to write a second book on the "body-soul" to accompany his anatomy text. True to his word, *De anima brutorum . . .* (Two Discourses on the Soul of Brutes which is the Vital and Sensitive Soul of Man) appeared in 1672, and was more decidedly clinically oriented.[9] *De anima brutorum* came out just a few years after Willis moved to London.

His presence in England's largest city was requested by the Archbishop of Canterbury, who lived there and was both friend and patient. Indeed, it was to Gilbert Sheldon, the Archbishop, that Willis dedicated his books on the nervous system. The Archbishop's call for help stemmed from the fact that many physicians had fled London during the Great Plague of 1665, while a large number of those who remained died from it. Willis responded to the need for medical help in 1667. Nevertheless, he kept his Sedleian Professorship and hence retained some ties with Oxford after he settled in London, now as much a center for new science as Oxford.

Willis died from pneumonia only a few years after *De anima brutorum* was published and before it could be translated from Latin into English. He passed away at age fifty-four and was buried in Westminster Abbey in 1675. His first wife, Mary, had died in 1670, and he was survived by Elizabeth, his second wife of only three years.

The Virtuosi: A Team Effort

Cerebri anatome resulted from a team effort, even though only one name can be found on its cover page. Within the text, Willis gratefully acknowledged the roles played by certain other *Virtuosi*, including Richard Lower, the talented experimentalist who performed the first blood transfusion. More than anyone else, Lower assisted Willis (some say instructed him) with his anatomy and physiology. The two men were inseparable, and Lower even accompanied Willis and his family when they moved to London.

Willis also expressed a debt of gratitude to Robert Boyle, the renowned physicist and chemist. Today Boyle is best remembered for his laws of gases and is often hailed as a founder of modern chemistry. Another contributor was Thomas Millington, the physician who would succeed Willis as Sedleian Professor at Oxford.

Christopher Wren, later to be knighted, became the best-known member of this illustrious group. He helped with the extensive dissections, injected blood vessels with ink and other substances to see them better, and was responsible for many of the magnificent anatomical plates that graced *Cerebri anatome* (Fig. 7.2). As Willis wrote in the preface to this book, "Doctor Wren was pleased . . . to delineate with his most skillful hands many figures of the Brain and Skull whereby the work might be more exact." Wren went on to greater fame as the architect of St. Paul's Cathedral and other buildings in London. He even drew up an ambitious plan to reconstruct vast stretches of London that had been reduced to rubble and ashes by the Great Fire of 1666.

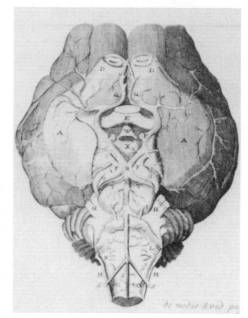

FIGURE 7.2.

A plate showing the underside of a sheep's brain. Drawn by Christopher Wren, this plate appeared in Willis' Cerebri anatome *of 1664.*

Willis, Wren, Lower, and others in this circle of talented men not only worked together, but strove for a new spirit of cooperation. They felt it important not to compete and preached the sharing of thoughts and skills. The ultimate goal, all agreed, had to be the advancement of science. By collaborating, they believed, researchers could more ably advance the cause of science.

Not surprisingly, this group played an important role in founding the Royal Society of London. This famed "commonwealth of philosophers" grew out of the "experimental philosophicall clubbe" (or, in Boyle's words, the "Invisible College"), that first met at the Oxford lodgings of Dr. Richard Petty. Willis was one of the original forty names on the roll of the Royal Society when it was founded in London in 1660 to promote "Natural Knowledge." He became a fellow of the Royal Society three years later, one year before he became a fellow of the Royal College of Physicians.

The Doctor, the Church, and the Soul

Willis believed in the teachings of the Anglican Church and had no intention of offending the ecclesiastics when he wrote *Cerebri anatome* or *De anima brutorum*. Especially when it came to issues pertaining to the soul, he chose his words carefully. It is significant that in *De anima brutorum* he distinguished between the immortal human soul and the "soul of brutes," which was also called the "corporeal soul," the "body soul," and the "material soul."

He stated that only humans possess an immortal soul, and it makes us special. But our immortal soul cannot be studied anatomically or physiologically. Although it provides fitting subject matter for the clergy and the philosophers, it is not something tangible that can be explored by natural scientists.

In contrast to the immortal soul, the corporeal soul is a material entity that we share with dogs, cats, cows, and other animals. It is an "inferior" soul in humans, but it is the highest controlling force in animals. Willis held that since animals have at least the rudiments of perception, cognition, and memory, these functions must be features of the corporeal soul. And since this soul can be tied to the brain, its functions can be studied scientifically, both in animals and humans.

Thus, Thomas Willis skillfully rejected the idea that only humans can think and act rationally—a belief central to the thinking of René Descartes. Instead, he found himself having much in common with the French philosopher Pierre Gassendi, whose ideas were discussed by the *Virtuosi*.[10] A bright and effective critic of Descartes, Gassendi adopted the commonsense belief that beasts have perception, cognition, and memory, just not at our level. To the satisfaction of the Church, he further maintained that the soul of brutes is material, and therefore is incapable of surviving the body.

With his faith in God and dedication to the Church of England uncompromised, and the mind-body problem that plagued Descartes circumvented, Willis felt free to put his thoughts about brain function and what we would now call psychiatry on paper. He even discussed his ideas with the Archbishop of Canterbury. To his relief, his friend the Archbishop was willing to accept his premise that matter (the corporeal soul) is capable of simple cognition.

The Language of Willis

Willis was an absolute master of flowery language and an innovator of terminology. From his Latin texts and the English translations made by Samuel Pordage not long after his books were written, there emerged a wealth of new words that are still in use today.

One such word is *neurology*, which appeared in the English edition of *Cerebri anatome*, dated 1681. *Neurologie* referred to the "Doctrine of the Nerves," a reference to the cranial, spinal, and autonomic nerves, as differentiated from the brain and spinal cord. *Neurology* will eventually assume a much broader definition than this—one signifying the study of the entire nervous system in health and disease—but this change would not take place in the seventeenth century.

Interestingly, it is also in the writings of Willis that we find an early use of the word *psychology*, quite possibly the first appearance of the word in English. Constructed from Greek roots (*psyche* = "soul"; *logos* = "study of"), the word *psychology* was first introduced as *psychologia* by Rudolf Goeckel in 1590.[11] In *De anima brutorum*, Willis defined it as the "Doctrine of the Soul."

Willis also coined a host of anatomical terms. They include *lobe, hemisphere, pyramid, peduncle,* and *corpus striatum*. And on the physiological front, it was Willis and not Descartes who coined the descriptive word *reflexion*. It is from this word that we get the word *reflex,* a term used to describe the very rapid, automatic, input-output actions of the nervous system.

For a reader who has taken some college courses in biology or perhaps psychology, the texts written by Willis may not seem especially difficult to understand. Nevertheless, they contain hidden pitfalls that can lead to serious interpretative errors. The difficulty relates to the fact that several of our common anatomical terms had different meanings to Willis. By looking at just three such words, we shall see how voyaging without a good map into the world of Willis, or into other works from this period, can be a most hazardous affair.

The first example of a word that is apt to prove confusing is *cerebel,* the old word for the cerebellum. For Willis, this word often referred not just to the spherical "little brain" (the literal translation of the word *cerebellum*) located behind the cerebral hemispheres, but also to specific regions of the midbrain and pons just below it. More specifically, he considered these lower-brainstem structures as important processes, or appendices, of the cerebellum proper. Today, of course, we do not put them under the cerebellar umbrella.

The second term is *corpus callosum,* the subcortical band of nerve fibers connecting the left and right cerebral hemispheres to each other. As we shall see in Chapter 17, the corpus callosum made headline news during the 1960s, when Ronald Myers and Nobel Prize–winner Roger Sperry showed that surgically cutting these connections prevents one hemisphere from knowing what the other is perceiving, doing, or commanding. But the twentieth-century definition of the corpus callosum was not what Willis had in mind. For Willis, the corpus callosum included the entire white core of the hemispheres.

The third and final example concerns the term *medulla.* Modern texts define the medulla as the most posterior part of the brainstem (medulla oblongata). This medulla is located just above the spinal cord and behind the pons. In the writings of Willis, however, the Y-shaped medulla, or medullary substance, is much larger. It extends from the spinal cord all the way up to his broadly defined corpus callosum.

With this better understanding of the language of Willis, we can now examine some of his thoughts about the functions of different parts of the brain. We can also see what led him to his conclusions. To begin this expedition, we shall start with the top of the brain and then work down.

The Cerebrum Replaces the Ventricles

Throughout the Renaissance, many scientists continued to maintain that memory is housed in the posterior ventricle. To some, however, it involved the nearby cerebellum itself. The idea that the cerebellum may function in memory was briefly mentioned by Andreas Vesalius in his *Fabrica.* In the first half of the seventeenth century, Johann Vesling, a German anatomist, and Nicolaas Tulp, a Dutch physician (who was featured in Rembrandt's famous painting "The Anatomy Lesson"), also considered the cerebellum the likely seat of memory functions.[12]

Willis initially entertained the notion that the cerebellum may have something to do with memory. He then rejected this relationship in favor of the more modern idea that memory for such things as words and ideas must reside in the "outmost banks" of the cerebral hemispheres.[13] As he saw it, our superior memory capabilities have to do with images being impressed on the cortex and being accessed when needed. He drew attention to the large number of deep grooves characterizing the human cerebrum. Dogs, cats, birds, and fish all have much smoother cerebral cortexes than we do. They not only lead simpler lives, he wrote, but can learn only by imitation.

The assignment of memory to the "cortical spires" was not just based on comparative anatomy. It was also grounded in clinical cases. With his interest in neurology, Willis knew that severe wounds to the cerebral hemispheres could affect memory. He also presented the brains of two congenitally impaired individuals to support his position.[14] One came from an adult who had a weak intellect and a second came from a mentally deficient child. Postmortems revealed the bulk of the cerebrum to be very small in both cases. Here, he reasoned, was further reason to associate memory functions with the roof of the brain.

Although this was a strong start, Willis showed that he could also be misled. He looked for an everyday action that would reveal the link between memory and the uppermost part of the brain. He was sure he had found one when he observed people rubbing their temples or foreheads while endeavoring to remember a fact. Today we realize that such actions reveal nothing at all about the underlying brain.

Willis' attention to a mundane act such as rubbing the head shows, however, that he was looking for anything that might tie structures and functions together. Today we would say that he was searching for "converging lines of evidence." Modern scientists do precisely the same thing when they attempt to bring anatomical, physiological, and behavioral evidence together to establish a point about parts of the brain mediating specific functions.

Willis made a sharp distinction between the gray matter forming the roof of the brain and his underlying corpus callosum. Because he saw a great number of blood vessels supplying the gray shell, he theorized that animal spirits are generated and stored in the cerebral cortex. He looked upon the underlying white matter as forming roads or paths for the "dispensation" of the spirits toward and away from the cortex. Said somewhat differently, he conceived of the cerebral white matter as important for the execution of voluntary or willed actions and other higher functions.

The ventricles, long considered critical for higher functions and given a prominent place in the pineal theory proposed by Descartes, were left out in the cold by Willis. He looked upon the ventricles as "vacuities" and compared himself to an astronomer who has little interest in empty spaces. No longer would the ventricles play the leading part or even a viable part in the localization story. Scientists would now move away from the chambers of the brain and devote their attention to the gray and white matter of the brain itself.

So how accurate was Willis in his assessment of the cerebrum? Overall, he was more right than wrong. He was accurate in rejecting the ventricles in favor of the cerebrum as the center for memory, cognition, volition, and imagination. He was also right to distinguish between the functions of the gray and white matter of the cerebral hemispheres. In contrast, his idea that the brain's crust, with its abundance of blood vessels, plays a key role in the formation of ethereal spirits was something less than a major leap forward.

The Corpus Striatum

In his *Cerebri anatome,* Willis described the corpus striatum adequately for the first time.[15] He wrote that this subcortical structure buried in the hemispheres near the front of the brain is streaked with alternating bands of gray and white matter (Fig. 7.3). He further noted that this sort of "chamfering," or streaking, cannot be

FIGURE 7.3.

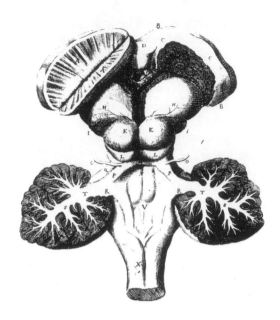

*Another plate
made by
Christopher Wren
for Willis' Cerebri
anatome of 1664.
The cerebral
hemispheres have
been removed from
this sheep's brain
and a piece of the
left corpus striatum
(A) has been cut
away to show its
striped (chamfered)
appearance.
The cerebellum (T)
has been cut
through to show it
in cross-section.*

found in any other part of the brain. Based on what he was seeing, he gave the striated corpus its name.

He proposed that this part of the brain plays a role in movement. He theorized that the corpus striatum contains channels through which animal spirits released at higher levels by the will traveled to the medullary substance, and from there to the peripheral nerves and the musculature.

To support his contention, he described a few patients who showed degeneration of the corpus striatum at autopsy. Prior to death, these individuals suffered from types of paralyses that prevented them from moving their arms and legs voluntarily.

> For as often as I have opened the bodies of those who dyed of a long Palsie, and most grevious resolution of the Nerves, I have always found these bodies less firm than others in the Brain, discoloured like filth or dirt, and many chamferings obliterated.[16]

Again Willis looked for converging evidence to support his theory. This time he found it when dissecting the brains of very young animals. He astutely observed that the striations are absent in newborn dogs. Moreover, movements of the limbs are limited in very young puppies. As he put it:

> Further, in Whelps newly littered, that want their fight, and hardly perform the other faculties of motion and sense, these streaks or chamferings, being scarcely wholly formed, appear only rude.[17]

Willis not only associated the corpus striatum to movement but, as can be seen in his statement about puppies, postulated that this part of the brain also functions in sensation. He suggested that impressions from the sensory organs project to the corpus striatum. These sensory inputs, he maintained, can trigger motor outputs that could lead a man to brush away a bothersome fly without so much as thinking about it. If some object is headed toward the face, the projections to and from the corpus striatum will also allow us to move to the side and blink involuntarily.

Thus, Willis viewed the corpus striatum as more than a simple replacement for the first ventricle, the common seat of all sensation (sensorium commune) to the Church fathers and many others prior to the seventeenth century. Instead, he saw the corpus striatum as a structure involved with both sensation and movement, a veritable "mart" for the distribution of the spirits to higher and lower brain centers.

Today we know that the corpus striatum is, in fact, involved in movement. This has been shown by scientists who have examined the brains of people who cannot stop their limbs from twitching, flailing, and making jerking movements. Patients with Huntington's disease, for example, show significant neuronal losses in this part of the brain, whereas those with Parkinson's disease exhibit a loss of inputs to this region.

Moreover, the major pathway for voluntary movements (the corticospinal tract) descends from the cerebral cortex through the corpus striatum. This descending pathway is responsible for the white stripes and bands in the region. Its presence explains why strokes or injuries to this region can affect voluntary (cortical) as well as involuntary actions.

In contrast to movement, sensation is no longer considered a function of the corpus striatum. In this domain, Willis was wrong, but not by much. The corpus striatum is located close to a very important part of the brain for sensation, the thalamus. This upper-brainstem structure receives inputs from all of our sensory systems, with the exception of olfaction, and plays a major role in processing sensory data and relaying some of it to the cerebral cortex. Had Willis paired the corpus striatum with movement and associated the nearby thalamus with sensation, his sensory-motor relationships would have been even more accurate.

The Cerebellum and Involuntary Functions

Willis postulated that the spirits produced in the cerebrum are responsible for willed actions, whereas those generated in the "Cerebel" govern internal activities or visceral actions which can take place without conscious awareness. He poetically dubbed the cerebellum his "Mistress of the involuntary Function." His evidence for this, once again, came from a variety of sources.[18]

As usual, he relied heavily on comparative anatomy. His brain dissections showed that the cerebellum varied little across species. Thus, it made sense to give it a function that humans share fairly equally with lower vertebrates. This effectively took higher functions, such as thinking and memory, out of the picture. It caused him to focus on more primitive acts—behaviors essential for all vertebrate life.

The notion that the cerebellum may have something to do with involuntary functions was strengthened by clinical and experimental observations. In particular, he noted that severe injuries of the cerebellum could affect the heartbeat, perhaps the most basic function of all. He also discovered that jerking the head in the direction of the cerebellum could cause changes in respiration.

Willis postulated that the nerves of the autonomic nervous system, which control the smooth muscles of the internal organs of the body, originate in the cerebellum. From its location in the back of the brain, the cerebellum "unperceivedly" regulates the motions of the heart and lungs, actions necessary for digestion in the gut, and other basic processes.

Willis never claimed that these functions cannot be affected by messages from the cerebrum. Instead, he maintained that especially strong spirits from the viscera could travel all the way up to the cerebrum to be appreciated in consciousness. He also postulated that spirits from the cerebrum can pass down to the cerebellum to permit conscious regulation of typically involuntary actions, such as changing the depth and rate of our breathing when asked by a physician to do so. To the extent that Willis postulated both higher and lower levels of control over the viscera, as well as the potential for interactions, his thinking represented a marked advance over earlier ideas.

Had Willis associated the cerebellum proper with involuntary movements of the skeletal muscles, rather than with movements of the smooth muscles of the internal organs, he would have been even more impressive. Today we know that the cerebellum plays an important role in allowing us to walk without thinking about how to move one leg in front of the other, and in allowing us to drink from a cup without consciously plotting how far to tip it as it approaches our lips.

Alternatively, had he included parts of the medulla (as we know it today) in his broad definition of the cerebellum, he also would have been more accurate. Early in the 1800s the role of the medulla in governing primitive visceral functions became clearer. In defense of Willis, however, we should recognize that some of his so-called cerebellar lesions could easily have damaged or disrupted the functions of the nearby medulla. Even today there is a tendency to overlook some of the secondary or distal effects of acute brain damage.[19]

Willis as a Psychiatrist

During the Middle Ages and through the Renaissance, so-called mental disorders, or afflictions of the mind, were treated largely by the clergy. Only diseases of the body were routinely cared for by physicians. Willis effectively called for a different approach in *De anima brutorum*. In his work, he approached behavioral problems as disorders of the corporeal soul. By doing so, he helped to make the study of mental diseases, or psychiatry, an earthly or secular science and a fitting subject for medical men.

In the text of *De anima brutorum,* Willis points to a relationship between mania and melancholia (depression). He mentions how one can change into the other and provides his readers with the first good description of manic-depressive illness, or what many psychiatrists now refer to as bipolar disorder:

> After Melancholy, Madness is next to be treated of, both which are so much akin, that these Distempers often change, and pass from one into the other.[20]

Willis wrote that this cyclic disorder is like a burning object, one that can produce smoke or flame. Continuing the analogy, he added that smoke is likely to follow a flame, just as a flame can burst forth from what had been smoke.

He also dealt with "stupidity" and "foolishness."[21] In *De anima brutorum* he described four degrees of stupidity. He commented that "Fools beget Fools," suggesting that some forms of low intelligence are inherited. He then commented that foolishness can also be acquired as well. He cited head injuries as one cause of acquired foolishness. Overindulging in wine was mentioned as another. Today, chronic alcoholism is associated with Korsakoff's disease, an acquired brain disorder characterized by severe memory and attentive problems.

Sleep disorders, including sleepwalking, constituted another aspect of psychiatry that fascinated Willis. One of his most vivid descriptions is of narcolepsy, the irresistible tendency to lapse into short episodes of sleep at inopportune moments. He describes this strange and dangerous sleep disorder as

> a sleepy disposition—they eat and drink well, go abroad, take care well enough of their domestick affairs, yet whilst talking, or walking, or eating, yea their mouthes being full of meat, they shall nod, and unless roused up by others, fall fast asleep.[22]

As for epilepsy, the feared "sacred disease" of old, Willis hypothesized a fault with the brain or its blood supply.[23] Although his iatrochemical theory of epilepsy is based on animal spirits and is fanciful, what he wrote is significant because he presented epilepsy as a natural disease, not the work of the devil and a sure sign of possession. At a time when witch hunting had reached epidemic proportions, it took great courage to take the supernatural out of epilepsy and present it as a physical or chemical disorder without mystical overtones. Willis not only had the fortitude to do so, but chided those still associating epilepsy with witchcraft.

Lastly, we should mention hysteria. In ancient times, hysteria was thought to be a physical disease of females, one thought to be caused by a wandering womb in unfulfilled women. Willis rejected this theory for several reasons. Among other things, he observed that men can have hysteria. In addition, autopsies on women who suffered spasms, fits, motions in the belly region, unusual respiratory symptoms, and "distempers" usually revealed perfectly normal wombs. Some of the same observations about hysteria will be reiterated by Jean-Martin Charcot in the nineteenth century (see Chapter 12).

Willis thought that the cause of hysterical disorders could be found in the back of the brain, where the nerves of the autonomic nervous system originate.[24] These nerves, he reasoned, go to the internal organs of the body that seem most affected. Hence, Willis did not look upon hysteria as a psychological disorder, or what we would now call a neurosis. What he did do, and it was significant, was to throw the wandering-womb theory into doubt, while recognizing that strong emotions can trigger the symptoms of this medical disorder.

Steno Criticizes Willis

Cerebri anatome was well received in many quarters, as is evidenced by the mostly positive book reviews and comments that followed its publication in 1664. A very favorable review appeared in *Journal des Sçavans*, the first scientific journal, in its inaugural year of 1665. As the author of a twentieth-century book on the history of neurology expressed it, *Cerebri anatome* was by far "the most complete and accurate account of the nervous system which had appeared to that date."[25]

A mixed review, however, came from Nicolaus Steno. As we read in the last chapter, he discussed the ideas of both Willis and Descartes in a speech delivered at the

Paris residence of Monsieur Thévenot in 1665.[26] Steno praised the brain diagrams in *Cerebri anatome* for their accuracy and overwhelming superiority over previous drawings. The Danish anatomist stated: "The best Figures of the Brain are those of Willis."[27] There were some faults with the diagrams, he continued, but they were relatively minor.

Nevertheless, Steno felt that Willis, like Descartes before him, was entirely too speculative.

> Willis is the Author of a very singular Hypothesis. He houses common Sense in the Corpora Striata, the Imagination in the Corpus Callosum, and the Memory in the cortical Substance. . . . How can he be so sure that these three Operations are performed in the three Bodies which he pitches upon? Who is able to tell us whether the nervous Fibers begin in the Corpora Striata, or if they pass through the Corpus Callosum all the way to the cortical substance? We know so little of the true Structure of the Corpus Callosum, that a man of tolerable Genius may say about it, whatever he pleases.[28]

Steno's audience was greatly impressed by his thoughtful lecture. He was critical but not personal or emotional, and he made his points clearly and succinctly. He could not have been more polite. There was none of the venomous language used a century earlier by Jacobus Sylvius when he responded with rage to what Vesalius had written about Galenic anatomy (see Chapter 5).

For his part, Willis recognized that his love of speculation was a fault, albeit one difficult to overcome. He therefore accepted Steno's comments about scientific objectivity and imaginary animal spirits in a positive spirit, without anger or hostility. True to his nature, Willis even praised "the most learned" Steno when he wrote his *De anima brutorum* just a few years after the Danish anatomist had criticized the speculative excesses in *Cerebri anatome.*

To his dying day, however, Willis was unable to veer from his "more pleasant Speculations" about brain functions or from his fanciful iatrochemistry. As Hansruedi Isler, Willis' twentieth-century biographer, put it:

> Willis' way of integrating observed facts by making use of every suitable idea within reach is characteristic for the whole species of pre-Newtonian scientists, the earlier *Virtuosi*, whose boundless scientific optimism gave them the necessary impetus to overcome their Scholastic adversaries, and made them nearly unable to realize the limits of their projects.[29]

The Lasting Contributions

In retrospect, Thomas Willis deserves great credit for refocusing attention on the brain itself. True, he was a starry-eyed optimist who made his share of errors. But his shortcomings appear small when we look at the scope of his work and the bigger picture.

It was Willis who correctly differentiated between cerebral functions, such as memory and volition, and lower-brainstem functions, such as control over respiration and heartbeat. It was also Willis who correctly associated the corpus striatum with motor functions and first discussed different levels of neural control. Moreover,

Willis should be recognized for the first good descriptions of a host of neurological and psychiatric disorders. In addition to bipolar disorder and narcolepsy, he probably gave the first medical description of myasthenia gravis.[30] He also described migraine headaches and tremors, most likely those of Parkinson's disease. When we add the best illustrations of brain dissections to date, a classification system for the cranial nerves, and a wealth of new terminology to this list, we can begin to appreciate the magnitude of his contributions.

Perhaps the most significant part of Willis' legacy, however, was his ability to stimulate others to test and improve his ideas. His work resulted in more human autopsies, a wealth of revealing experiments on animals, better clinical observations, and new ideas about how the nervous system works. When it came to making the brain exciting to other scientists, Willis was head and shoulders above Descartes, the philosopher who loved mechanics but lacked clinical experience and a real understanding of the medical sciences.

Many of Willis' achievements would not have been possible had he separated the mind and body in the manner of Descartes. Willis, of course, agreed with Descartes that only humans possess an immortal soul. But Willis did not accept the Cartesian notion that animals cannot perceive, think, or remember. Further, he saw these higher functions as products of the brain. By clearly distinguishing between the corporeal soul and the immortal soul, and by refusing to bog himself down with Cartesian metaphysics, Willis opened the door for scientists to study higher functions of mind in the clinic and with animals.

Willis' approach to the mind also had an influence on the philosophers of the day. One of Willis' most distinguished students at Oxford was John Locke, whose writings contain many important philosophical concepts traceable at least in part to Willis. Indeed, the very notion of the mind being a *tabula rasa* (blank slate) at the time of birth, a concept now considered synonymous with Locke, can be found in Willis' writings. Like Locke, Willis maintained that nothing is in the brain that is not first in the senses.[31]

Sir Charles Sherrington, winner of the 1932 Nobel Prize whom we shall meet in Chapter 14, wrote of his debt to Thomas Willis. As he saw it: "Thomas Willis of Oxford practically refounded anatomy and physiology of the brain and nerves."[32] Sherrington continued by saying that it was Willis who "put the brain and nervous system on their modern footing." Today, more than ever before, scholars are agreeing with Sherrington's assessment of the talented Oxford physician who elevated the brain sciences to a new level during the second half of the seventeenth century.

8

Luigi Galvani

Electricity and the Nerves

From the things that have been ascertained and investigated thus far, I believe it has been sufficiently well established that there is present in animals an electricity which we . . . are wont to designate with the general term "animal." . . . It is seen most clearly . . . in the muscles and nerves.

—*Luigi Galvani, 1791*

Nerve Spirits and Juices

The seventeenth century was a time of great accomplishments in the brain sciences. Thanks in part to Thomas Willis and his associates in England, the advances included a much-improved neuroanatomy, a host of new terms to describe parts of the brain, and fresh attempts to associate brain structures with behavioral functions. Additional changes included better clinical descriptions of neurological diseases and a greater willingness to intervene surgically. Nevertheless, as was true during the Renaissance, scientific advancements did not proceed equally along all fronts. In particular, investigators living in the opening decades of the eighteenth century still could not boast about knowing how the nerves work.

Three different theories of nerve function were debated as the seventeenth century drew to a close.[1] First, many scientists still adhered to the notion of spirits running through the hollow nerves to contract the muscles or convey impressions to the brain. This was an ancient idea, but René Descartes, among others, embraced it.

A second theory held that the nerves secrete droplets of fluid onto the muscles to activate them. Thomas Willis, for example, thought that when nerve fluids mixed with blood and fermented, they could cause minute "explosions," which would result in muscular contractions.

The third theory was the idea that the nerves transmit information by vibration. This idea found its champion in Sir Isaac Newton, who attributed color perception to different waves of light causing corresponding vibration patterns in the nerves from the eye to the brain.[2]

Each of these notions of nerve action—ethereal spirits, fluids, and vibrations—had serious problems. Not only was each hard to test, but each seemed to be at odds with laboratory findings. For instance, in 1677 Francis Glisson, a physics professor at Cambridge University, took a glass tube closed at one end, put his arm in it, and then filled it with water. He hypothesized that more water should be displaced by the inflow of spirits into his arm when he flexed his muscles. Yet he found no more displacement when he did this.

At about the same time, an Italian scientist by the name of Giovanni Borelli decided to test the theory of spirits in a slightly different way. In 1680 he immersed the limb of an animal in a tub of water and cut some of its muscles. To his surprise, no bubbling ferment came out. Based on what he failed to see, he too questioned whether a case could be made for nerve spirits inflating the muscles.

Experiments in which nerve bundles were tied off also left many scientists scratching their heads. In theory, the buildup of spirits or juices should cause the nerves to swell just behind the knot. Not only were the experimental results inconsistent with this expectation, but when the nerves were then cut, drops of fluid failed to materialize. Findings of this sort these forced Albrecht von Haller, the leading physiologist of the mid-eighteenth century, to assert that if the nerves secrete some sort of a vital juice, it cannot be anything like water.[3] Moreover, he doubted whether any fluid encountered in everyday life could move fast enough to account for nerve actions.

As for vibrations, the third idea, this notion seemed wrong to most scientists from the start. The immediate problem was that the nerves appeared soft and pulpy. They were not pulled tight like the strings of a bow, which vibrate when struck. In addition, they did not spring back when cut. For these reasons, the idea of nerve vibrations was criticized by Herman Boerhaave. In a lecture given in 1743, the esteemed Leiden physician referred to vibration theory as just another "repugnant" idea.[4]

Could the use of newly constructed microscopes shed some light on the problem? Anton Van Leeuwenhoek, the pioneering microscopist who published some of his findings in 1674, was distressed when he failed to observe an opening in the optic nerve of a cow.[5] After all, Galen had written that such an opening could be seen by simply holding the severed optic nerve up to the sun. Other fledgling microscopists searched high and low for hollow openings in the nerves, a prerequisite for theories involving nerve spirits or juices. To their dismay, such openings could not be found at all or demonstrated in a manner that inspired confidence.

Steno, the Dane who was critical of Descartes and Willis for their fanciful theories of brain function, did not mince words when it came to expressing his own doubts about existing theories of nerve action, especially those involving mysterious animal spirits. In 1667, after having conducted some experiments of his own, he complained:

> We are still more uncertain about what relates to the Animal Spirits. Are they Blood, or a particular Substance separated from the Chyle by the Glands of the Mesentery? Or are they derived from a Lymphatic Serum? . . . Our common Dissections cannot clear up any of these difficulties.[6]

Physiologists in the opening decades of the eighteenth century consequently had many good reasons to be skeptical of existing theories, whether based on animal spirits, fluids, or vibrations. Obviously, there was a need for a more plausible mechanism of nerve action—a theory based on an agent that could work rapidly, be manipulated, and perhaps even be measured. In short, there was a demand for a fresh start.

The Spark

During this period of rising frustration with existing nerve theories, some scientists began to wonder whether the nervous system may work by electricity, a term coined by the British physicist William Gilbert around 1600. But was there a single figure dominating the study of electricity at this time? Because so many people were involved with electricity, and because knowledge about electricity led to developments in so many different scientific fields, some broad-minded historians say no. When discussions narrow to just the history of neurophysiology, however, one name stands above the rest. As put by historian Mary Brazier: "In the history of the electrical activity of the nervous system the outstanding figure is, without rival, that of Galvani."[7]

Luigi Galvani (Fig. 8.1) worked in the last quarter of the eighteenth century. But, as Brazier herself admits, given the Zeitgeist, or spirit of the time, Galvani's experiments and theories were not as revolutionary as they were evolutionary. Galvani studied electricity with animals because this was one of the most exciting things an aspiring scientist could do, and he was the recipient of a wealth of experimental findings and new ideas. The favorable climate that existed when he worked can best be appreciated by looking at some of the phenomena that attracted him to the field.

The Instruments

Let us begin by looking at the use of electricity in the healing arts. As we have already seen, fish with electric organs were known to the ancients (see Chapter 4).[8] Scribonius Largus and Galen used shocks from the electric ray to treat headache, gout, paralyses and various other disorders in ancient Rome. They sometimes had patients stand barefoot on a live ray or more than one fish to increase the effect. The therapeutic use of these fish continued in the Middle East after this time and was known to the Europeans.

Nevertheless, finding electric fish for therapy or experimentation was not always

FIGURE 8.1.

Luigi Galvani (1737–1798).

easy. Special machines for creating electricity on demand were needed. These appeared as "friction machines," but not until the eighteenth century. They were made
of globes of sulfur, porcelain, or glass that were rubbed by hand.[9] Once electrically
charged, an individual could produce a spark by touching an object or even another
person. Inventors soon found a way of generating "electrical fire" simply by touching revolving glass globes, cylinders, or plates.

Once electrically charged by a friction machine, a man could set a glass of brandy
ablaze by throwing a spark from his tongue. Johann Winkler, a professor at Leipzig,
often demonstrated this unexpected phenomenon at social gatherings to the delight
of his guests.[10] For the nobility who were looking for new ways to amuse themselves, spectacles of this sort were the talk of the town.

Johann Krüger was of a more serious frame of mind when he learned about the
new frictional machines. He envisioned a role for them in mainstream medicine. In
1744 he wrote:

> But what is the usefulness of electricity, for all things must have a usefulness,
> that is certain. We have seen it cannot be looked for either in theology or in ju-
> risprudence, and therefore nothing is left but medicine. The best effect would
> be found in paralyzed limbs to restore sensation and reestablish the power of
> motion.[11]

With the advent of the Leyden jar in 1745, storing and releasing electrical charges
on demand became much easier. This device, named for the Dutch city south of Amsterdam in which it was invented (now spelled Leiden), allowed electricity to be
stored until needed. The often-cited man of the hour was Petrus van Musschen-
broek, a very talented instrument maker and a pupil of Herman Boerhaave (A similar device for conserving and releasing electricity was constructed at about the same
time in the north of Germany.[12])

The typical Leyden jar has a narrow neck with a rubber stopper. A metal foil
coating over the glass (except at the top) serves as an outer conductor, whereas a liq-
uid (or, later, metal) inside the jar serves as an inner conductor. After a frictional machine is connected to the jar, the separation of the inner and outer conductors by the
glass allows the charge to be conserved. The charge can be released when wires from
both conductors touch a person or an animal. As the early researchers quickly found
out, the discharge could throw a grown man off his feet and even kill a small animal.

In the hands of Abbé Jean Antoine Nollet, the man who coined the term "Leyden
jar," the device made for some truly spectacular demonstrations.[13] He made a nine-
hundred-foot line of Carthusian monks jump en masse when they held hands and
completed a circuit to a fully charged Leyden jar. The instrument, however, was destined to be used for more than just entertainment. Nollet himself stressed that it
should be utilized therapeutically. He reported some success with one in a military
hospital in Paris, and even wrote a book about how to apply electricity to paralyzed
body parts.

Although some cautious physicians felt that electricity was overrated as a cure-all,
Leyden jars and electrical machines now made their way into Europe's more progressive hospitals and clinics. Between 1750 and 1780 more than twenty-five articles
on curing paralyses with electricity appeared in just one French periodical, the *Journal de Médecine*. But many scientists still did not view electricity as the mysterious

fluid of the nerves. Instead, they held tight to the idea that electricity worked its wonders by increasing the flow of sluggish juices in the nerve canals.

A Most Unusual Trio

Electrotherapy became so popular in Europe and North America that many individuals who were not formally trained as physicians began to practice it. Some hoped to make their fortunes by administering shocks to sick and injured patients. Other medical "outsiders" were drawn to it by a wish to help humanity or by a desire to become famous.

Three people who are much better known in other domains illustrate just how alluring electrotherapy had become. The first is John Wesley, the English religious reformer and founder of the Methodist Church. Wesley did not formally attend medical school or pass any kind of certifying examination, but this did not dampen his enthusiasm when it came to applying the new technology.[14]

In his *Primitive Remedies,* which first appeared in 1747, Wesley listed 288 medical conditions that he thought could be prevented or healed by electricity.[15] He was convinced that no remedy from nature was as good as an electrical machine for disorders of the nervous system. He suggested fifty to a hundred weak shocks for most medical conditions. For paralyses, he recommended electrifying the malfunctioning limb once per day for three months. Wesley had been using his "electrical fire" therapeutically for many years before he published *The Desideratum or, Electricity Made Plain and Simple by a Lover of Mankind and Common Sense.*[16] This pamphlet appeared in 1759 and was so popular that it quickly went through several editions.

A second nonphysician who had something to say about medical electricity was Benjamin Franklin. He worked on his electrical science in Philadelphia from 1747 to 1755. His experiments with kites and lightening rods are well known. In 1751 he published a series of letters he had written to friends in London as *Experiments and Observations on Electricity, Made at Philadelphia in America.*[17] But in 1759, when the medical use of electricity was already widespread, he penned a letter to John Pringle of the Royal Society of London expressing his doubts about electricity as an effective treatment for paralyses:

> I never knew any advantage from electricity in palsies that was permanent. And how far the apparent temporary advantage might arise from the exercise in the patients' journey, and coming daily to my house, or from the spirits given by the hope of success, enabling them to exert more strength in moving their limbs, I will not pretend to say.[18]

The last member of this most unusual trio was Jean-Paul Marat, who treated patients in England and France.[19] Marat aspired to be a respected healer and reported successes with electricity for a variety of ailments, including paralyses and pain. He even recommended it for children who were failing to develop properly. But like Franklin, Marat saw limits to electrotherapy. He mentioned that it cannot cure epilepsy and cannot cause malignant tumors to undergo remission. Marat published his most important works on electricity in 1782 and 1784.[20] The latter was awarded first prize in a French competition.

Today Marat is much better remembered for his politics during the French Revolution and for his fanaticism during the Reign of Terror. A self-proclaimed "friend of the people," Marat was instrumental in sending many innocent men and women to the guillotine. Charlotte Corday, a member of the opposing Girondist revolutionary party, finally took a knife hidden under her dress and killed him while he was writing on a desk that covered his medicinal bath. Marat's dramatic death in 1793 was immortalized by his friend Jacques-Louis David, the leading French painter of the Neoclassical period (Fig. 8.2).

In contrast to Marat's fanaticism, Franklin's diplomacy during and after the American Revolution are remembered more positively. Even the Europeans viewed Franklin as a talented statesman, journalist, and scientist in the culture-starved American colonies. As for John Wesley, he is best remembered today as the founder of the Methodist Church. By purchasing electrical machines for the practice of medicine and by turning to God, he strove to save the largest number of bodies and souls in what he claimed would be the most economical way.

The Nature of Animal Electricity

Those individuals who first thought that electricity might be the "fluid" of the nervous system listened carefully to what the electrotherapists were saying. Even after eliminating the outright frauds from the picture, the claims for electrotherapy were impressive and supportive of the idea that electricity might be the key to understanding nerve action. After all, what other force was known to have such an effect on the vital processes of the body? Was anything else more effective in eliciting muscular contractions? And was there any other agent that could travel as fast as electricity?

Stephen Gray, Stephen Hales, and Alexander Monro (Primus), three early

FIGURE 8.2.

The painting by Jacques-Louis David entitled "The Death of Marat" (1793). (Reproduced with permission of the Musée Royal des Beaux Arts, Brussels.)

eighteenth-century scientists in Great Britain, were among the first to write that electricity could be the mysterious fluid of the nerves.[21] Others followed with stronger statements. Still, it was a long jump from the observation that people could be electrically charged to the claim that electricity underlies normal nerve action.

One of the biggest problems was the idea that it may not be possible to confine electricity to just the nervous system. What would prevent electricity from leaking out of the nerves onto surrounding tissue? This was a legitimate concern, and it was an issue that kept many researchers a respectable arm's length away from the idea that electricity must be the mysterious agent of nerve action.

Back to Electric Fish

Some of the mysteries surrounding electricity were cleared up during the 1770s, when a few scientists decided to take a fresh look at the specialized fish that had so amazed the ancients. Unlike their predecessors, this new generation of explorers had better instruments for studying these fish. They also had well-thought-out questions they wanted answered. At the top of most "wish lists" was the most crucial questions of all: How could researchers be sure that the shocks released by electric fish are really electrical in nature?

The task of understanding how these specialized fish may function began in earnest in the second half of the seventeenth century. An Italian scientist by the name of Francesco Redi described the painful shocks of the electric ray (also called the torpedo fish) in 1666, and his student, Stefano Lorenzini, examined the electric organ itself just a few years later.[22] In 1678, Lorenzini wrote:

> The Cramp-Fish hath not this stupefying Quality in all the Parts of his whole Body, but only in one particular Part, and this determin'd or particular Part is those two hooked Muscles . . . which unless they are immediately touched with the bare Flesh, produce no Effect at all; and besides, in touching those Parts, it is necessary that the Fibres of those said Muscles be contracted, to produce the Effect on the naked Part of those who touch them.[23]

John Walsh arrived on the scene about a hundred years later and conducted a series of landmark studies on torpedo fish caught off the French coast.[24] In 1772, he presented some of his findings at a meeting of a scientific academy in France. He also penned several letters to Benjamin Franklin that were published a year later. A report by Walsh on the larger English ray followed in 1774.

Walsh described how an electric ray could discharge fifty or more shocks in a minute and a half. He also showed how its electricity could be transmitted through wires. He noted that it was important to have the wires connected to the upper and lower surfaces of a healthy ray to experience the full effect, which he found indistinguishable from the shock produced with a Leyden jar.

Walsh hypothesized that the shock from the fish is due to the buildup and release of compressed electrical fluid. The electrical organs draw their charge from the nerves and, like a Leyden jar, store it for eventual release. To Walsh, the only real difference between the fish and a Leyden jar was that the ray could decide whether or not to release its own shocks. This, he said, is evidenced by the fact that it closes its eyes when it begins to release electricity.

Having satisfied himself that these fish can generate electrical shocks, Walsh enlisted the aid of Dr. John Hunter, an English surgeon skilled in anatomy, to study the fine structure of the ray's electrical organs.[25] Hunter found that these specialized organs make up about half of the ray's body and have a hefty nerve supply. They are composed of a large number of perpendicular columns, each made of many hexagonal discs separated from each other by a thin layer of fluid (Fig. 8.3). A small ray may possess about five hundred columns; a large one almost twelve hundred.

Knowing that the shocks originate only from this highly specialized organ was of considerable importance to Walsh. It meant that the charge somehow does not dissipate into surrounding tissues. The fact that the fish do not electrocute themselves meant that there must be some sort of insulation around the electric organs and the nerves associated with these specialized structures.

The Royal Society awarded the Copley Medal to Walsh for his outstanding work, but conservatives in the scientific establishment still expressed their doubts about the specialized fish. What these fish discharge may act like electricity and feel like electricity, they admitted, but is it really electricity?

The trouble was that the shocks from these creatures were not accompanied by flashes of light or popping sounds. Without a flash and a crackle, or at least something akin to lightning and thunder, many scientists remained unconvinced that these fish were really producing electricity. The fact that a charged ray or eel would fail to deflect pith balls hanging on strings or thin gold leaf (as do other electrically charged objects) made the doubting Thomases even more suspicious.

Ever hopeful of quieting his critics, John Walsh continued his experiments. His idea was to form a circuit from the fish to some onlookers using wire and tinfoil, and then to make a thin cut in the foil. At first, the electric spark did not jump the gap, in contrast to what Walsh had found when he used a fully charged Leyden jar. Still, he persevered and, by making a thinner cut in the tinfoil and working in a dark room, he showed that a discharge from an electric eel can produce a spark capable of jumping a small gap.

Unfortunately, Walsh died before he was able to publish the results of his crucial experiment. But in 1795, the year in which Walsh passed away, his most important experiment was described by Tiberius Cavallo, who witnessed the demonstration:

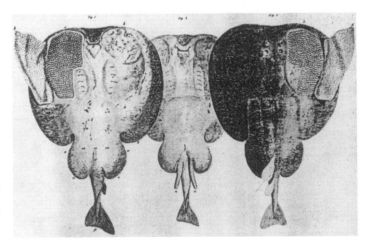

FIGURE 8.3.

The torpedo fish, also known as the electric ray, and the location of the electric organ.

The strongest shocks of the *gymnotus* will pass a very short interruption of continuity in the circuit. . . . When the interruption is formed by the incision made by a penknife on a slip of tin-foil that is pasted on glass, and that slip is put into the circuit, the shock in passing through that interruption, will shew a small but vivid spark, plainly visible in a dark room.[26]

There were fewer doubters about fish electricity once this experiment became known. Yet it was one thing to say that electricity from a specialized fish can produce an electrical shock, and quite another to conclude that frogs, barnyard animals, and even humans can also generate electricity. The ability to generalize from some unusual fish to other life forms was one of the issues foremost on people's minds when Luigi Galvani stepped out of obscurity.

Galvani and His *Commentary*

Aloisio Luigi Galvani was born in 1737.[27] He was a mild-tempered man, a good husband, and a kind doctor, who shied from publicity. He was also an honest, modest person, quite content to spend almost his whole life in one location, the northern Italian city of Bologna. Inclined at first to study theology, Galvani found anatomy, physiology, and medicine more to his liking. Before turning to the study of electricity, he worked on skeletal development, on the comparative anatomy of the ear, and in the field of obstetrics.

Galvani probably spent some time studying with Leopoldo Caldani, a Bolognese scientist who looked upon electricity as the strongest possible stimulus for animal tissue. He obtained degrees in medicine and philosophy and eventually took Caldani's place as professor of anatomy at the University of Bologna. Much of his own research, however, took place at his home. There Galvani kept many electrical devices, including machines for producing electricity by friction, primitive condensers, and Leyden jars. His primary subject was the frog, but he went on to experiment on sheep and even people. Among his assistants was his talented wife, Lucia Galeazzi, the daughter of his anatomy teacher.

From his notebooks, we know that Galvani began to experiment with the *fluido elettrico* (electrical fluid) in the 1770s. Although he wrote four memoirs on his electrophysiological research during the 1780s, for unknown reasons he decided not to publish his findings at the time.[28] The treatise that made Galvani famous was published in 1791. It was entitled *De viribus electricitatis in motu musculari commentarius* (Commentary on the Effects of Electricity on Muscular Motion).[29]

In his *Commentary*, Galvani traced his experiments in chronological order, explaining the logic behind each study and elaborating on the results. Here one can find the young scientist expressing surprise when an unexpected finding emerged and showing enthusiasm for new evidence to suggest that "animal electric fluid" does, in fact exist. Throughout the work Galvani strives for synthesis, never losing sight of the whole while assembling the parts.

Galvani tried to establish four points when he packaged his many experiments together. The first is that a frog muscle can be made to twitch when touched by a metal scalpel held by a person near an electrical machine shooting sparks. This strange event first occurred in a room where his assistants were amusing themselves with

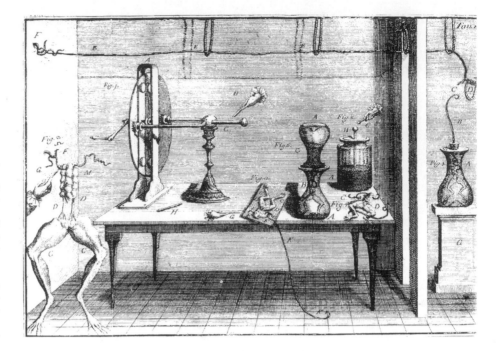

FIGURE 8.4.

The interior of Galvani's laboratory, showing some of the devices for producing electricity, including a Leyden jar (far right corner of table) and a frog muscle-nerve preparation (left).

one of his devices (Fig. 8.4). The apparatus threw a spark at precisely the same time that a scalpel in his hand touched a nerve leading to a frog's leg muscle. It caused a quick contraction of the leg muscles that caught everyone off guard.

Galvani repeated the contraction-at-a-distance experiment and found that touching the nerve with a scalpel in the absence of a spark did nothing. These experiments showed that electricity could be conveyed through the human body (hardly a new finding) and that the exposed nerve of the frog could be activated with electricity to produce a seemingly natural muscle contraction (again, not particularly new). Galvani, however, went a step further. He emphasized that the muscle contraction occurred without the usual circuit involving direct contact with wires and machines. Without the closing of a circuit, he argued, the spark must have triggered natural electricity in the nerves themselves!

Galvani, impressed with his serendipitous finding, set forth to vary his basic experiment even more. The result was his second contention—a long wire can take the place of a person in the experiment. Moreover, it did not matter if living frogs were substituted for isolated nerves and muscles, or if sheep and chickens were substituted for frogs. The experiments failed only if his assistant cut the long wire or if the researchers tried to use nonconductors, such as silk or glass.

In Part II of his *Commentary,* Galvani described experiments with atmospheric electricity, some dating back to 1780. Studies with lightning rods attached to frog limbs led to his third major point—the muscles can be made to twitch even without man-made generators. Functionally, there is no difference between Nature's own lightning and shocks from electrical machines; both are capable of stimulating frog legs. Energized by Benjamin Franklin's famous kite experiments, Galvani seemed oblivious to the fact that lightning had killed a number of other foolhardy experimenters who set forth to study or capture it.

Galvani's last finding was that two dissimilar metals making contact with a nerve attached to a muscle can make the muscle contract. He observed this when he used brass hooks to hang fresh frog legs and live frogs from the iron railings outside his house. His intent was to study how changes in atmospheric conditions could affect his preparations. He noticed that the frog legs moved not only when there were thunderstorms but also when the sky was cloudless. The contractions occurred more reliably when he deliberately pressed the brass hooks holding his specimens against the iron railings. Comparable experiments conducted inside his house had precisely the same effect, showing that it did not have to do with atmospheric electricity. Galvani then tried different metals and found that some combinations worked better than others, whereas using just a single metal had no effect at all.

Not all of the experiments described by Galvani were successful. In contrast to his work on the peripheral nerves and musculature, he was unable to get muscles to contract by stimulating the brain itself. In this he was not alone. But by 1803 Giovanni Aldini, his devoted nephew and a tireless coworker, had completed a number of successful experiments on the exposed brains of oxen.[30] Aldini showed that he could produce movements of the eyelids, lips, and eyes. On an even more surrealistic note, he also collected fresh human heads at the base of the guillotine and found that he could evoke grimaces, jaw movements, and eye openings by passing current through the brain (Fig. 8.5).[31]

Galvani probably began his work with few preconceived notions, but he clearly developed a thesis as he progressed. In Part IV of his *Commentary,* he proposed that animal electricity is secreted by the brain and distributed through the inner core of the nerves to the muscles. The nerves, he explained, must have a fatty or oily covering, which prevents leakage of the electricity to surrounding tissues. His "proof" of this insulation came when he "distilled" a few nerves and obtained some droplets of oil. He further hypothesized that the nerves lose their insulation where they come in contact with the muscles. The muscles, in turn, receive and store the electricity like Leyden jars. Only when there is a proper trigger will there be a discharge.

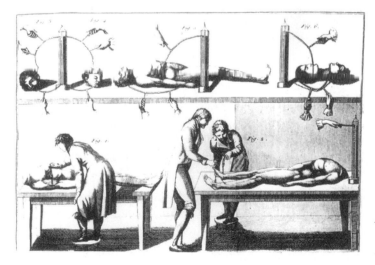

FIGURE 8.5.

A plate from Aldini showing experiments with electricity being performed on humans after decapitation.

Impact

Galvani's landmark treatise of 1791 had a major impact on other scientists, even though some of his experiments and the very idea of animal electricity were not completely original. In retrospect, it was really Galvani's planning, execution, and interpretation of a large number of experiments, rather than any one preparation, that commanded the attention of the scientific community. No one before him had brought such an array of experimental material into the arena. No one's research was able to serve as such a stimulus for other physiologists and physicians to conduct their own studies. And no one had made a stronger case for the concept of intrinsic animal electricity in birds, tortoises, sheep, and even humans than did Luigi Galvani in 1791.

The theoretical jump to mammals now seemed complete. It began with the ancients discovering electric fish. It turned to frictional machines for producing electricity in the seventeenth century, was extended to atmospheric electricity by Franklin, and then was led back to electric fish by Walsh. Now, because of Galvani, many scientists saw animals without specialized electrical organs for shocking their prey operating by a subtle electrical force—one that could travel with incredible speed from the brain through insulated nerves.

Theories based on animal spirits, nerve fluids, or the idea of nerve vibrations were now discarded. Electricity, proclaimed by Galvani to be the true agent of nervous action, was by far a more plausible force. With electricity flowing through the nerves, scientists finally had an animating power that was natural, and one they could see, manipulate, and try to measure.

Johann Friedrich Blumenbach, a leading physiologist and anthropologist who worked in Göttingen, Germany, quickly recognized the importance of Galvani's experiments and ideas. He contrasted them to the highly speculative ideas of nerve action proposed by Descartes and Willis in the previous century. In 1795 Blumenbach wrote the following about Galvani:

> By the combined labours of experimental physiologists in different parts of the world, this branch of science was at length matured for giving birth to another discovery, which will probably be found of equal importance, in explaining the phenomena, and in removing the diseases of the animal system, with that which consigned to immortality the name of the illustrious Harvey. The discovery to which I wish at present to direct the attention of the reader is that of, what is usually called "animal electricity," or, the existence and operation of a fluid extremely similar to electricity in the living animal system. For the fortunate Galvani, professor of anatomy at Bologna, was referred the honour of lighting by accident on this beautiful and divine discovery—a discovery which entitles its author to be ranked with the great promoters [of] science and the essential benefactors of man.[32]

Emil du Bois-Reymond, the leading neurophysiologist in the mid-nineteenth century, wrote that Galvani created a scientific storm, equaled only by the political upheavals occurring in Europe at the end of the eighteenth century.[33] Tongue in cheek, he went on to express dread for the future of European frogs. With thousands of zealous scientists out to catch, dismember, and electrify them, these once common and truly lovely creatures, he feared, would soon be headed for extinction.

As for practicing physicians, they increasingly believed that electricity was indeed the long-sought magic pill (The new theriac) that could cure just about any ailment. Galvani himself strongly advocated electrotherapy and, with his concept of animal electricity, found a way to explain seizures and other disorders of the nervous system.

Unfortunately, Galvani enjoyed little of the fame that followed the publication of his *Commentary*. He never recovered emotionally from his wife's death in 1791, the year in which the *Commentary* appeared. In addition, after Napoleon created the Cisalpine Republic in Italy, Galvani refused to take an oath of allegiance to the new state. As a result, he lost his appointments and affiliations in 1798. To make matters worse, another Italian, a man considered much more knowledgeable about physics and electricity than Galvani, was now assailing him and what he had to say.

The Fight with Volta

Alessandro Volta (Fig. 8.6) was born in the beautiful northern Italian town of Como in 1745. Attracted to the electrical sciences even before his eighteenth birthday, he was a skillful and imaginative physicist who paid attention to even the smallest details. Holding a chair at the University of Pavia, he was the acknowledged authority on electrical matters during the closing decades of the eighteenth century.

Volta replicated many of Galvani's experiments and at first concluded that Galvani was a hero worthy of praise for his scientific acumen and conclusions. In 1792 he called his experiments "great" and "brilliant." Soon afterward, however, he began to wonder whether Galvani was correct in his interpretations. The more Volta thought about it, the more he became convinced that *elettricità metallica* (electricity from two different metals) could explain practically all of Galvani's findings.

By 1793 Volta's short period of admiration for his compatriot had come to an end.[34] He now voiced the opinion that Galvani's experiments did not prove that *elet-*

FIGURE 8.6.

*Alessandro Volta
(1745–1826).*

tricità animale (intrinsic animal electricity) really existed. Electricity from the torpedo and other specialized fish was one thing. But electricity generated by frogs and mammals? This was unproved and probably nonsense. As far as he was concerned, Galvani and Aldini were only able to show that electricity can be a very powerful stimulus for nerves and muscles. They did not prove that muscles contract under natural conditions because of electricity intrinsic to the body.

Volta was correct in pointing out that many of Galvani's preparations were flawed. Galvani, who hated controversy, understood what Volta was saying about experimental artifacts and did not deny that some of his findings could have resulted from his use of different metals. He therefore set forth to demonstrate the existence of animal electricity without using different metals. With Aldini to help, he designed several experiments to demonstrate that muscle contractions could be triggered with just one metal or, even better, no metals at all.[35]

In one such experiment, they dipped the end of a nerve and its detached leg muscle in mercury. They found that the nerve was still able to stimulate the isolated muscle, even though only a single metal was involved. In this case, the mercury could be no more than a conductor of electricity from one tissue to another.

Another important experiment was published as a supplement to a treatise that appeared in 1794. The name of the author was not given, but authorities are convinced that both the treatise and its supplement were written by Galvani, probably with the help of Aldini.[36] The experimenter exposed a frog muscle and then cut the spinal cord. When the cut end of the spinal cord touched the muscle, there was a reliable twitch. No machines, scalpels, wire, mercury pools, or brass hooks were even involved.

In 1797 Galvani followed up on this experiment by showing that the nerve from the leg of one frog can stimulate a nerve in another frog's leg and make a muscle contract. Here, he argued, was the best evidence yet for the thesis that animal electricity exists in more than specialized fish.

After Galvani died in 1798, Aldini, now a professor of physics at the University of Bologna, continued the fight to defend his uncle's theory of animal electricity against Volta's continued onslaughts.[37] Given the evidence now favoring animal electricity, Volta should have thrown in the towel and not returned to the ring to fight another round. Unfortunately, his sharply worded criticisms, based on such things as imperceptible differences in the mercury pool or humidity factors, continued almost unabated. In all, Volta wrote twenty memoirs and many letters between 1793 and 1800, always insisting that frogs, sheep, and humans were not generators of electricity. For reasons of his own, he remained absolutely fixated on the idea that metals or extrinsic factors could account for what others were calling intrinsic animal electricity.

Well after the once-healthy scientific debate had degenerated into a heated brawl, Alexander von Humboldt, the dean of German science, entered the fray as an unbiased, independent judge. The originator of the term *galvanism* repeated many of Galvani's experiments in Berlin, including the one in which he let a nerve drop on a muscle. He also designed several new studies. Humboldt's conclusion, published in 1797, was that *both* animal electricity and bimetallic electricity are real phenomena.[38] With his pronouncement, and with Volta now showing more willingness to compromise, the fight seemed over.

After inventing the "pile" (wet-cell battery), however, Volta returned to argue effectively against Galvani's assertion that electricity originates in the brain, travels through the nerves, and is stored in muscles.[39] Without question, Volta's understanding of the production of electricity by dissimilar conductors (such as those in the specialized organs of electric fish) making contact was significant. But most important, and in terms of the bigger picture, after the Galvani-Volta debates had ended the life sciences would no longer be driven by the animal-spirits paradigm of the past.

From Therapy to Gothic Horror

Electrotherapy now advanced even more rapidly, stimulated by new experiments performed throughout Europe, claims of successful clinical trials, and new instruments. The logic was simple: Nerves are electrically excitable, and nervous energy is electrical. From these two premises, nervous diseases can be explained and treated as electrical breakdowns.

Predictably, many clinicians made exaggerated claims for the therapeutic benefits of electricity. Given what some scientists who worked with laboratory animals were telling them, this is somewhat understandable. At times, even the so-called scientific literature resembled a veritable fantasy land.

Perhaps the most bizarre example of this exuberance in the post-Galvani era can be found in the work of a German physician-scientist by the name of Karl August Weinhold.[40] He was among the many experimenters who maintained that the brain was like a battery with attached wires. But unlike most of his learned contemporaries, he was driven to prove this hypothesis.

In 1817, Weinhold described his work on kittens. In one experiment, he removed the cerebrum and cerebellum of a kitten and claimed that he was able to revive the dead animal by filling the cranial cavity with different metals (bimetallic electricity):

> I removed with a small spoon, through an opening at the back of the head, the cerebrum and cerebellum, as well as, by means of a screw probe, the spinal cord. After this, the animal lost all life, all sensory functions, voluntary muscle movement, and eventually its pulse. Afterward, I filled both cavities with the aforementioned amalgam [zinc and silver]. For almost 20 minutes, the animal got into such a life-tension that it raised its head, opened its eyes, stared for a time, tried to get into a crawling position, sank down again several times, nevertheless finally got up with obvious effort, hopped around, and then sank down exhausted. The heartbeat and the pulse, as well as the circulation, were quite active during these observations. . . . Also, body temperature was fully restored.[41]

Weinhold's exaggerated description of this unfortunate kitten, and several related experiments with metals replacing brain, could easily have come from an imaginative novelist. In fact, his narrative is in many ways like the famous gothic horror story written by nineteen-year-old Mary Shelley at almost the same time. Mary, Percy Shelley, Lord Byron, and Dr. John Polidori each agreed to make up a ghost story while spending some time near Geneva, Switzerland. Their nightly conversations covered terror, theories of the origin of life, and galvanism.

The Prometheus legends especially intrigued the members of this group. Indeed,

Byron had written his celebrated poem "Prometheus" in 1816. In the original Greek myth by Aeschelus, Prometheus gave mankind fire from the sun, for which he was severely punished by Zeus. In a second version, which was more popular in Rome, Prometheus re-created people by giving life to a clay figure. The two ideas were fused in the second or third century A.D. The result was a story about how Prometheus stole fire from the sun and then used it to give life to inanimate human forms.

Mary Shelley, who enjoyed the Prometheus myths, was also familiar with some of the medical developments involving galvanism.[42] She and Percy read and discussed medical science books, and she had attended lectures by leading scientists. She also had extensive conversations with Dr. Polidori, who had recently obtained his medical degree from Edinburgh. Percy Shelley himself had studied chemistry before being expelled from Oxford. He once tried to cure his sister's skin disorder with electricity and, to his dismay, managed to electrocute the family cat in the process.

Mary Shelley published her novel in 1818.[43] Because electricity replaced the sun as the life-giving source in *Frankenstein,* she gave it the subtitle *The Modern Prometheus.* Throughout the first edition, and even more in the second edition, dated 1831, there were allusions to electrical machines, lightning, and the remarkable powers of galvanism (Fig. 8.7).

In the preface to a revised version of *Frankenstein,* she told how the group's earlier conversations about galvanism helped her construct the story:

> Many and long were the conversations between Lord Byron and [Percy] Shelley, to which I was a devout but nearly silent listener. During one of these, var-

FIGURE 8.7.

Victor Frankenstein and his creature. This plate appeared at the front of the 1831 edition of Mary Shelley's novel.

ious philosophical doctrines were discussed, and among others the nature of the principle of life, and whether there was any probability of its ever being discovered and communicated. They talked of the experiments of Dr. [Erasmus] Darwin. . . . Perhaps a corpse would be re-animated; galvanism had given a token of such things.[44]

Chapter 5 of her novel opens with the following chilling paragraph:

It was on a dreary night of November, that I beheld the accomplishment of my toils. With an anxiety that almost amounted to agony, I collected the instruments of life around me, that I might infuse the spark of being into the lifeless thing that lay at my feet. It was already one in the morning; the rain pattered dismally against the panes, and my candle was nearly burnt out, when, by the glimmer of the half-extinguished light, I saw the dull yellow eye of the creature open; it breathed hard, and a convulsive motion agitated its limbs.[45]

The difference between Mary Shelley and Karl August Weinhold is that Mary Shelley knew she was writing fiction about the powers of electricity. Weinhold, in contrast, was so caught up in the electrical frenzy that he actually believed he could replace the nervous system of a dead animal with dissimilar metals to do more than just elicit some movements; he was convinced he could restore life. In retrospect, it is clear that both writers were swept up in the spirit of the times and the excitement that followed in the wake of Galvani's *Commentary*.

Back to Science

Animal electricity effectively gave birth to the modern discipline of neurophysiology and also had a profound influence on electrotherapy. Yet in a very real way, the ideas championed by Galvani, Aldini, and even Volta preceded the measuring instruments needed to verify them. This is why the scientific community was so excited during the 1840s and 1850s, when Emil du Bois-Reymond began to build recording instruments (galvanometers) sensitive enough to detect electrical changes in nerves, and when his close friend Hermann Helmholtz indirectly, but correctly, estimated the speed of nerve conduction.

But this is jumping ahead of other developments. Even before the first quarter of the nineteenth century was over, another idea emerged that proved just as stimulating to the scientific and medical community as electricity. Some scientists began to assert that different mental "faculties"—such as language, mathematics, and even love of offspring—are controlled by distinctly different parts of the gray mantle of the brain, the cerebral cortex.

To appreciate how different functions of mind were assigned to specific cortical territories, we must now travel to Vienna to meet Franz Joseph Gall. En route we shall come upon Emanuel Swedenborg, a brilliant man who just might have been the first scientist to write in detail about cortical localization of function, but who is not often remembered for any of his thoughts about the brain.

9

Franz Joseph Gall
The Cerebral Organs of Mind

> The object of my researches is the *brain*. The cranium is only a faithful cast of the external surface of the brain, and is consequently but a minor part of the principal object.
>
> —*Franz Joseph Gall, 1796*

> The opinions of Drs. Gall and Spurzheim . . . are a collection of mere absurdities, without truth, connexion, or consistency.
>
> —*Anonymous Reviewer, 1815*

One Organ of Mind

The major achievement of the brain sciences in the eighteenth century was a better understanding of the mysterious agent by which messages are sent through the nerves. But in contrast to the excitement caused by scientists experimenting with electricity, new ideas about the organization of the cerebral cortex were not forthcoming. From 1664, the year in which Willis published his *Cerebri anatome,* to 1791, when Galvani argued for "animal electricity" in his *Commentary,* the dominant belief was that the outer mantle of the cerebrum constitutes a single unified structure, one not divisible into distinct functional parts.

During the period in which Bach and Handel were composing music and Voltaire was writing plays, scientists seemed to have no inkling of the fact that different parts of the cerebral cortex control different functions. Only one person saw the world of the brain differently. His name was Emanuel Swedenborg, and in medical circles he was virtually unknown.

Swedenborg's World

Emanuel Swedenborg (Fig. 9.1) was a handsome man blessed with exceptional intuitive powers and extraordinary energy.[1] Born in Stockholm in 1688 and educated in mathematics at Uppsala University, he admired Sir Isaac Newton and hoped to follow in his footsteps. In fact, he drew plans for flying machines and submarines and established himself as an authority in mathematics, astronomy, and mining before he turned to the life sciences in 1736.

Once Swedenborg decided to study anatomy and medicine, he left his post as director of Swedish mines and visited medical centers in France, Italy, and the Nether-

FIGURE 9.1.

Emanuel Swedenborg (1688–1772). (Courtesy of the Royal Swedish Academy of Sciences, Stockholm.)

lands. He learned all he could about the nervous system. His guiding belief was that by acquiring every bit of knowledge he could about the brain, he would be able to have a better understanding of the relationship between the body and the soul.

Swedenborg admitted that it was not his own anatomical investigations that led him to see the brain differently. Instead, it was his ability to synthesize what others were finding that contributed most to the development of his new ideas. In fact, Swedenborg stopped conducting his own experiments because he felt they limited the objectivity he needed. His strategy was to place every notable fact in front of him for careful reflection and rational analysis.

Swedenborg agreed with Thomas Willis and his own contemporaries that the large cerebral mass above the lowly brainstem must be associated with understanding, thinking, judging, and willing. Comparative anatomy and pathology made this abundantly clear. Between 1740 and 1745, however, he took this idea one giant step further. He wrote that different functions must be represented in different parts of the cortex, an idea missing from the writings of Willis and one far from everyone else's mind at the time.

Swedenborg saw specialized cortical territories as the best way to account for the ability of the cortex to function "without confusion or disorder," to use Steno's words from the previous century.[2] After all, we do not confuse what we hear with what we see, nor do we mix up taste and touch. Further, cortical localization seemed to be the only way to explain why higher functions (memory, cognition, perception) are not equally affected by diverse injuries of the roof brain.

By 1745 Swedenborg had localized muscle control in a cortical region that includes the back of today's frontal lobes. He also described how the muscles of the feet are controlled by the upper convolutions, those of the middle part of the body by the middle convolutions, and the head and neck muscles by the lower convolutions. He even maintained that the motor cortex is responsible only for voluntary movements. More autonomic and habitual movements, he concluded, must be mediated by more primitive motor control centers, such as those in the corpus striatum, cerebellum, and spinal cord.

Although Swedenborg's three cerebral convolutions encompassed a larger territory than is recognized for the motor cortex today, he was quite accurate in his statements. As we shall see in Chapter 11, his ideas about the location of the "voluntary" motor cortex and the upside-down representation of the body surface will be verified experimentally in the 1870s.[3]

Swedenborg was also well ahead of his time when it came to linking the frontal lobes with the intellectual functions of the brain. He wrote:

> If this anterior portion of the cerebrum is wounded, then the internal senses—imagination, memory, thought—suffer; the very will is weakened, and the power of determination blunted. . . . This is not the case if the injury is in the back of the cerebrum.[4]

Given that Swedenborg was so accurate in his statements about cortical organization, why is his name so rarely mentioned in contemporary books on the brain sciences? The answer to this question is complex. One problem is that his published work of 1740–1741 never caught the attention of the medical community.[5] Another is that a follow-up book was never completed.[6] Further, these books were really religious in nature.

Even more sadly, Swedenborg's brain treatises, which were more to the point, were not discovered until 1868 and not translated for decades.[7] The thoughts expressed in these papers could have made Swedenborg famous among brain scientists, but because they remained unknown for so long, they had no impact at all on the emergence of cortical localization theory.[8]

Swedenborg's ideas about cortical localization of function could also have become better known had he lectured from a university post, but he did not have an academic appointment. Additionally, it would have helped if he had stayed with the sciences and satisfied the image of a stable investigator.

In 1744, however, he began to experience mystical visions, many of which involved communicating with the dead. Dedicating himself to reinterpreting Christianity and rewriting Scripture, he fled to London, only to die there impoverished. The Swedenborgian (New Jerusalem) Church was founded in England in 1784, twelve years after the immensely talented man whose ideas could have put the brain sciences on the fast track was laid to rest.

The Background of Franz Joseph Gall

With Swedenborg never really in the picture, the monumental job of leading an all-out assault on the notion of a functionally indivisible cerebral cortex was left to Franz Joseph Gall (Fig. 9.2).[9] He was born in 1758 in the small southwest German town of Tiefenbrunn (Baden); his father was a merchant of Italian descent (the family name had once been Gallo) who hoped his son would one day become a priest. The pallid young man's interests, however, lay elsewhere—he was intrigued by the more earthly study of medicine. Youth prevailed, and Gall began his medical studies in Strasbourg.

Gall married in Strasbourg, after which he moved to Vienna to complete his med-

FIGURE 9.2.
Franz Joseph Gall
(1758–1828).
(From J. B.
Nacquart,
Traité sur la
nouvelle
physiologie du
cerveau.
Paris: Leopold
Collin, 1808.)

ical degree. There he proceeded to establish a name for himself as one of the most outstanding physicians in Austria. In 1794, when the physician responsible for the health of the Holy Roman Emperor Francis II retired, he was nominated to replace him. In order to preserve his independence, however, Gall declined the honor, saying he was the wrong man for the job; other physicians he knew of would be better suited for court life.

The Insight

During the final decades of the eighteenth century, many scientists were trying to associate bodily features with personality characteristics. In particular, physiognomy, the art of judging character from facial features, became very popular at this time. Gall agreed with a basic tenet of the physiognomists, who held that where there is variation in structure, there must also be variation in function. But unlike them, he chose to focus on the brain itself.

By 1792 he was convinced that the cerebral cortex must be composed of many different specialized organs. Four years later he was giving public courses promoting the idea that the development of the different cerebral organs is reflected in the pattern of bumps on the overlying skull. This was his *Schädellehre*, or "organology," the part of his theory best remembered, although less than positively, today.

Gall traced his theory of cortical localization to some vivid childhood observations. When he was only nine years old and living with an uncle in the Black Forest, he was intrigued by a classmate who was excellent at memorizing verbal material, his own ability to memorize passages being very poor. Individual differences in verbal memorization skills never left his mind during his university years. Suddenly it occurred to him that the outstanding memorizers had a physical feature in common—large, protruding eyes. This was a characteristic that he believed the poor memorizers did not possess. From this insight, he postulated that bulging eyes, or "ox-eyes," are the result of the excessive development of an underlying cortical area governing verbal memory.

Expanding the Theory

From this one localization, it was relatively simple for Gall to conceive of the cerebral cortex as a mosaic of many specialized organs. Each organ would control a particular mental function, but each would also be tied to others, including its twin on the opposite side of the brain, to assure cooperation. By the time he finished correlating skull features with mental functions, he had defined twenty-seven faculties of mind and located each on both the left and right hemispheres of the human brain (Fig. 9.3).

Nineteen of our faculties, he maintained, are shared with animals; only the remaining eight are distinctly human (Table 9.1). Among the faculties common to humans and beasts are "reproductive instinct," "love of one's offspring," "affection," "destructiveness or tendency to murder," and "desire to possess things." In contrast, some faculties unique to people are "wisdom," "poetic talent," and "religious sentiment."

Although he believed head size could be a good indicator of overall mental power, Gall was convinced that this measure revealed nothing about how the faculties of mind are organized. An ox may have more brain than a dog, but the ox is hardly more intelligent. To understand the specifics of one's character, he maintained, it is necessary to study the individual parts of the overlying skull, since each part reflects the size and shape of the brain below it.

Gall, like Swedenborg before him, further surmised that the frontal lobes must house man's greatest attainments and social characteristics. Functions that humans are more likely to share with the animals must be found more toward the back and underside of the brain. Gall based these ideas on the fact that the frontal lobes in particular are much smaller in animals than in humans. Most important, these thoughts seemed consistent with what he was able to determine by interviewing people with differently shaped crania.

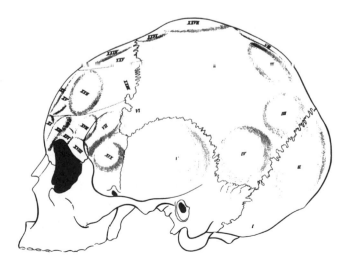

FIGURE 9.3.

Plate XCIX from Gall's atlas, showing the approximate locations of the various faculties of mind. (From F. J. Gall, Anatomie et physiologie du Systèm Nerveux.... *Paris: F. Schoell, 1810.)*

Table 9.1   *Gall's Faculties of Mind*

Faculties Shared by Humans and Animals

1. *Reproductive instinct*
2. *Love of one's offspring*
3. *Affection or friendship*
4. *Instinct of self-defense, or courage*
5. *Destructiveness, carnivorous instinct, or tendency to murder*
6. *Cunning*
7. *Desire to possess things*
8. *Pride*
9. *Vanity or ambition*
10. *Circumspection or forethought*
11. *Memory for facts and things*
12. *Sense of place*
13. *Memory for people*
14. *Memory for words*
15. *Sense of language*
16. *Sense of color*
17. *Sense of sounds, gift of music*
18. *Sense of numbers*
19. *Mechanical or architectural sense*

Distinctly Human Faculties

20. *Wisdom*
21. *Sense of metaphysics*
22. *Satire and wit*
23. *Poetic talent*
24. *Kindness and benevolence*
25. *Mimicry*
26. *Religious sentiment*
27. *Firmness of purpose*

Censorship in Austria

As Gall's public lectures became more and more popular, Dr. Stifft, the man he had personally recommended to be the emperor's physician, turned sharply against him. He called Gall's doctrines subversive, dangerous, and ill-founded, and took it upon himself to convince the authorities that they had to stop him from lecturing.

Although Gall tried to respond to the charges, the conservative Catholic authorities were not willing to listen. They were disturbed by the fact that he had been linking intellect to a material substance, the brain, rather than to the immaterial soul. They were also upset with his contention that character is basically fixed at birth and highly resistant to change. As they saw it, Gall was championing materialism, atheism, and fatalism bordering on heresy.

On December 24, 1801, Gall received a personal letter from Francis II. One part of the Hapsburg monarch's threatening "Christmas letter" read: "This doctrine concerning the head, which is talked about with enthusiasm, will perhaps cause a few to lose their heads, and it leads also to materialism."[10] Two weeks later Gall was ordered to cease lecturing by the Austrian government. The occasion was marked by a "General Regulation," which prohibited all public lecturing unless special permission could be obtained from the authorities.

Gall never bothered to apply for permission. He was not blind to where the emperor stood on the issue, the strength of his opponents, or the reasons for the General Regulation. Disgusted, he finished some projects and said good-bye to those friends who had supported him in Vienna. In 1805, at age forty-seven, he left the beautiful but overly conservative country where he was denied basic speaking rights, and he began to travel.

Travels

Gall's stated objective upon leaving Austria was to lecture in major cities and university towns of Europe's more liberal countries. He spoke tirelessly on crania, brains, and basic personality traits in the northern provinces of Germany during the summer of 1805. He even lectured before the royal family in Berlin, where two medals were minted in his honor. In addition, he traveled to Denmark, Holland, Switzerland, and France, where his lectures were also well received.

Gall gave demonstrations in dissection, hoping to prove to incredulous anatomists that different parts of the cerebral cortex are connected to different specialized structures in the brainstem. To accomplish his objective, he began his brain dissections with the spinal cord or lower brainstem, and proceeded to trace the nerves to successively higher levels.

Gall was seen as a first-rate anatomist by just about everyone who attended his dissections. In 1805 Johann Christian Reil, one of the most respected anatomists and physicians of the day, remarked: "I have seen in the anatomical demonstration of the brain made by Gall, more than I thought a man could discover in his whole lifetime."[11] Even Pierre Flourens, the Frenchman who would fight with Gall over the issue of distinct cerebral organs, praised his skill and methods of dissection. He described Gall as a great anatomist and stated that before he watched Gall do a dissection, he had never really seen or appreciated the detailed anatomy of the brain.[12]

When there were openings in his schedule, Gall seized the opportunity to collect new material for his files. He was especially interested in criminals and the insane, and visited prisons and asylums during his travels. Such side trips allowed him to study the heads of almost five hundred robbers as well as some murderers. He found that many thieves were repeat offenders, which he quickly attributed to an enlarged organ of "acquisitiveness." In contrast, he observed a massive organ of "destructiveness" and a small organ for "love of one's offspring, or philoprogenitiveness," in a woman found guilty of infanticide without remorse (Fig. 9.4).

Gall's Paris Years

Gall reached Paris in 1807 and planned to stay for about one year. To support himself, he built a practice, which grew to include diplomats from twelve embassies and

FIGURE 9.4.

Another plate from Gall's atlas showing a woman who loved children (top) and one who was indifferent to her children (bottom). The faculty for "love of one's offspring" was placed in the back of the cerebrum.

individuals from some of the finest families in Europe. With his wealthy clientele, he lived luxuriously in the French metropolis. The freedom to express himself that he found in Paris led him to remain there and seek French citizenship.

The year 1808 was a particularly important one in Gall's life. On March 14 he applied for membership in the Académie des Sciences and submitted a paper on the anatomy of the brain with his application. Aware of Napoleon's bitter hatred of foreigners, the committee, chaired by Georges Cuvier, rejected his application.[13] They downplayed the importance of his anatomical contributions and stated it was not within their province (being a mathematical and physical committee) to assess the idea that different parts of the cerebrum serve different functions.

Gall was outraged when he learned of this. He felt the committee had never intended to evaluate him fairly. After all, only a paper dealing with "ordinary" anatomy had been submitted with his application. This work was original, important, and did not step on sensitive toes. Yet the committee brought his controversial theory of cranial bumps into their deliberations, even though his organology was not among the material he chose to submit.

In the same year in which he experienced this setback, Gall began writing his monumental *Anatomie et physiologie du système nerveux en général et du cerveau en particulier* (Anatomy and Physiology of the Nervous System in General and the Cerebrum in Particular).[14] The first two volumes and the atlas of this massive four-volume work were published in 1810, and the second two volumes appeared in 1819. With its hundreds of pages of text and one hundred detailed copper-plate engravings, the cost of publication was enormous. The expense was reflected in its nearly prohibitive price of 1,000 French francs per copy.

Between 1822 and 1826 Gall published a revised edition of his multivolume work under the title *Sur les fonctions du cerveau*. In 1835 this appeared in English as *On the Functions of the Brain*.[15]

During this period, Gall's wife died. This ended an unhappy union that was made worse by the fact that Gall had a reputation for being a womanizer who fathered at least one illegitimate child.[16] Three years afterward, in 1828, he suffered a paralytic stroke. Later that same year, the man who saw himself as a misunderstood scientist dedicated to discovering the truth about the brain and behavior died at the age of seventy-one. His body was buried without a religious ceremony. His skull, however, was added to his own large collection of crania, as he had requested. Three years later, his second wife presented the entire collection to a Paris museum in return for an annual pension of 1,200 francs.[17]

A Closer Look at Gall's Methods

Gall's science was based primarily on studies of the cranial features of people from the extremes of society. These people included great writers, poets, statesmen, and musical or mathematical prodigies. They also included lunatics, criminals, the feeble-minded, and deaf, dumb, and blind individuals. The overriding belief was that because certain traits are exaggerated or markedly deficient in these unusual men and women, relationships between brain and behavior could be determined more easily by studying them.

When Gall found a person with a special talent, he examined the form of the head for a cranial prominence and tried to determine the approximate boundaries of the specialized organ. He also tried to talk to people with unusual crania to learn more about their abilities. Unfortunately, he was so firm in his belief about the cranium being a faithful cast of the underlying brain that he rarely looked at the brains themselves.

Gall collected skulls and made casts of those he could not acquire to help him associate cranial features with brain functions. By 1802 he had amassed over 300 skulls from individuals with well-documented mental traits ranging from literary talents to murderous tendencies. He also possessed 120 casts from living persons who had distinguished themselves in one way or another. These skulls, casts, and their associated records served as his most important reference library. After leaving the bulk of his first collection in Austria, he began a new collection in Paris that eventually contained over 600 items.[18]

Although Gall described the study of human skulls as his primary method, he supplemented it with findings from animals. He compared the skulls of different species and also kept an eye open for unusual pets. He was always eager to learn about an animal that would eat only "stolen" food or a lost dog able to find its way home from a distant place. He would then look at the cranial features of these animals and compare them to the skulls of nonexceptional animals of the same species. He also relied heavily on cross-species comparisons.

It is sometimes thought that Gall had absolutely no interest in neurological cases. This idea may have come from his stated conviction that "accidents of Nature" cannot be duplicated and are often associated with secondary effects, including infections, that can make testing difficult and clinical data impossible to interpret.

For these reasons, Gall never claimed to discover a cranial organ on the basis of clinical material. Similarly, he did not hesitate to explain away contradictions to his theories coming from clinical case studies. Nevertheless, he did show an interest in neurological patients whose behaviors confirmed the localizations he had already established by cranioscopic examinations on humans and animals.

Baron Dominique-Jean Larrey, the surgeon whom Napoleon Bonaparte called "the most virtuous man I have ever known," sent Gall some of his battlefield cases.[19] One was Edouard de Rampan, who sustained a fencing injury of the frontal lobe. He showed a loss of memory for words and exhibited a paralysis on the right side of his body. Although this man always recognized Baron Larrey, he was unable to say his doctor's name.

As we have seen, Gall assumed that each organ of mind is represented twice— once in each hemisphere. But if one side of the brain simply duplicates the other, why should a patient like Rampan lose his memory for words after a lesion of just one hemisphere? Gall and his followers were forced to address this neurological riddle. They concluded that an injury on one side of the brain can upset the balance between the two hemispheres, thus affecting the faculties on both sides. This, they explained, is why damage to just one hemisphere can cause dramatic changes in behavior.

The Portrait Gallery

To appreciate more fully how Gall chose, classified, and localized mental faculties, let us now turn to some specific examples from his *On the Functions of the Brain*. Although he emphasized human craniological examinations, we shall now see how he was willing to look at all sorts of corroborative evidence, ranging from naturalistic observations of animals to the study of ancient art, to construct and support his theories.

The faculty of "Destructiveness, carnivorous instinct, or tendency to murder" is a good place to start. The organ for this faculty was localized above the ears for several reasons. First, this area was found to be bigger in carnivores than in grass-eating animals. Second, it was huge in a successful businessman who gave up a good livelihood to become a butcher. Third, a large prominence was found here in a student who was so fond of torturing animals that he became a surgeon. Fourth, this region was discovered to be well developed in an apothecary who later became a public executioner. And fifth, it was found to be unusually large in sadistic tyrants who delighted in seeing others die by the sword or on the gallows.

"Desire to possess things," Gall's faculty to account for the tendency to steal, was located in an area on the side of the head that seemed unduly large in pickpockets. Gall first made this association after observing lower-class boys who ran errands for him. Some had a high sense of morals, whereas others bragged about their thefts or were simply indifferent to their crimes. Examining the heads of these boys, he found that only those who were prone to petty thievery had well-developed prominences in the upper temple region. This association was confirmed by examining the crania of others who stole repeatedly. They included inmates at prisons and in hospitals, as well as troublemakers at institutions for the deaf and dumb. One poor man was so incapable of resisting this temptation that he decided to become a tailor. He explained that this profession allowed him to keep his hands in other people's pockets without fearing the wrath of the law.

As for "Religious sentiment," the organ for this faculty was located toward the front of the top of the head because Gall noted that the most fervent worshipers at church services had rather distinct prominences in this location. He also found a bump here in one of his nine siblings, his exceedingly religious brother. This relationship was confirmed by studying inmates suffering from paranoia with religious themes, by visiting monasteries, and by examining religious art. The latter involved Western portraits of ecclesiastics and ancient art depicting high priests.

Gall's Disciple, Johann Spurzheim

Gall was assisted in his dissections, lecturing, and writing by Johann Spurzheim (Fig. 9.5), a tall man who was born in 1776 in the Mosel region of the lower Rhine.[20] Although his parents hoped he would secure a position in the church, Spurzheim, like Gall, opted to pursue a medical career. The two men first met in 1800, while Spurzheim was studying medicine in Vienna. Spurzheim began to help Gall in 1804, and the two men were inseparable for the next nine years.

Gall had already formulated his doctrine when Spurzheim began assisting him. He found Spurzheim so helpful that he included his name on the first two volumes of his *Anatomie et physiologie*.

Gall and Spurzheim parted company in 1813. The split arose when Spurzheim began to propose a craniological system that differed in several ways from that of his mentor. Although he agreed with Gall's basic premise that the form of the head reflects the development of the brain beneath, he maintained that there are no "bad" or "evil" faculties that have to be suppressed, such as those for "murder" and "theft." These acts, he contended, are due either to the overpowering actions and abuses of other faculties or to the underdevelopment of the moral faculties. Further, Spurzheim took it upon himself to add some new organs, such as "hope" and

FIGURE 9.5.

Johann Spurzheim (1776–1832), Gall's assistant for nine years, who went on to develop his own phrenological system.
(From Capen, Reminiscences of Dr. Spurzheim and George Combe. *New York: Fowler and Wells, 1881.)*

"right" (moral sense), to Gall's original list.[21] By the time he finished, the number of cerebral organs was increased from twenty-seven to thirty-three.

Spurzheim emphasized the importance of training and education more than did Gall, who was pessimistic about changing human nature. He was also responsible for popularizing the term *phrenology*. With the Greek word *phren* (mind) as its prefix and a form of *logos* (discourse) as its suffix, the term was intended to mean "mental science." This word might have been used first in 1805 by the Philadelphia physician Benjamin Rush.[22] Ten years later Thomas Foster applied the term to the organology of Gall and Spurzheim.[23] *Phrenology* was adopted by Spurzheim soon after this, and in 1818 it was incorporated in the title of one of his books, *Observations sur la phrénologie*.[24]

Gall never used the word *phrenology*. It pertained to mind and his emphasis was on the brain. Moreover, once Spurzheim claimed the word as his own, Gall felt its use would have signified his acceptance of Spurzheim's system. Hence, he continued to use the term *organology* or sometimes *zoonomy* to describe his own scheme. As for his methods, he preferred *organoscopy* and *cranioscopy* to *phrenological examination*. Nevertheless, *phrenology* became a rallying word, and before long it was used to describe the ideas of anyone who tried to correlate cranial features with behavior, Gall included.

Gall had little good to say about Spurzheim once his second-in-command left Paris for Great Britain to seek greater glory. In the preface to the third volume of his 1810–1819 magnum opus, he politely criticized Spurzheim's logic and "innovations." But in private, Gall was less restrained and called his one-time partner a plagiarist and a quack.[25]

Although he did not have Gall's blessing, Spurzheim's crusade in England, Scotland, and Ireland proved to be very popular. Stopping first in Vienna to receive a medical degree, he now commenced writing his own books. Within a few years he was made an honorary member of the Royal Irish Academy and was considered for a chair at University College, London.

With the "evil" faculties eliminated from his system, Spurzheim talked optimistically about mankind. Unlike Gall, whose fundamental interest was to learn about the functional anatomy of the brain, his "practical phrenology" was aimed at the masses in a way that brought social reform and individual betterment to the fore. Individuals could now learn more about themselves and live happier lives by turning to science.

The progressive shift into character analysis made "practical phrenology" extremely appealing to criminologists hoping for a better classification of convicts—one based on individual propensities as revealed by skull features rather than on just the act committed. It also appealed to educators, who previously felt compelled to teach all pupils in precisely the same way. Social reformers and psychiatrists who believed that mental illness must have a physical basis also found the system attractive. Based on Spurzheim's teachings, insanity was akin to brain disease, and mental problems could be treated by exercising the moral faculties. In short, phrenology began to play a role as an important liberalizing force in a variety of social and reform movements.

The Number-Three Man, Combe

George Combe, who at first rejected the new doctrine, became an follower of Spurzheim in 1816. For Combe, the turning point came when he was invited by a friend to watch Spurzheim dissect a brain from the brainstem up.[26] He then attended more of Spurzheim's lectures, visited him, and tested his theory on his own. Years later he penned a letter to Spurzheim stating: "Chance brought me first into your presence; but the day when I met you was the fortunate one of my life."[27]

Combe, who was born near Edinburgh Castle, had no formal medical training. He was a barrister and a philosopher who, after meeting Spurzheim, believed that people could promote the growth of the most admirable faculties of mind by venerating God and living morally. Within the context of phrenology, he actively crusaded for charity, worship, temperance, and hard work. Instead of seeing phrenology as a materialistic threat to religion, he argued that phrenology could be beneficial to it.

Along with his brother Andrew and several other learned citizens of Edinburgh, George Combe founded the Edinburgh Phrenological Society in 1820. He then watched its membership swell to over six hundred as it became the leading phrenological association in the world. Three years after its formation, the society began publishing the *Phrenological Journal and Miscellany* under Combe's capable editorship.

Combe became the leading spokesman for phrenology after Spurzheim died in 1832. By this time there were twenty-nine phrenological societies for the well-educated in Great Britain alone. Some of Combe's lectures drew over twelve hundred paying customers.

Combe also distinguished himself with his writing. His *Constitution of Man*, which first appeared in 1827, sold more than seventy thousand copies in Great Britain and the United States by 1838, and well over a hundred thousand copies before the presses stopped running.[28] The Bible and John Bunyan's *Pilgrim's Progress*, which was first published in 1678, were the only two books more likely to be found on the shelves in English-speaking homes at this time.[29]

Combe was even consulted by the British royal family.[30] Prince Albert asked him to examine his son, since the young Prince of Wales seemed to be a slow learner. Combe also evaluated Queen Victoria, but only at a distance. He saw her at an opera and noted that her broad and high forehead ensured the British people a leader possessing firmness of purpose. He then diplomatically added that her cranial features also correlated with great self control and awareness of moral and political principles. To say the least, Combe was both a dedicated phrenologist and a particularly adroit politician.

Phrenology Across the Atlantic

The ideas proposed by Gall, Spurzheim, and Combe were not embraced just in Western European circles. Three Americans, Drs. John C. Warren, John Bell, and Charles Caldwell, all visited Europe and attended lectures on phrenology. They returned to the United States around 1822 to spread the word in their own lectures and by writing books and articles.[31]

Warren, who would later perform Spurzheim's autopsy, expanded his developing research program at Harvard, which involved correlating the skulls of different animals with their behavioral patterns. Caldwell and Bell helped to launch the Central Phrenological Society in Philadelphia, the first of many such societies in the United States.[32] Caldwell also published the first American textbook on phrenology.[33]

Spurzheim visited the United States during the summer of 1832 to rekindle interest in phrenology, which was now on the wane, and to study the natives and citizens of the young republic.[34] He arrived in Boston late in August, expecting to give two series of lectures. Unfortunately, he fell ill and died before his lectures were completed. In accord with his wishes, his skull was detached and delivered to the Harvard Medical School for study.

George Combe also made an American tour. It began in 1838 and ended in 1840. Although he met three American presidents—Martin Van Buren, William Henry Harrison, and John Quincy Adams—his subject matter was no longer new, and he was dismayed by the fact that fewer than the expected number of physicians attended his lectures.

Reflective of the change in sentiment, the *American Journal of Phrenology*, which was founded in the same year as Combe's visit, flopped as a magazine for professionals. In addition, the once-enthusiastic Boston Phrenological Society shut its doors. By the 1840s phrenology as a science was dead; what remained was little more than a mixture of popular entertainment, fortune-telling, and a way for sweet-talking itinerants to make a living as they wandered across the countryside.[35]

The Case Against Phrenology

At the peak of their popularity, the systems proposed by Gall and his disciples were welcomed in many quarters. In fact, as one historian put it, "No scientific discovery, save Darwinism, and, later, Freudian psychology, ever aroused so wide and immediate an interest."[36] Phrenology's advantages were that it was easy to understand, exciting, and practical.

But while many people were swayed by the logic of the phrenologists and impressed with their well-chosen words about science and careful measurement, most members of the scientific community soon followed Professor Thomas Sewall of Washington, D.C., who completely rejected cranioscopy as a method for understanding the physiology of the brain.[37] Sewall pointed out that brain injuries rarely affect the faculties in a manner consistent with phrenological thinking. He also argued that it is not possible to measure brains accurately from the skulls themselves.

A strong reaction against phrenology also occurred back in France, where Gall was perceived by Napoleon and the scientific elite as a threat to French culture and science. Pierre Flourens was Gall's most powerful scientific adversary (Fig. 9.6).[38] In 1822 he was commissioned by the Académie Française to test Gall's theory by conducting brain lesion and stimulation experiments on animals. His resulting message could not have been clearer—Gall's concept of a mosaic of cortical organs was all wrong. As he saw it, all parts of the cortex are responsible for intelligence, voluntary actions, and perception. Citing numerous animal studies in which varying amounts of cerebrum were destroyed, he proclaimed that when one cortical function is affected, so are all the others, and if one function recovers, so do all the rest.

Flourens presented his experimental findings and conclusions in a book first pub-

FIGURE 9.6.

Marie-Jean-Pierre
Flourens
(1794–1867),
the French
experimentalist
who opposed the
phrenologists.

lished in 1824.[39] His theories opposing phrenology were also summarized several years later in his exposé *Phrenology Examined*. In this work he even took a moment to lament: "Descartes goes off to die in Sweden, and Gall comes to reign in France."[40]

The failure of Flourens to find behavioral differences following lesions in different parts of the cerebral cortex can in part be explained by the fact that most of his work was done on animals with poorly developed cerebral hemispheres, notably hens, ducks, pigeons, and frogs. As Gall and his supporters saw it, generalizing from birds or frogs to humans was lamentable, if not laughable. Moreover, Flourens examined little more than sleep and wakefulness, the ability to move without bumping into things, and the capacity to eat and drink typically soon after brain damage.

The battle that took place between the proponents and opponents of phrenology did not just center on new laboratory findings. Flourens also tried to portray Gall as a madman—a person driven beyond reason to build up a sizable collection of skulls and a man to be feared. He was successful in this more emotional endeavor, especially when addressing people who did not know Gall personally. To enhance the desired negative perception of Gall, Flourens included the following passage in his *Phrenology Examined*. It was taken from a letter written in 1802 by Charles Villers to Georges Cuvier:

> At one time every body in Vienna was trembling for his head, and fearing that after his death it would be put in requisition to enrich Dr. Gall's cabinet. He announced his impatience as to the skulls of extraordinary persons—such as were distinguished by certain great qualities or by great talents—which was still greater cause for the general terror. Too many people were led to suppose themselves the objects of the doctor's regards, and imagined their heads to be especially longed for by him, as a specimen of the utmost importance to the

success of his experiments. Some very curious stories are told on this point. Old M. Denis, the Emperor's librarian, inserted a special clause in his will, intended to save his cranium from M. Gall's scalpel.[41]

In addition, Flourens tried to shatter Spurzheim's scientific reputation. He recounted the following story in one of his books:

> The famous physiologist, Magendie, preserved with veneration the brain of Laplace [a leading French naturalist]. Spurzheim had the very natural wish to see the brain of a great man. To test the science of phrenology, Mr. Magendie showed him, instead of the brain of Laplace, that of an imbecile. Spurzheim, who had already worked up his enthusiasm, admired the brain of the imbecile as he would have admired that of Laplace.[42]

The opponents of phrenology also made their opinions known in Great Britain. In an article published in the 1815 *Edinburgh Review,* an anonymous author included the following caustic remark on his very first page:

> We look upon the whole doctrine taught by these two modern peripatetics, anatomical, physiological, and physiognomical, as a piece of *thorough quackery* from beginning to end.[43]

The fire-and-brimstone paper ended forty-one pages later with this long sentence:

> The writings of Drs. Gall and Spurzheim have not added one fact to the stock of our knowledge, respecting either the structure or the functions of man; but consist of such a mixture of gross errors, extravagant absurdities, downright misstatements, and unmeaning quotations from Scripture, as can leave no doubt, we apprehend, in the minds of honest and intelligent men, as to the real ignorance, the real hypocrisy, and the real empiricism of the authors.[44]

Several other British critics followed the lead of the writer for the *Edinburgh Review.* Among their comments were "This bubble! What an outrage on common sense, on natural laws, on scientific facts"; "such ignorant and interested quacks as the craniologist, Dr. Gall"; "these infernal idiots, the phrenologists"; and "We have already said that in our opinion 'fool' and 'phrenologist' are terms as nearly synonymous as can well be found in any language."[45]

Many politicians, writers, dramatists, and artists picked up on the growing negative response to phrenology. The result was a tidal wave of farces, ballads, editorials, plays, and jokes lambasting phrenology and equating it with astrology, palmistry, numerology, and the occult.

One badly written poem targeting Spurzheim was called "Boston Notions" and appeared in *New England Magazine* in 1832. The first verse of eight, with the word *noodles* spelled *noddles,* goes as follows:

> Great man of skulls! I must let loose
> My pen against you;—more's the pity,
> For surely you have played the deuce
> Among the noddles of the city.

> I won't malignantly assail
> Your fame, and say you mean to joke us;
> But faith, I can't make head or tail
> Of all this mystic hocus pocus.[46]

Despite the fact that phrenology was indirectly endorsed by the writings of such literary greats as Honoré de Balzac, Charles-Pierre Baudelaire, Gustave Flaubert, George Eliot, Charlotte Brontë, Walt Whitman, and Edgar Allan Poe, the opinions of its critics prevailed.

As the negative feelings promoted by the sarcastic editorial writers and cartoonists gained momentum, some revisionist physicians who had once praised phrenology now denied ever having supporting such a foolish, unscientific doctrine. Astonishingly, when the Americans Bell, Caldwell, and Warren died, their long obituaries did not even mention what they had done to promote the phrenological movement in the United States.[47]

Gall's Place in History

At an invited lecture presented at the University of Edinburgh almost one hundred years after Gall's death, Grafton Elliot Smith, a leading early-twentieth-century anatomist, spoke kindly of Gall. In contrast to his contemporaries, who tended to portray Gall as a quack, Smith, an Australian, said the following in his 1923 speech:

> The time has arrived for a juster appreciation of the important part played by Gall and a more adequate recognition of his achievements than has been made in the past. . . . His contributions to the anatomy of the central nervous system are of far reaching importance, and to the physiology of the brain and to psychological theory he gave a new orientation and a new inspiration.[48]

Gall's anatomy was exceptional, but his most significant contribution had to be the concept of cortical localization of function. In contrast to Swedenborg, whose neurology was essentially unknown in his lifetime, Gall was the first scientist to make a strong case in public for specialized cerebral organs. Although he was ridiculed by most scientists for his craniometry, his work forced people to entertain the possibility that the cerebral cortex may be composed of a myriad of distinct functional organs—an idea that, from the early 1800s on, would continue to surface at scientific meetings.

Unfortunately, Gall was too biased and unquestioning when it came to collecting data for his theory, as were Spurzheim and the others who also believed in the new doctrine. These men sought only cases that confirmed their theories, and each new case further convinced them that they had the right idea. In the same vein, Gall and his followers were too quick to discard material that contradicted their theories. They always had an excuse for why a person showing a particular skull feature lacked the expected character trait, or for why an individual showing a great talent did not exhibit the right bump.

One example of discarding evidence involved a cast of the right side of Napoleon's skull made a few hours after the emperor passed away. This cast revealed a "miserably small" forehead and a tiny organ for "sense of numbers and mathematics." These two features were completely inconsistent with Napoleon's known personality characteristics. The phrenologists first responded that the cast was a poor rendering of the skull. Then they shot back that the unavailable left part of the skull probably would have been more in accord with Napoleon's personality.[49]

A second celebrated case was that of René Descartes (see Chapter 6), who had a small, depressed forehead, allegedly signifying poorly developed organs of reasoning. Spurzheim himself pondered the dilemma posed by the philosopher's skull before replying: "Descartes was not so great a thinker as he was held to be!"[50] By no means was it mere coincidence that Flourens dedicated his *Phrenology Examined*, his most biting condemnation of phrenology, to Descartes.

If Gall and Spurzheim had been more objective in the way they collected data and more interested in the effects of brain damage, they probably would have been able to see that their localization idea was reasonable, but that their sampling procedures and cranioscopic assumptions were faulty. Both men, however, were oblivious to the standards for data collection that more modern scientists were beginning to demand.

Today it is fair to think of Gall as a visionary with the right idea about cortical localization, but also as a man led astray by poor methodology. In contrast, Pierre Flourens had a better method, but drew the wrong conclusions. Although both men were important pioneers in the study of the brain, each was blind to the other's meaningful contribution.

The bias against phrenology grew so strong after Flourens' condemnation of Gall's system that the parallel idea of using brain-damaged patients or animals to argue for cortical localization of function was shunned by most scientists for decades. But, as we shall now see, the fundamental notion of specialized cortical areas was never really allowed to die, at least not in Paris. In the 1860s Paul Broca will step forth to argue for cortical localization, citing clinical cases to make his points.

Paul Broca

Cortical Localization and Cerebral Dominance

Our observation thus confirms the opinion of Monsieur Bouillaud who places in these [frontal] lobes the seat of the faculty of articulate language.

—*Paul Broca, 1861*

The Shift to Pathology

After arousing considerable interest early in the 1800s, the phrenological movement was shunned by most scientists and practicing physicians. Most of the intellectual elite saw little good in Franz Joseph Gall's cranioscopic methods or in the idea of cortical localization. Yet some individuals were more open-minded. They rejected Gall's system of bumps, but were still willing to consider the idea that the cerebral cortex may be composed of distinct functional areas.

If measuring the skulls of people with remarkable talents or deficiencies could no longer be considered a valid method for learning about the brain, where did these more open-minded scientists look for evidence for the localization idea? The answer to this question is that they studied human patients with brain damage.

Without question, these minority clinician-scientists were more interested in some functions than others. They were especially intrigued by the so-called higher functions of mind. Of these functions, one alone took center stage—spoken language.

Jean-Baptiste Bouillaud

Jean-Baptiste Bouillaud (Fig. 10.1) was the most ardent supporter of speech localization in the cerebral cortex during the second quarter of the nineteenth century.[1] He studied medicine with two famous French physicians, Guillaume Dupuytren and François Magendie. After receiving his medical degree in 1823, he was elected to the French Académie de Médecine and appointed chair of clinical medicine at the Paris Medical School. He also joined the staff of the Charité, one of the famous Paris hospitals. Later he became dean of the Faculty of Medicine in Paris, president of the Académie de Médecine, and commander of the Legion of Honor. Politically,

FIGURE 10.1.

Jean-Baptiste Bouillaud (1796–1881), the French clinician who localized speech in the anterior lobes in 1825.

Bouillaud was far to the left of center, and he was not afraid to champion unpopular positions.

Bouillaud was convinced that diseases could serve as natural experiments to uncover great physiological mysteries. To him, the clinical examination, especially when followed by autopsy, assumed new importance. Moreover, unlike those physicians who were content to report single case studies, he analyzed data from large numbers of brain-damaged patients. In fact, his "library" of clinical cases grew to well over seven hundred in his lifetime. Before Bouillaud and some of his contemporaries adopted the use of large samples and showed how informative such a method can be, this research strategy, which is so important for clinical studies today, had attracted few supporters.

In 1825 Bouillaud published an article and a book, both of which described patients with brain lesions who had lost the ability to communicate effectively with words.[2] He noted that impaired speech was the only notable mental symptom in some of these men and women, most of whom could still use their tongues. This observation suggested that speech may have its own distinct territory in the brain. In fact, after assessing the pathological anatomy, he concluded that speech must be an "anterior" lobe function, just as Gall had surmised some years earlier.

Actually, Bouillaud postulated the existence of two distinct speech areas. He maintained that the intellectual center for word memory, or the executive organ of speech, is housed in the anterior cerebral cortex. The white matter below this anterior gray shell, he theorized, is responsible for executing the movements needed to produce words.

Loss of speech after anterior lobe injuries remained a central theme in many of Bouillaud's publications over the next few decades. He repeatedly defended his basic thesis, citing case after case to anyone who would listen.[3] In addition to his positive cases, he also cited negative cases to support his theory. These cases involved individuals with normal speech functions who had brain damage that spared the anterior lobes.

Bouillaud was seen as a sincere investigator by all who knew him; his integrity was never an issue. Nevertheless, most scientists ignored what he had to say about localization. What he promoted sounded like phrenology. In addition, he had trained with Gall, still defended Gall as a misunderstood genius, and was a founding member of the Société Phrénologique.

But even more important, the medical literature was filled with patients with anterior-lobe damage who still possessed normal speech. There simply was no recognition of cerebral dominance—the idea that the left hemisphere is considerably more important than the right hemisphere for speech. Interestingly, Bouillaud's own data would have shown this left-hemisphere specialization.[4] In 1825, for example, 73 percent of his cases with damage to the left hemisphere had speech defects, in contrast to only 29 percent of his cases with right-hemisphere damage. Bouillaud, however, was obsessed with front-back differences, and did not have the mind-set to analyze his cases in a right-versus-left way as well.

A related problem was that most clinicians had encountered patients with damage sparing the anterior lobes who still had speech problems. For example, in 1840 Gabriel Andral discussed fourteen cases of speech loss with damage that spared the anterior lobes.[5] Because Andral was an established authority, his decidedly negative opinions about localization carried considerable weight, especially in his native France.

Bouillaud, who was not one to be dismayed by even the most formidable of opponents, responded by presenting even more clinical cases linking speech to the anterior lobes. The heated debates that took place in the French academies led him to make what may be the most famous bet in the history of the brain sciences. Late in the 1840s, after collecting hundreds of clinical cases that supported his position, he announced that he would give a substantial prize to anyone who could provide him with an example of a deep lesion of the anterior lobes that did not affect speech.

Many years passed before Bouillaud's prize money was finally awarded.[6] The prize was finally given to a surgeon named Velpeau for a case he had presented back in 1843. His sixty-six-year-old patient had a massive tumor that, Velpeau believed, had "taken the place" of the two anterior lobes. The fact that this patient not only spoke fluently but proved to be loquacious ("a greater chatterer never existed") was not contested when Velpeau claimed the money. Instead, it was argued that the prize money should not be awarded because there was too much sparing in the front of the brain.

The Battle Lines Are Drawn

In February 1861 the respected anatomist Pierre Gratiolet addressed the Société d'Anthropologie in Paris. He presented his observations on the skull of an "Indian" brought back from Mexico by the occupying forces of Napoleon III. Because it was unusually large relative to contemporary French skulls, an argument erupted about whether brain volume was really a meaningful correlate of intelligence. Indeed, the possibility that some "savages" may have bigger brains and more intelligence than the erudite Frenchmen passing judgment on them proved to be a matter of considerable concern.[7]

The idea that intelligence is directly related to brain size quickly turned into a debate over cortical localization. The localizationists did not mince words. They argued that there is a more meaningful correlate of intelligence than overall brain size—namely, anterior lobe development. It is in the front of the brain, they insisted, that our highest human functions (like speech) reside. The Indian skull under consideration, they went on to say, did not suggest tremendous intellect because its large size stemmed from disproportionate growth in the more pedestrian back of the brain, not in the intellectual front.

The leading spokesman for the localizationist position at these debates was young Simon Alexandre Ernest Aubertin, who was married to Bouillaud's daughter Elise.[8] Aubertin, *chef de clinique* at the Charité, agreed with his father-in-law that it would take just one agreed-upon localization to destroy the reigning idea of a single, indivisible cortical organ. With this in mind, he became Bouillaud's right-hand man, and he too proceeded to cite cases from the literature to show that speech is an anterior-lobe function.[9]

At a meeting that took place early in April 1861, Aubertin described a patient who had been at the Hôpital St. Louis in Paris. This individual had deliberately shot himself in the head, exposing his brain. For a while, the dying man seemed to possess both speech and intellect, making him an ideal candidate for a simple neurological experiment.

Aubertin told his audience what happened when light pressure was applied to the wounded man's anterior cortex. In his own words:

> During the interrogation the blade of a large spatula was placed on the anterior lobes; by means of light pressure speech was suddenly stopped; a word that had been commenced was cut in two. The faculty of speech reappeared as soon as the compression ceased.[10]

Aubertin acknowledged that some people with anterior-lobe injuries may exhibit normal speech. He thought this was possible because not all of the anterior cortex need be involved with speech. Moreover, one side of the brain may have the ability to compensate for the loss of the other after some injuries. He therefore maintained that only a massive lesion of both anterior lobes that would leave speech well preserved would force him to reconsider his theory.

He also told his audience that he would publicly renounce his localizationist ideas if a dying patient named Bache, who was at that moment exhibiting severe speech problems, did not show extensive anterior lobe damage on autopsy. With this remark, he hoped to bring the discussions of the Société d'Anthropologie to an end.

But Pierre Gratiolet stood up and fired back:

> I do not hesitate to conclude that all attempts at localization, which up to now have been tried, lack any foundation. They are no doubt great efforts, titanic efforts! But when one attempts to grasp the truth at the height of these babbles, the edifice crumbles.[11]

Paul Broca

As secretary of the fledgling Société d'Anthropologie, which he had founded in 1859, Paul Broca (Fig. 10.2) was present at these emotionally charged meetings. In March 1861 Broca even delivered a paper of his own on the relationship between

FIGURE 10.2.

*Paul Broca
(1824–1880).
(Courtesy of
L'Académie de
Médecine, Paris.)*

brain size and intellect.[12] The more he thought about it, the less willing he was to accept the idea that all parts of the cerebral hemispheres function in the same way.

Although more research was obviously needed, Broca felt the localization position promoted by Bouillaud and Aubertin had good clinical support. Additionally, his own microscopic studies revealed cellular variation in the cerebral cortex, also suggestive of functional differences. A third consideration was that localization was consistent with comparative anatomy; comparisons across species suggested a strong relationship between frontal-lobe size and what he construed as intelligence.

But who was Paul Broca, the man who was leaning more and more toward the localizationist philosophy during these debates? The records show that he was born to Protestant parents on June 29, 1824. His birthplace was Sainte-Foy-la-Grande, the same small town near Bordeaux in which Pierre Gratiolet was born. Following in the footsteps of his father—who had served with distinction as an army surgeon and was present at the Battle of Waterloo (as was Bouillaud)—he opted to study medicine. This decision took him to Paris, where he excelled in his medical studies and completed his degree in 1848.[13]

In a surprisingly short time, the talented young surgeon who was skilled in languages, painting, and music rose to become a respected figure in French science. He first made a name for himself by showing that cancer cells can be spread through the blood, and with his discoveries about muscular dystrophy and rickets.[14] Broca's early discoveries in pathology, coupled with his strong belief that laboratory and clinic must join forces to advance medicine, helped him to secure several very desirable appointments in the Paris hospital system.

Scientific societies, where ideas could be presented and debated by interested parties, were popular in Paris at this time. Broca joined the Société Anatomique

(Anatomical Society) in 1847 and, a few years later, became a member of the Société de Chirurgie (Surgical Society). Not satisfied with spending several afternoons a week at the meetings of these societies, he founded the world's first anthropological society. It was at the meetings of his Société d'Anthropologie that scientists tried to account for racial, gender, and social inequalities by looking for differences in brain size, shape, and organization.

Broca's Protestant upbringing in predominantly Catholic France made him particularly sensitive to minority positions in all walks of life. More than just liberal in his politics and science, he was in many ways a crusader like Bouillaud—a man who enjoyed skillfully defending new or controversial issues against the well-entrenched establishment. Nevertheless, Broca was not the sort of person who was quick to take a stand. He preferred to analyze issues from every conceivable perspective before opening his mouth. His cautious approach to language and the frontal cortex exemplified how he dealt with perplexing issues. He could be quiet and pensive, but then, as he became more sure of himself, turn animated and even passionate in stating his convictions.

The Landmark Case

On April 12, 1861, only eight days after Aubertin stirred up the Société d'Anthropologie with his localization challenge, a fifty-one-year-old man named Leborgne was transferred to Broca's surgical service at the Bicêtre hospital for men. Described as mean and vindictive by the other patients, Leborgne had suffered from epilepsy since youth and was hospitalized at age thirty-one, after losing his power to speak. He developed a paralysis on the right side of his body with loss of sensitivity on the same side about ten years after his speech became affected. He now faced an additional problem, cellulitis with gangrene of his paralyzed right leg—the reason he was sent to Broca's surgical service.

Because Leborgne was not expected to live much longer, Broca took the opportunity to test the concept of cortical localization of function in his patient. He invited Aubertin to join him in an examination of the dying man.

> Since M. Aubertin had declared a few days previously that he would renounce it [cerebral localization] if we could show him a single case of well characterized aphemia [aphasia] without lesions in the frontal lobes, I invited him to see my patient, first of all to know his diagnosis. Was this one of the cases which he would admit as conclusive evidence? . . . My friend found the actual state . . . sufficiently clear to affirm, without hesitation, that the lesion should have begun in one of the frontal lobes.[15]

As expected, Leborgne died less than a week later. An autopsy was immediately performed, and his brain was found to show a "chronic and progressive softening," which was centered in the third frontal convolution of the left hemisphere (Fig. 10.3). The brain was presented to the Société d'Anthropologie one day later. Broca issued only a brief statement about the lesion and the loss of articulate speech at the time. Nevertheless, he promised a more detailed report in the near future.[16]

True to his word, Broca presented a more complete description of the case of Leborgne at a Société Anatamotique meeting four months later.[17] At the meeting and

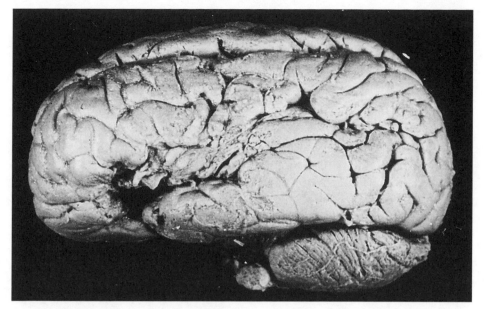

FIGURE 10.3.

The brain of Leborgne (nicknamed "Tan"), perhaps the most famous brain in the history of the brain sciences. (Courtesy of the Musée Dupuytren; courtesy of Assistance publique, Hôpitaux de Paris.)

in his second published report on speech in 1861, he described the epilepsy that had plagued Leborgne since childhood, his subsequent loss of speech, and his paralysis of the right side. He also described the growth of the lesion from its probable point of origin in the third frontal convolution to other parts of the brain.

Broca called Leborgne's inability to speak *aphémie* (in English *aphemia*, meaning "without speech"). He distinguished this disorder from an inability to hear or understand words, as well as from loss of speech due to paralysis of the mouth. Now firmly committing himself to the localizationist idea, he congratulated Bouillaud for amassing considerable clinical evidence in support of the role of the anterior lobes in fluent speech. He also praised Aubertin for seeing merit in his father-in-law's position, for skillfully defending it, and for accurately predicting the site of Leborgne's lesion.

Broca's presentation was received by an enthusiastic audience. The case of "Tan" (the nickname given to Leborgne because he tended to make this sound when he tried to give his name) became a turning point in the history of the brain sciences. It was the landmark case that persuaded many learned men to accept what they and others had previously called blasphemy—namely, cortical localization of function.

Why did Broca's case have such a tremendous effect on neurological thinking? And why did Broca's predecessors, notably Aubertin and Bouillaud, not draw a greater following given the plethora of cases they presented? There are no simple answers to these questions, but a number of important factors obviously converged in the case of Tan.

First, Broca provided more information about his case than could be found in other case reports at this time. There was detail in the case history, in the emphasis on articulate speech (as opposed to other speech defects), and in the search for the precise locus of this one faculty of mind. As a result, most of the salient facts about Leborgne were put on the table.

Second, Broca went out of his way to show that his speech area was different from the frontal lobe localization proposed by the phrenologists, whose ghosts were still haunting the halls of science. Gall had localized speech behind the eye orbits; Broca, in contrast, pointed to the third frontal convolution, a more posterior location off to the side of the head.

Third, the spirit of the times had changed since the deaths of Gall and Spurzheim. The scientific community was now more willing to distinguish between Gall's discredited system of "bumps" and the newer proposal of a neurologically based cortical localization of function.

A fourth factor that made Broca's paper a landmark was his own credibility and reputation. By 1861 he was viewed as a highly respected scientist, a distinguished surgeon, the founder of a prestigious society, and a man of vision. He was also seen as a cautious and fair-minded individual who spoke only when sure of himself. Thus, when Broca stepped to the podium to support the localization movement, the time was ripe and he had amassed a sizeable following willing to listen.

The Broader Role of the Frontal Lobes

Broca did more than argue for a cortical center for articulate language in the frontal lobes. He raised the possibility that the frontal lobes may serve other executive functions, including judgment, reflection, and abstraction. Indeed, in Tan's final years, while his lesion was spreading throughout his frontal lobes, he showed signs of losing his intellect.

The broader role for the frontal lobes was warmly welcomed by Broca's French followers. Among other things, it explained why some individuals with very large skulls, such as Gratiolet's earlier case from Mexico, may not be very bright, whereas others with smaller skulls, namely those doing the evaluations, could be geniuses. In the same way, Broca was now able to downplay the fact that many recently unearthed Cro-Magnon specimens from central France had cranial capacities that far exceeded those of the French who were evaluating them.[18] The great size of the Cro-Magnon cranium, he pointed out, is due to the excessive development of the posterior brain, and is not an indicator of intelligence.

These arguments were soon embraced by most Europeans, who were convinced that Caucasians were superior to the other races, even if they did not have the biggest skulls.[19] In England, Sir Humphry Davy Rolleston later asked whether there are material differences between the brain of an educated man and that of an "animal-like" savage.[20] Rolleston answered his own question with reference to the brain of an Australian aborigine:

> If the convolutions of this Australian brain be compared with those of an average European brain the simplicity of the former is at once thrown into relief. . . . The simplicity of the frontal region is a point of importance.[21]

The Left Hemisphere Is Special

Just months after his victory with Tan, Paul Broca presented the case of another person who had lost his ability to speak fluently.[22] An old man named Lelong had collapsed from a stroke some months before admission to Broca's surgical service.

When examined, he could say only a few simple words, such as *oui, non,* and *tou-jours* (yes, no, always), in addition to his own name.

Because Lelong still seemed to show good comprehension, Broca concluded that his intelligence was not too seriously affected. He also was certain that paralysis of the mouth was not responsible for his speech deficit. In short, Lelong, like Tan, seemed to display a higher-order deficit—an inability to communicate effectively with words.

When Lelong died, his brain was removed and examined. It showed a well-circumscribed lesion in the posterior part of the left frontal lobe. With a case thought to be considerably less complex and less fraught with uncertainty than Tan's, Broca rejoiced over the realization that he could confirm his previous localization so quickly.

Following Tan and Lelong, other cases of speech loss were brought to Broca's attention. By April 2, 1863, he was able to talk about eight cases of *aphémie,* a word that would soon be replaced by Armand Trousseau's French word *aphasie* or aphasia in English[23] Broca was surprised to find that all eight cases showed lesions of the left hemisphere. Temporarily at a loss for words himself, he could tell members of the Société d'Anthropologie only that this was a truly "remarkable" finding. He felt that more cases were needed before he could make a definitive statement about hemispheric differences in function, an unexpected phenomenon that must have struck him as very strange indeed.

The left-hemisphere phenomenon did not evaporate into thin air as additional cases were brought to Broca's attention. Instead, the idea that he was dealing with an unusual chance effect, like a balanced coin repeatedly coming up heads, seemed less and less likely as his sample size continued to increase. New cases exhibiting lesions of the right hemisphere without impaired speech further suggested that the left hemisphere must be special.

Broca's clearest statements and most important thoughts about what we now call cerebral dominance appeared in 1865 in an article published in the *Bulletin de la Société d'Anthropologie.*[24] In it he theorized that the left hemisphere is special or dominant for language because it matures faster than the right hemisphere. Broca reasoned that the two hemispheres, which look so much alike, are probably not very different in innate capacity, but one clearly takes the lead in the case of speech and then dominates over the other.

But earlier in 1865 two other papers on cerebral dominance saw the light of day. One was by Marc Dax and the other was by his surviving son, Gustave Dax, both of whom hailed from southern France.[25] The result was a fight over who should be recognized for discovering cerebral dominance.[26]

The Priority Fight

In March 1863, while Broca was slowly warming to the possibility of cerebral dominance, an obscure country physician by the name of Gustave Dax sent a manuscript to the Académie de Médecine in Paris. He hoped the document would show that his long deceased father, Marc Dax, had recognized the special importance of the left hemisphere for speech before anyone else. His father's revolutionary ideas, he claimed, were presented in 1836 at a congress on southern French cultural achievements held in Montpellier. In addition to including his father's report as a part of his own lengthy paper, the younger Dax presented many more case reports that he had collected to support his father's contention.

The theme of the original Marc Dax paper was that speech defects can be associated with lesions of the left hemisphere alone, even if both hemispheres sustain damage. This association holds for patients who can no longer speak fluently as well as for those who remain fluent but no longer use the correct words. Translated from the French, the title of Marc Dax's 1836 paper was "Lesions of the Left Half of the Brain Coincident with the Forgetting of the Signs of Thought."

Marc Dax based his insight on a large number of patients, some with sword wounds of the skull, others with strokes, and still others with brain tumors. Collected over a twenty-year period, forty cases came from the medical literature and an equal number were derived from his own clinical practice. He did not present autopsy material or explain why one hemisphere should be more important than the other for speech. Nevertheless, he was probably the first person to recognize and discuss cerebral dominance; from all indications, his son was the second.

But could evidence be found to support Gustave Dax's claim that his father's material had been presented in Montpellier in 1836? The local newspaper had published a list of subjects covered at the congress and no mention was made of anything even suggestive of Marc Dax's paper. Acting on Broca's behalf, the kindly town librarian even interviewed twenty physicians who had attended the meetings. Yet no one could remember such a paper. As Broca saw it, Marc Dax might have recognized the importance of the left hemisphere for speech well before he did, but Dax then must have lost the courage to present his surprising findings to the public.

The exact date that Broca learned about the Daxes may never be known for sure. Historians can say only three things. The first is that Broca expressed great surprise about the location of the lesion in his own eight cases just ten days *after* the Dax manuscript arrived in Paris. The second is that Gustave Dax's lengthy paper, which included his father's insightful *mémoire*, was immediately given to a select committee of the Académie de Médecine for secret review, following the established rules of that organization. Nevertheless, receipt of Gustave's 1863 submission was immediately published, and the lengthy title, "Observations Seeming to Prove the Coincidence of Speech Disorders with a Lesion in the Left Hemisphere of the Brain," pretty much told the whole story. Further, there were leaks, and details of the so-called secret material were known well before the three-man committee finally reported back to the society at the end of 1864. And third, by 1865, when Broca published his own celebrated paper on cerebral dominance, he was no longer tentative about the phenomenon.

Because some of Broca's fair-minded contemporaries thought that both Marc Dax and Paul Broca should be recognized for the recognition of cerebral dominance, hemispheric specialization was initially tied to the names of both men. Over the years, however, the elder Dax's name slowly disappeared from the literature, furthering the impression in some quarters that cerebral dominance was first recognized by Broca.

Not surprisingly, this unfairness raised the blood pressure of the vitriolic younger Dax, who fought for his father's priority and his own fame to his dying day. In addition to seeing himself as the second person to recognize cerebral dominance, Gustave claimed that, based on a diagram initially prepared by his father, he was the first to recognize that the left temporal lobe is especially important for speech.[27]

Today, after being ignored for years, the two Daxes are again receiving some recognition for their insights, thanks to a growing number of historians who feel that these two nineteenth-century explorers of the brain have for too long been denied a rightful place in the pantheon of science.

Exceptions and Therapy

In his 1865 paper on cerebral dominance, Broca was forced to deal with some recently surfaced exceptions to the idea that the center for articulate language resides in the third frontal convolution of the left hemisphere. One such case involved an epileptic woman at the Salpêtrière, the large Paris hospice where Jean-Martin Charcot was already leaving his mark (see Chapter 12). An autopsy showed that the woman was probably born without Broca's critical region. Nevertheless, she was able to read, speak fairly well, and express her ideas without difficulty. As a result, Broca hypothesized that her healthy right hemisphere had taken over for her compromised left hemisphere. He also postulated that a small percentage of healthy people may be born "right-brained."

Broca then asked himself the same question that many others raised before he did: Why do we not see more sparing and recovery following unilateral lesions if, in fact, one hemisphere has the capacity to compensate for the other? He postulated that one limiting factor may be that most aphasics also suffer from intellectual deficiencies, limiting their ability to relearn. This was something he had repeatedly seen in patients with large frontal lesions.

Broca also lamented that professionals were doing little to retrain their aphasic patients. He suggested teaching aphasics to speak in the same way that a child learns to speak. Therapy should begin with sounds of the alphabet, then words, then phrases, and eventually sentences. By doing this sort of thing, the right hemisphere may find it easier to take control over from its injured counterpart on the left side.

Speech therapy based on this sort of progression had been tried before, but not within Broca's theoretical framework. For example, one of the first systematic attempts at reeducation was described by Jonathan Osborne in 1833.[28] His twenty-six-year-old patient had a different sort of speech problem: He was fluent but unable to repeat words. When first seen after his stroke, he was asked to repeat a passage from the by-laws of the College of Physicians. Osborne read, "It shall be in the power of the College to examine or not to examine any Licentiate, previous to his admission to a Fellowship, as they shall think fit"; his patient responded, "An the be what in temother of the trothotodoo to majorum or that emidrate ein einkrastrai mestreit to ketra totombreidei to ra fromtreido asthat kekritest."[29] Osborne wrote that he taught his patient "to speak like a child, repeating the first letters of the alphabet, and subsequently words after another person."[30] The end product, the good doctor reported, was most satisfactory.

In his 1865 report Broca wrote that he tried speech therapy with one of his own cases. The man was successful in relearning the alphabet and in working with syllables, but did not do well when it came to constructing longer words. Broca was optimistic, however, and expressed the hope that although he had been able to give his patient only a few minutes now and then because of other demands, others would be able to devote more time to this sort of therapy. He thought more good would come from longer, regular sessions.

The Barlow Case

Broca's theoretical statements seemed logical to his contemporaries and to scientists later in the nineteenth century. One of his widely accepted ideas was that the right hemisphere has the ability to take over for the left hemisphere, and is especially likely to do this successfully after birth defects or injuries in childhood. An important case study supporting this possibility came a little more than a decade later. It was presented in 1877 in an article written by Thomas Barlow for the *British Medical Journal*.[31]

The Barlow case had three notable features. First, it involved a boy who had a lesion that resulted in a loss of speech. Second, the impairment was followed by some recovery. And third, the speech problem emerged again after a second brain lesion.

The boy was ten years old when he first suffered a loss of fluent speech. At the time, he also exhibited a paralysis of the right side of his body. People who knew him rejoiced ten days later when he began to regain his ability to speak clearly. The boy seemed fully recovered after a month, but again became aphasic three months later. This time his ability to speak did not return, although he died not very long after the second attack.

An autopsy on the child's brain revealed calcium deposits and evidence of blocked blood vessels in the third frontal convolution of each hemisphere. The critical feature was that the right hemisphere appeared to have been damaged weeks after the left hemisphere (Fig. 10.4). This characteristic suggested that the right hemisphere had taken over speech functions when Broca's area in the left hemisphere was damaged, only to lose its newfound ability following the second lesion.

The Barlow case was cited by many scientists as evidence for takeover of function.[32] Charlton Henry Bastian, a British neurologist, seemed to be one of the few who disagreed with this popular interpretation, and for good reason.[33] In a statement that was all but ignored, he argued that ten days was too short a time for complete transference of function. He also pointed out that Barlow never told his readers whether the boy was strongly right-handed. He wondered: Could this boy's right hemisphere have been the leading hemisphere for speech from early childhood?

During the twentieth century, it was discovered that most left-handed people still have speech localized in the left hemisphere; a smaller number of left-handed individuals have speech more evenly distributed between the two hemispheres (a phenomenon now called mixed dominance), and even fewer have true right-hemisphere speech. In recognition of these findings, scientists now distinguish between "normal," or "familial," left-handers, a group genetically destined to have speech localized in the left hemisphere, and "pathological" left-handers, who seem to have been forced into their left-handedness by problems in brain development. People in the latter category are more likely to rely on the right hemisphere or both hemispheres for spoken language. Individuals in this group also tend to exhibit more learning disabilities, attentional problems, and "hard" neurological signs, including epilepsy. To his credit, Broca was right when he warned others not to

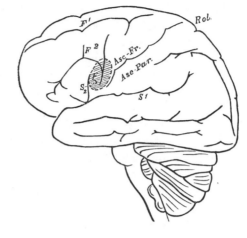

assume that the intellectual side of the brain can be determined with confidence just by knowing handedness.[34]

Jackson and the Right Hemisphere

During the summer of 1868 Broca traveled across the English Channel to the city of Norwich. He was invited to present his thoughts on language and the brain to the British Association for the Advancement of Science. John Hughlings Jackson (Fig. 10.5), the dean of British neurology, was also invited to speak at the session. Because Jackson was not as convinced as Broca that cerebral functions could be localized so discretely, the meeting was billed as a "debate" involving the two best neurological minds of the period.

Broca spoke about different forms of speech afflictions in his presentation.[35] He was treated warmly by Jackson, a shy man who had no intention of engaging such an important guest in any type of debate. To Jackson, the Frenchman was the man of the hour—a visiting hero worthy of praise for his exciting new discoveries.

Jackson had begun to comment on cerebral differences some four years before this meeting, after he learned what Broca had been saying before the scientific societies of Paris.[36] Broca's reports stimulated him to examine some seventy cases with losses or defects of speech. In all but one, the paralysis was on the right side, indicating left-hemispheric injury. As a result, he concluded that Broca was probably correct in associating aphasia with lesions of the left hemisphere.

In 1868, the year of the Norwich conference, Jackson wrote that his aphasic patients usually did reasonably well when examined on perceptual tasks. He then astutely recognized, probably for the first time, that damage to the right hemisphere is more likely to impair spatial abilities than damage to the left hemisphere.[37] Speech and spatial functions, Jackson believed, must be mediated by different brain areas.

Some of Jackson's right-hemisphere injury cases are especially noteworthy. For example, in 1872 he described a man with paralysis of the left side who could not recognize people, including his own wife.[38] This man also had difficulty when it came to recognizing places and things, even though his vision seemed adequate.

A few years late, he presented the case of Elisa P., who had lost her sense of direction.[39] He wrote:

> She was going from her own house to Victoria Park, a short distance and over roads that she knows quite well, as she has lived in the same house for 30 years, and has had frequent occasion to go to the park; on this occasion, however, she could not find her way there, and after making several mistakes she had to ask her way, although the park gates were just in front of her.[40]

Elisa P. died three weeks after she was seen. A large malignant tumor was found in the posterior part of her right temporal lobe.

Jackson coined the term "imperception" to describe the defect of memory for persons, objects, and places.[41] He associated imperception with lesions of the posterior part of the right hemisphere. In his mind, the back of the right hemisphere was every bit as special as the front of the left hemisphere.[42] The right hemisphere may not think in words, but its importance in getting from one place to another, in recognizing people, and even in getting dressed, was undeniable.

The Beast Within

On the basis of clinical material, Dax, Broca, and now Jackson saw the two hemispheres as decidedly unequal. Agreeing with them was Carl Wernicke, who in 1874 wrote that damage to the left temporal lobe is responsible for a disorder in which speech remains fluent but not meaningful.[43] This speech disorder, which had been described countless times in the past but had not been associated with its own specific cortical territory, is now called sensory aphasia or Wernicke's aphasia in his honor.

Wernicke's work added to the growing idea that the left hemisphere is more involved with intellectual functions than the right hemisphere. The right hemisphere, in turn, increasingly was thought to govern skills shared by both humans and beasts, such as locating an object or finding one's way home. With this dichotomy simplifying matters, more and more scientists began to envision the "civilized" left hemisphere as the guardian of right hemispheric and lower structures, that can act in beastly ways if unleashed.[44]

This was the Zeitgeist in the 1870s, the very period in which Robert Louis Stevenson began to work with William Ernest Henley on a play about a man with two very different personalities.[45] The subject of the play was a real person who had lived in Edinburgh a hundred years earlier. He was a respected cabinetmaker during the day and a nefarious burglar at night. *Deacon Brodie; or the Double Life; A Melodrama, Founded on the Facts, in Four Acts and Ten Tableaux* was completed in 1880.[46] The spellbinding play based on opposing personalities within a single body then underwent several revisions.

The idea of developing the theme of dual minds even further came to Stevenson in a terrifying nightmare. The hypnotic influence of Edgar Allan Poe's short story "William Wilson" might also have influenced Stevenson to write his decidedly more famous *Strange Case of Dr. Jekyll and Mr. Hyde*.[47]

The novel itself deals with a respected citizen, Dr. Jekyll, who no longer wants to contend with the war between the two opposing parts of his mind. By drinking a potion, he finds that one part of himself is able to break away in the guise of Mr. Hyde. This evil-looking figure is "more express and single, than the imperfect and divided countenance" of the good doctor, who remains a mix of two opposing minds. Mr. Hyde does not represent violence or lust in the pure sense. Rather, he is an unrestrained, primitive soul who cannot abide by the notion that certain acts, including murder, are not permissible in civilized society.

The Jekyll and Hyde story, which first sold as "a shilling shocker," was an immediate success. Within a few years, every educated person in Great Britain either had read the book or at least knew about the terrifying story from other sources. But let us take a moment to pause and think about the Jekyll and Hyde characters and ponder whether Stevenson, who was clearly more interested in morality than in science, might have been at least partly affected by the science of his day. Was the respected Dr. Jekyll not largely the personification of the cultivated left hemisphere? And was not Mr. Hyde, the doctor's morally defective self, the growing personification of the "primitive" right hemisphere?

The Jekyll and Hyde story could well have affected how some clinicians subsequently viewed their cases. In 1895, nine years after Stevenson's book first appeared, a Scottish psychiatrist by the name of Lewis Bruce published a case study of a real man who seemed to have two distinct consciousnesses.[48] One consciousness of patient H.P. was demented and spoke in gibberish with some Welsh. It could not understand English and was shy and suspicious. When it predominated, the man's left hand (read right hemisphere) was used for writing. His other consciousness showed fluency in English and used the right hand (read "educated" left hemisphere) to write more legibly. Neither had any memory of anything that occurred when the other consciousness was in control.

Bruce hypothesized that the two distinct consciousnesses, which he named the Welsh and the English, were due to the different actions of the two cerebral hemispheres. The more primitive Welsh consciousness was thought to reflect the functioning of the right hemisphere, whereas the English consciousness was believed to reflect the normal actions of the more civilized left hemisphere. But what could account for the switching back and forth?

Two years later Bruce again presented the case of H.P. and followed it by mentioning two other cases from his asylum in Scotland.[49] He noted that his two new patients suffered from epilepsy. Moreover, their personalities changed immediately after a seizure. Bruce now had his mechanism—unilateral seizures could "paralyze" one hemisphere, allowing the other hemisphere to dominate until the first hemisphere recovered. If the seizures paralyzed the civilized left hemisphere, the uneducated right hemisphere could break loose and reveal its more bestial self, much like the educated Dr. Jekyll giving way to the beastly Mr. Hyde.

The ideas expressed by Bruce had potential legal ramifications. Can a right hemisphere (right mind) be found guilty of a crime if it had only been behaving in a natural way? Indeed, no one would want to say that a wild animal is guilty for stealing or killing its prey. And can the left hemisphere (left mind), if diseased, be brought to trial for not being its brother's keeper? However one may choose to answer these questions, one thing was clear in Victorian circles and in the reformist social milieu of the times. Civilized people had to learn to suppress or change the dark right side of the mind.

Movements to Educate the Two Hemispheres

Among some freethinking individuals, there was a growing belief that the inferior right hemisphere could be educated, perhaps even as much as the "intelligent" left hemisphere. In particular, this was presumed to be true early in life. The presumption was that early "bilateral training" could produce a better person, a superior intellect, and a more civilized being.[50]

FIGURE 10.6.

Charles-Edouard Brown-Séquard (1817–1894), who suggested trying to educate the "uncivilized" right hemisphere.

Charles-Edouard Brown-Séquard (Fig. 10.6), a physician close to both Broca and Jackson, stood among those who were convinced that existing educational systems failed to train the right hemisphere as well as the left. In the 1870s he argued that the primitive right hemisphere may become better developed and more comparable to the civilized left in function if the left side of the body could be used more.[51] He wrote:

> I think, therefore, the important point should be to try to make every child, as early as possible, exercise the two sides of the body equally—to make use of them alternately. One day or one week it would be one arm which would be employed for certain things, such as writing, cutting meat, or putting a fork or spoon in the mouth, or in any of the other various duties in which both the hands and the feet are employed.[52]

Such notions were put into practice in some schools in the United States and Great Britain. Further, "ambidextrous culture societies" were spawned to train adults to do two things at once, such as playing a piano with one hand while writing a letter with the other.

The movement to educate the two hemispheres lasted into the twentieth century and was seen by many scholars as a viable solution to the "problem" of the right hemisphere. Nevertheless, ambidextrous training also had its share of sharp, vocal critics. One such person was James Crichton-Browne, a British physician who ran psychiatric asylums and was interested in medical policy (see Chapter 11). He looked on the bilateral training movement as a cult, writing:

> In this present movement in favour of ambidexterity I fancy I detect the old taint of faddism. Some of those who promote it are addicted to vegetarianism, hatlessness, or anti-vaccination, and other aberrant forms of belief; but it must be allowed that beyond that it has the support of a large number of intelligent and reasonable people, and of some men of light and leading.[53]

To say the least, there were two sides to this issue. Although the reformers promised great things, not everyone was willing to risk tampering with Mother Nature.

Broca's Later Years

Although he once stated that science was for the young, Paul Broca remained active during the 1870s, after he turned fifty. Hence, he was able to see some of the ramifications triggered by his own neurological discoveries. It was during the 1870s that John Hughlings Jackson presented his most important cases of "imperception" after lesions of the posterior right hemisphere and Charles-Edouard Brown-Séquard started his push for bilateral training. It was also in this decade that Carl Wernicke argued for a second speech area and Thomas Barlow drew new attention to the phenomenon of recovery of function.

Broca, however, did not write any more papers about speech and the brain after 1877. By this time he was more interested in the remains of early humans, especially those unearthed in his native France.[54] In addition, he had become intrigued by the limbic lobe, a collection of brain parts then thought to be associated with olfaction, but today looked upon as particularly important for emotion.[55]

In 1880 Broca died from heart disease, leaving a wife and three children behind. He had been a hard-driving perfectionist who published over five hundred books and articles during his intense scientific career. He was also a heavy cigar smoker and a man who enjoyed eating rich foods. He passed away only months after he was elected to the French Senate to represent science and medicine, an indication of the esteem in which he was held by his contemporaries.

Broca was a giant in many fields. He contributed significantly to human physical anthropology and wrote important papers on the anatomy of the brain. But most people remember him best for his cortical localization of speech and for his recognition of cerebral dominance just a few years later. In these domains, his acumen as a clinician was instrumental in giving birth to the modern neurosciences, a set of disciplines largely guided by careful neurological observations and a firm belief in localization of function.

David Ferrier and Eduard Hitzig

The Experimentalists Map the Cerebral Cortex

A part of the convexity of the hemisphere of the brain of the dog is motor . . . another part is not motor.

    —*Gustav Fritsch and Eduard Hitzig, 1870*

Experiments on animals, under conditions selected and varied at the will of the experimenter, are alone capable of furnishing precise data for sound inductions as to the functions of the brain and its various parts.

    —*David Ferrier, 1876*

London, 1881

The Seventh International Medical Congress took place in London late in the summer of 1881. More than 120,000 people traveled to Britain's largest city for this huge Victorian spectacle, including scientists, physicians, and royalty from all over Europe.[1] The railroads and steamship lines arranged special travel packages with banquets and parties for the event. There was even a huge reception at the Crystal Palace, a monumental glass edifice. To everyone's delight, the reception was followed by "a pyrotechnic display, the original feature of which consisted in the fire portraits of Sir James Paget, M. Charcot, and Professor Langenbeck"—great men of medicine from England, France, and Germany, respectively. A medal showing Queen Victoria and Hippocrates was issued as a special souvenir of the congress.

The congress had over one hundred sections, each with several presenters. In the session on physiology, which met during the morning of the fourth of August, two individuals with contrasting ideas about brain physiology were scheduled to present their experimental findings and theories.

Friedrich Goltz (Fig. 11.1), a forty-seven-year-old physiologist from Strassburg, Germany, was one of the scientists. He was well known for questioning the view that the cerebral cortex is made up of specialized organs. A heavyset man with a walruslike mustache, he was an engaging speaker with a flair for the dramatic.

David Ferrier (Fig. 11.2), a thirty-eight-year-old physiologist from King's College Hospital, London, was expected to take the opposing position. Unlike Professor Goltz, he believed in specialized sensory and motor areas of the cerebral cortex. With his handlebar mustache, the short, thin, and dapper Scot looked more like a military officer than a scientist who worked with animals.

The well-attended session at the Royal Institute on Albemarle Street met expectations. It opened with Professor Goltz, who told his audience that he had conducted many experiments on dogs in which large amounts of the cerebral cortex had been destroyed. He stated that although the dogs with large cortical lesions became demented, they did not become paralyzed, deaf, or blind. If there really were specialized cortical organs for movement and the different sensory systems, he for one could not find them. To emphasize his point, Goltz showed the audience the surgically damaged skull of a dog and a small remnant of its brain. This dog, he explained, had undergone four brain operations, yet it had still moved about and was not robbed of a single sense before it was sacrificed in the name of science.

Goltz then added that he had even brought one of his experimental dogs with him. This animal, he explained, had both its parietal and occipital lobes removed; like the dog just described, it did not show the specific losses predicted by the localizationists. He then left the podium to a nice round of applause.

David Ferrier now made his way to the front of the room and began from a defensive position, politely criticizing Goltz's experiments and conclusions. Ferrier believed that vision, hearing, smell, touch, and taste each has its own cortical territory, with still another part of the cerebral cortex governing voluntary movements. He felt that Goltz's dogs performed well for three reasons: first, critical brain areas for sensation and movement had not been completely destroyed; second, dogs have relatively small cerebrums and are not dependent on the cortex for much of what they do; and third, the German professor had failed to ask the right questions.

Ferrier then went on to describe some monkeys he was studying together with Gerald Yeo, his surgical colleague at King's College Hospital. One still-living monkey, he explained, could not voluntarily move the limbs on the right side of its body, even though it could see, feel, and hear. This animal, he emphasized, had had its left motor cortex removed seven months earlier. He then mentioned a second monkey that was still alive after undergoing bilateral temporal lobe surgery six weeks earlier. It was completely deaf, but had no trouble using its other senses or moving about. The picture of cortical specificity painted by Ferrier could not have contrasted more with what Goltz had just described.

Before the morning discussions ended, the chairman asked both participants if their animals could be examined later that day.[2] Both Goltz and Ferrier readily agreed, each confident of victory. At three o'clock about seventy-five attendees walked over to King's College to see their animals.

Examination of Goltz's dog did not reveal any specific sensory or motor losses. The dog ran around and reacted to a wide variety of stimuli. For example, when a black and white flag was put on the floor, the dog nimbly stepped around it, and when cigar smoke was blown into its face, it turned away. Here, Goltz explained, was convincing proof that the increasingly popular theory of cortical localization should not be taken too seriously.

Next the two monkeys were brought into the room. One animal dangled an arm and hobbled about in such a way that Jean-Martin Charcot, the neurologist whom we shall meet in the next chapter, gasped, "*C'est un malade*" (It is a patient).[3] The hemiplegic animal used only its left hand to take food from the demonstrator and showed absolutely no ability to utilize its right limbs at will. In all other respects, however, it seemed normal.

In contrast to this animal, Ferrier's second monkey, with damage to its temporal (auditory) cortex, moved about with ease. Yet when Ferrier fired a percussion cap not far from the animal's head, it did not cringe—it behaved as if it were completely deaf.

The onlookers agreed that the dog and two monkeys had been accurately described during the morning session, at least in terms of how they behaved. The pressing question now had to do with pathological anatomy—were the lesions located where they were said to be? From the beginning, Ferrier doubted that the lesions made by Goltz were as complete as the Strassburg professor was maintaining. He felt that significant parts of the dog's cerebral cortex must have been spared.

To determine whether the lesions were placed as described, Ferrier and Goltz were asked if they would object to having the dog and one of the monkeys sacrificed. Again, both men agreed. Subsequently, the two animals were sacrificed and their brains removed and turned over to a panel of experts.[4]

The panel found that the lesion sustained by the hemiplegic monkey was exactly where Ferrier and Yeo said it would be (Fig. 11.3). It occupied the left cortical motor area, which at the time was broadly defined to include both frontal and parietal cortex in the region of the Rolandic fissure (today, only the frontal region is really considered motor). In contrast, just as Ferrier and Yeo suspected, Goltz's dog was found to have more motor and sensory cortex spared than the professor had thought. Recognizing that spared cortex could account for the dog's good performance, the localizationists claimed the victory as theirs. Ferrier was congratulated and that evening toasts were made in his honor.

Unfortunately, Ferrier did not remain happy very long. Three months later he received a summons to appear at the Bow Street police station in response to a criminal charge filed against him by a militant group of animal-rights activists.[5] The Victoria Street Society for the Protection of Animals from Vivisection, well aware of his ex-

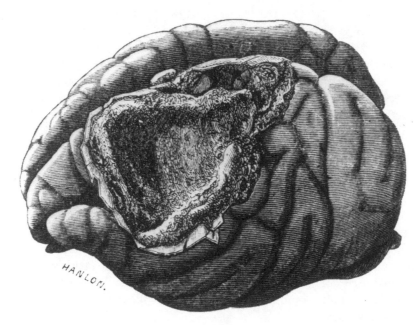

FIGURE 11.3.

The brain of the hemiplegic monkey exhibited by Ferrier at the International Medical Congress of 1881. (From D. Ferrier, The Functions of the Brain. New York: G. P. Putnam's Sons, 1886.)

periments, was determined to put an end to animal research. Claiming that he violated the Cruelty to Animals Act of 1876, they planned to make a spectacle of the celebrated physiologist and to send a message to others of his kind.

The Experimentalists Demand More

To appreciate the events that led up to the summons, as well as the fury of the antivivisectionists in Britain, we must begin with what transpired after the clinical cases of Leborgne ("Tan") and Lelong were presented by Paul Broca in 1861 (see Chapter 10). Although many clinicians accepted what Broca had to say, the hard-core experimentalists wanted more proof than some isolated clinical case studies to convince them of the reality of the localizationist doctrine. With the failed efforts of Pierre Flourens to find cortical localization in animals still fresh in their minds, these flag bearers for "real science" were not quite ready to back the revolutionaries.[6]

The experimentalists did not deny that Broca's case studies cast suspicion on the time-honored concept of a single, multipurpose cortical organ. But, they emphasized, clinical observations tend to be unreliable and considerably less informative than well-controlled experiments on animals. Only by working with animals could scientists confine brain lesions to specific territories, study behavior before and after surgery, and conduct autopsies at will. This point was not lost on Gustav Fritsch and Eduard Hitzig when they went searching for the motor cortex in dogs.

Fritsch and Hitzig: The Motor Cortex

If Broca's 1861 report can be singled out as the most important clinical paper in the history of cortical localization, the discovery of the cortical motor area by Fritsch and Hitzig must be regarded as the most significant laboratory discovery. From a series of experiments published in 1870, these two experimentalists drew four important conclusions.[7] First, cortical localization does not pertain just to speech. Second, even dogs have specialized organs in the cerebral cortex. Third, a part of the cerebral cortex is electrically excitable. And fourth, the experimental methods of stimulation and ablation (removing tissue) can shed light on the organization of the brain.

Fritsch and Hitzig (Fig. 11.4) were both born in Germany in 1838.[8] Hitzig, the more important of the two men in the history of experimental brain research, came from a distinguished Jewish family living in Berlin. He first wanted to practice law, but then realized that being a physician was more his calling. After a stay in Würzburg, he returned to Berlin to receive his doctorate in 1862 and then settled down to practice medicine.

Hitzig became interested in cortical motor functions late in the 1860s, when he noticed that applying electrical currents to the back of the head or to the ears caused eye movements in humans.[9] He then turned to rabbits and, in a preliminary experiment, again obtained encouraging results with electrical stimulation.

Gustav Fritsch was known for his anthropological and geographical studies in South Africa before he joined Hitzig. His interest in the motor functions of the cerebrum was aroused during the Prusso-Danish War. When called upon to clean and dress head wounds, he noticed that accidentally irritating exposed brain caused an injured soldier to twitch on the opposite side of the body.[10]

Both men knew their initial observations did not prove that a motor cortex exists. The stimuli they used could have affected the cerebellum or the corpus striatum, two more primitive structures long associated with movement. This, in fact, was the criticism Pierre Flourens had leveled against Luigi Rolando after the Italian scientist electrically stimulated the cerebrum in some animals and observed movements.[11] Fritsch and Hitzig were also cautious because other respected scientists were unable to elicit movements when they applied very mild electrical current to the cerebral cortex. Negative findings had been reported not only by Flourens, but by a host of other world-class investigators.

Fritsch and Hitzig knew nothing about Swedenborg's insights (see Chapter 10). Nor did they notice some observations made in 1849 by Robert Bentley Todd, who electrically stimulated the cerebral cortex in rabbits and noticed twitching movements of the facial muscles.[12] Sadly, Todd did not have the mental set or readiness to realize what the twitching meant. He was interested only in eliciting full-blown epileptic seizures and, like others at the time, believed that epilepsy was not a cortical phenomenon.

It was in this Zeitgeist that the two men decided to join forces to conduct their series of experiments on dogs. The use of dogs was in itself a major advance over the pigeons and hens favored by Flourens. Their work, however, would not take place in a new university laboratory or in the operating room of a major hospital. Although Hitzig had an affiliation with the Physiological Institute in Berlin, it had no facilities for conducting experimental brain research on mammals. The two investigators therefore decided to undertake their study in Hitzig's house.

What Frau Hitzig had to say about a part of her house being turned into a laboratory is unknown. We can only guess that if she stayed around long enough, she would have been horrified to see that the first few dogs were not even anesthetized for the experiments.

The surgical procedure involved exposing and electrically stimulating the cerebral cortex. Knowing that current could spread, the investigators used very low levels of electrical stimulation. They determined the strength of the stimulus by first placing the electrode on one of their own tongues. Once the current was set, they stimulated many different cortical regions, searching for telltale twitches, like those that had been exhibited by Fritsch's injured soldier.

They eventually discovered that stimulating one cortical area near the front of the cerebrum led to an immediate movement of the forepaw on the opposite side of the body. Nearby, they found a corresponding hind-paw region, followed by face and neck territories (Fig. 11.5). Low-level stimulation in other regions failed to produce any movements at all.

Fritsch and Hitzig now felt sure they had discovered a specialized motor center. They also recognized that this region is made up of smaller areas corresponding to various parts of the body. Today we would say that the motor cortex is somatotopically organized, meaning organized according to body part.

How could their predecessors have failed to see this? The two researchers had an explanation. They reasoned that the motor cortex is difficult to expose, especially in dogs. Evidently previous experimenters stimulated the brain in more-accessible places and, after failing to elicit movements, simply gave up. In contrast, Fritsch and Hitzig were not biased by prevailing theories, and they continued their search after failing to elicit responses from easily exposed parts of the brain.

Fritsch and Hitzig next set out to see if dogs would exhibit movement problems

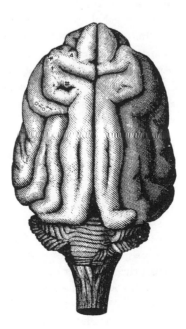

FIGURE 11.5.

Diagram of the dog's motor cortex published by Fritsch and Hitzig in 1870. (Hatch mark = opposite hind leg; cross = opposite front leg; diamond = face; triangle = neck.)

after destruction of the paw area. They used the handle of a scalpel to scoop out some of the forepaw region on the left side of the brain in two dogs and watched them a few days later. Unlike humans with severe strokes or head injuries, these dogs did not show severe paralyses. Nevertheless, they did have problems. Each seemed to exhibit little conscious awareness of its right forepaw; this limb repeatedly slid out from under them when they ran or sat.

These findings led Fritsch and Hitzig to think about the function of the newly discovered motor cortex. Because their dogs could still scamper about, they concluded that the circuits for basic movements must be situated below the cortical mantle. Everything suggested that the excitable cortex must be involved with a decidedly higher function, one more closely related to conscious awareness. The most likely function of the excitable cortex, they eventually concluded, is to allow the organism to perform voluntary acts—to "will" precisely what it wants to do when it wants to do it.

The Germans ended their monumental paper of 1870 by encouraging other researchers to search for sensory cortical areas and for regions that could account for even higher functions. As they put it, "certainly some psychological functions, and perhaps all of them . . . need circumscribed centers of the cortex."[13]

Enter David Ferrier

Fritsch and Hitzig's paper was read by James Crichton-Browne, the progressive Scottish supervisor of a Yorkshire institution called the West Riding Lunatic Asylum.[14] The term *Riding*, in this instance, has nothing to do with horses. Yorkshire was divided into three parts, and *Riding* was derived from *thriding*, an archaic word simply meaning "a third part of a region."

Crichton-Browne was a friend of David Ferrier, also a Scot now living in England.[15] Born near Aberdeen in 1843, Ferrier first studied psychology with the free-thinking psychologist-philosopher Alexander Bain. Then, after a brief period in Heidelberg, he returned to Scotland to claim top honors in medicine in 1868. He next briefly assisted Thomas Laycock, who looked upon brain functions as being sensory or motor. After joining a general practice in a small English town, he left for London, where he accepted an appointment at King's College Hospital and rapidly ascended the ranks. Simultaneously, he secured a position at the National Hospital for the Paralysed and Epileptic, Queen Square, the first British hospital built specifically for people with diseases of the nervous system.[16] There he befriended John Hughlings Jackson (see Chapter 10), who had also studied with Laycock.

The turning point in Ferrier's life occurred when he decided to visit Crichton-Browne in Yorkshire. The two physicians were eager to see each other and talked about many things, including the Fritsch and Hitzig experiments. Both Scots saw the work of the Germans on the motor cortex as a landmark, and both also saw several reasons to follow it up. Crichton-Browne then made Ferrier an offer he could not refuse. He told him that he would give him everything he needed if he would conduct the same sort of experimentation at his West Riding institution. He hoped to show the world that good brain research could be conducted at a psychiatric hospital, or what was then called a lunatic asylum.

As for Ferrier, he was quick to seize the opportunity for two main reasons. First, he wanted to test Jackson's theory that epileptic seizures can be triggered from the cortex. Second, he felt that he could add new facts by continuing this type of research.

True to his word, Crichton-Browne provided Ferrier with good laboratory space and plenty of birds, guinea pigs, rabbits, cats, and dogs. In 1873 Ferrier eagerly began the studies that would soon make his name well known among scientists and physicians, as well as animal-rights advocates. He first explored the brains of rabbits, cats, and dogs with mild alternating current and discovered many more excitable areas of the cortex than the Germans had reported. Moreover, the movements seemed natural, integrated, and directed toward an end. His animals walked, grabbed, scratched, blinked, and even flexed their digits.

These stimulation experiments supported Jackson's theory of cortical epilepsy. The hand, leg, and face movements of certain epileptics could be mimicked by stimulating select cortical areas. In addition, more massive seizures could be induced by increasing the strength of the current.

Ferrier next set forth to study the effects of damaging the motor cortex, just as Fritsch and Hitzig had done. Although he operated on only a few animals and did not keep them alive for more than a few days, he also observed movement disorders. Unlike the Germans, however, he showed that animals higher on the phylogenetic scale are more impaired by motor-cortex lesions than their lower cousins.

These findings led Ferrier to agree with Fritsch and Hitzig that the excitable cortex must have something to do with voluntary, goal-oriented movements, as opposed to instinctive or unconscious motor acts. The more the movements seemed to depend upon the will, the more they tended to be affected by cortical damage. Ferrier, however, parted company with the Germans when it came to the specifics of his motor-cortex theory. First, he argued that the motor cortex is strictly an organ of motion. In contrast, the Berliners brought sensory feedback from the muscles into the picture and suggested that the motor cortex may have both sensory and motor functions. And second, unlike Fritsch and Hitzig, he saw no reason to drag the soul into the picture to describe how the excitable cortex may mediate voluntary movement.

One of Crichton-Browne's innovations at West Riding was to have special monthly gatherings for scientists to present their new findings and ideas. The invited talks took place in the magnificent main hall of the asylum and were accompanied by various demonstrations, exhibits, photographic displays, and discussions. For ambiance, fine music was played and refreshments were served. Scientists and physicians coveted invitations to these gatherings, as did successful people from other walks of life.

David Ferrier first presented his findings at a West Riding gathering late in 1873. They began to appear in print later that year.[17] The forum for Ferrier's first major publication on cortical localization was, appropriately enough, Crichton-Browne's *West Riding Lunatic Asylum Medical Reports*. This strange-sounding journal was short-lived, but while it lasted it was known for very high-quality papers.

Of Jackson and Monkeys

Soon after starting his work at West Riding, Ferrier applied to the Royal Society for a grant to study monkeys. Before 1873 had ended, he found himself busy collecting data on these primates. His stated objective was to show that experimental work on the monkey would shed even more light on human cases, and in this endeavor he was quite successful. Some of his findings on the monkey motor cortex also appeared in Crichton-Browne's journal.[18]

Ferrier compared his monkey and dog findings in a presentation before the Royal Society in 1874.[19] He ran into a problem, however, when he tried to publish this material in a scholarly journal.[20] The referees assigned to review his submission felt that he did not adequately acknowledge the pioneering work of Fritsch and Hitzig. He made some changes, but his paper was still deemed unacceptable. Ferrier then made the decision to eliminate all of the dog experiments from his paper. Rather than having to say more about the dog work of the Berliners, he would confine himself to his new findings with monkeys.[21]

Why Ferrier was so cool to the German experimentalists in this and related papers is not known with certainty, but two theories are worth considering. One is that he was young and hoped to enhance his own reputation by not glorifying others. A second, much more likely possibility is that he was upset because Fritsch and Hitzig failed to mention the case studies and brilliant insights of his spiritual mentor, John Hughlings Jackson, in their famous paper of 1870.

Jackson's thoughts on the motor cortex stemmed from his early interest in epilepsy. In 1861 he presented his first report on seizures that cause involuntary movements on just one side of the body.[22] After Jackson moved to the National Hospital for the Paralyzed and Epileptic, he had more opportunity to study these so-called partial seizures in patients with cerebral syphilis.[23] He observed that the seizure tends to "march" over the body in a very predictable way. Convulsions starting with the foot spread up the leg to the torso and then to the arm before finally affecting the hand and face. In contrast, those starting in the hand go up the arm and then spread to the face and the hip before affecting the leg and its toes.

For several years, however, Jackson allowed himself to be influenced by the long-standing belief that epileptic movements are orchestrated below the cerebral cortex. Nevertheless, as he continued to study his own patients with seizures or paralyses, and as he witnessed autopsies, he recognized that the cerebral cortex must be playing a significant role in these seizures.

In 1870 he wrote that a cortical area organized according to body part could best account for the march of the seizures across muscle groups.[24] This was also the best way to explain why small cortical lesions can affect just one body part and not others. He then correctly deduced that the motor cortex controlling the right side of the body has to be located near Broca's area in the left hemisphere. This, he explained, is why speech is almost always arrested when the right side of the body goes into seizure, and why Broca's aphasia is usually accompanied by paralysis of the right side of the body.

Jackson's insights made perfect sense to Ferrier, who felt that Fritsch and Hitzig should have cited Jackson in their 1870 paper. By not citing the Germans, he might have felt that he could now even the score.

The ice that existed between Ferrier and his German counterparts only slowly thawed as Ferrier realized that the Berliners never really intended to slight Jackson. From all indications, they had been unaware of Jackson's insights, much as they knew nothing about Swedenborg, when they wrote their landmark paper. As a result, Ferrier slowly began to give more credit to the Germans for their pioneering work.[25]

Sensory Cortical Areas

Ferrier said little about the sensory systems in his 1873 report. Nevertheless, he found that some of his electrical explorations well outside the motor region elicited movements. Sites in the parietal and temporal lobes clearly drew more of Ferrier's attention when he began to study monkeys in 1874. He now emphasized that the movements elicited from some of these locations are like the reactions shown by normal animals to sights, sounds, and other sensory stimuli. Perking the opposite ear and turning the head to mild electrical stimulation of the superior temporal lobe suggested a role in hearing, whereas certain eye and head movements indicated that another cortical area was probably visual in nature.[26] As Ferrier liked to point out: "The mere fact of motion following stimulation of a given area does not necessarily signify a motor region."[27]

Given the inability of electrical stimulation to provide good information about the sensory cortex in organisms that cannot speak, Ferrier tried damaging these electrically responsive areas. He then tested to see if a monkey would still sniff an apple before tasting it, turn around when called by name, or in true British fashion sip some tea from a cup.

His conclusions proved to be accurate for some sensory systems. Today we localize the main center for hearing at the top of the temporal lobes (superior temporal gyrus), just as he maintained. He was also on the mark when he localized smell lower in the temporal lobes. Electrical stimulation of his suspected smell area caused his animals to close a nostril, raise a lip, and grimace as if they were smelling a disagreeable odor. Animals with lesions of the same region, he reported, do not even sniff their food.[28]

In contrast to hearing and smell, Ferrier did not do as well when it came to localizing the sense of sight. He first maintained that vision is controlled by a region in the back of the parietal lobe called the angular gyrus.[29] Only slowly did he recognize the primary role played by the occipital lobe, a discovery made by Hermann Munk late in the 1870s.[30]

Ferrier also had difficulty localizing the skin senses (touch, temperature, pain) and taste. He erroneously localized touch and taste in the temporal lobes, rather than in the parietal region just above it. Other scientists were equally baffled when it came to the location of these senses.[31] No one in the nineteenth century correctly pointed to the anterior parietal lobe as the primary site for the skin senses or, for that matter, the sense of taste.[32]

The "Silent" Frontal Cortex

During the 1870s Ferrier also tried to understand the functions of the cortex in front of the motor area. When he damaged this frontal region on both sides in some monkeys, they acted as if their intellectual powers were impaired; all were inattentive to what was going on, taking little notice of the animals in cages next to their own.[33] They were also apathetic, dull, and listless, but occasionally exhibited periods of restlessness characterized by moving back and forth and wandering aimlessly. Ferrier believed that these animals were suffering from attentional deficit disorders.[34]

Interestingly, Hitzig also began to study the "silent" frontal cortex at this time.[35] He found that dogs who knew how to search for food lost this ability after bilateral lesions of this frontal brain region. They even appeared to forget about pieces of food that were just shown to them. These same dogs, however, quickly ran to food and ate it when it was placed directly in front of their eyes, so loss of hunger was not an adequate explanation. Although Hitzig believed that intelligence cannot be confined to any one part of the brain, he argued that the anterior frontal cortex plays a special role in abstract thought.

Who was correct, Ferrier or Hitzig? Are the frontal lobes important for attention or do they mediate abstract thought? As it turns out, we now know that frontal lobe damage can lead to problems in both domains, and cause other difficulties as well. A dedicated Italian experimentalist by the name of Luigi Bianchi, who went so far as to live with some of his monkeys, wrote that frontal-lobe lesions could also cause defects in socialization, emotion, and personality.[36]

Actually, before Ferrier, Hitzig, or Bianchi even began their studies, there was a good clinical description of the myriad of changes that can follow severe frontal-lobe damage.[37] It involved a man by the name of Phineas Gage, who worked for a railroad

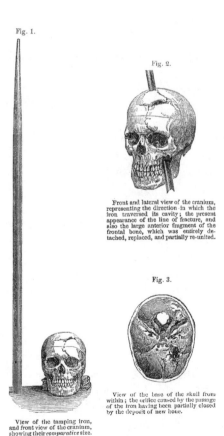

Fig. 1.

Fig. 2.

Front and lateral view of the cranium, representing the direction in which the iron traversed its cavity; the present appearance of the line of fracture, and also the large anterior fragment of the frontal bone, which was entirely detached, replaced, and partially re-united.

Fig. 3.

View of the base of the skull from within; the orifice caused by the passage of the iron having been partially closed by the deposit of new bone.

View of the tamping iron, and front view of the cranium, showing their *comparative* size.

FIGURE 11.6.

The skull of Phineas Gage and the tamping iron that shot through it in 1848. (From Harlow, 1868.)

in New England and was responsible for setting and firing the explosives needed for leveling the rocky terrain.

In 1848 an accidental explosion shot Gage's large tamping rod point-first through the left side of his jaw and out the top of his skull (Fig. 11.6). Over the next few weeks he slipped in and out of a coma and battled meningitis. But miraculously and against all odds, he survived and a few months later even tried to get his old job back.

Gage was turned down by the railroad because his intellect and emotional stability had changed. The "new" Gage used profanity, acted impulsively and in a childlike manner, and was unable to adapt to the social setting "because the equilibrium or balance . . . between his intellectual faculties and animal propensities, seems to have been destroyed."[38] The man the railroad officials were interviewing was no longer the one they had known—as friends and acquaintances put it, Gage was no longer Gage.

In 1852 Gage left for Chile to drive horses for a coach line. After several years in South America, he sailed back to California, where he died. An autopsy was not performed, leading to endless speculations about the exact brain regions damaged.[39] From his doctors' reports, however, there is no doubt that his frontal cortex was severely affected. Ferrier himself cited the Gage case during the 1870s, when discussing frontal-lobe lesion effects and functions.[40]

Some Books and a Journal

The year 1876 was a banner year for Ferrier. He was honored by being elected a Fellow of the Royal Society, and his most important book, *The Functions of the Brain,* was published.[41] In this work he presented all of his early research findings and theories with clarity and conviction. Appropriately, he dedicated his book to John Hughlings Jackson with "esteem and admiration."

By now, Ferrier was so confident that his experimental findings with monkeys could be applied to humans that he even transposed functional maps of the monkey cortex, sketched by London landscape painter Ernest Albert Waterlow, onto Ecker diagrams of the human brain (Fig. 11.7). The boldness of the transfer was reminiscent of Leonardo da Vinci sketching ox ventricles onto drawings of the human brain some 370 years earlier (Chapter 5).

Although some critics wrote that *The Functions of the Brain* did not adequately present contradictory data, the book was well received.[42] In fact, it was so popular that Ferrier decided to follow it with a second book, *Localization of Cerebral Disease.* This work was published in 1878 and was considerably more clinical and applied in nature.[43]

Now in full stride, Ferrier joined forces with his friends Crichton-Browne, Jackson, and Bucknell to launch a new scientific journal, one aptly called *Brain.* This specialized journal soon became one of the most prestigious periodicals for the publication of brain research. Today, it remains one of the premier journals in the field.

Ten years after *The Functions of the Brain* first appeared, an updated edition came off the press.[44] In the 1886 edition Ferrier reexamined his earlier ideas and provided new findings and insights. His sections on the sensory areas alone were three times longer than those of the 1876 edition. He also wrote considerably more about higher brain functions, such as how ideas may be formed.

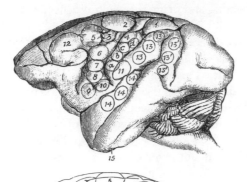

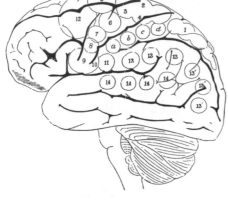

FIGURE 11.7.

*Ferrier's maps of
the monkey brain
(top) and the
human brain
(bottom). The map
of the human brain
was based on
findings with
monkeys. (From
D. Ferrier,*
The Functions of
the Brain. *New
York: G. P.
Putnam's Sons,
1886.)*

The Matter of Recovery

A major stimulus for the updated edition of *The Functions of the Brain* was that some of Ferrier's earlier experiments and conclusions had been criticized by others in the scientific community. Their challenges concerned the locations and boundaries of his functional areas. Closely tied to this issue was the question of recovery. To the critics, recovery suggested that the relevant centers were larger than Ferrier thought, or perhaps even located elsewhere.

For his part, Ferrier did not believe that there was any recovery following destruction of the main functional areas in primates. He felt, for example, that the movement disorders he observed in his monkeys were every bit as permanent as those seen in people with severe injuries of the motor cortex. His critics maintained, however, that he would have come to a very different conclusion if he had kept his monkeys alive for more than just a few days.[45] Ferrier certainly wished that he could have kept his animals alive longer, but he had to sacrifice them before they developed meningitis from the surgery.

His ability to keep animals alive for longer periods improved greatly after the first edition of *The Functions of the Brain* appeared in 1876. The welcome change for the better was made possible by the work of Joseph Lister, whose crusade to minimize infections by using carbolic acid as an antiseptic had begun in the mid-1860s. Prior to Lister, the acts of opening the skull, cutting the meninges, and exposing the brain to the elements was a virtual death sentence.

Although Lister's ideas were initially met with some skepticism, Ferrier was en-

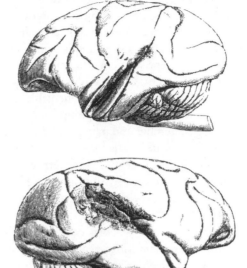

FIGURE 11.8.

*The brain of the
deaf monkey
exhibited by
Ferrier at the 1881
International
Medical Congress.
(From D. Ferrier,*
The Function of the
Brain. *New York:
G. P. Putnam's
Sons, 1886.)*

thused and recognized that the principle of antisepsis also applied to surgery on animals. By 1881 Ferrier and his surgical colleague Gerald Yeo were able to study healthy monkeys that had undergone antiseptic surgery months earlier. One monkey still exhibited a severe paralysis on the opposite side of its body seven months later. This monkey convinced Ferrier that the deficits following large motor-cortex lesions are, in fact, permanent in primates.

Ferrier and Yeo also had one animal, Monkey F, with lesions of the auditory cortex that remained in good health well after its surgery (Fig. 11.8). The monkey showed no evidence of recovering from its deafness.[46] Ferrier looked upon Monkey F as being just as special as his paralyzed monkey. These two healthy animals, he thought, should silence the critics of his theories.

Ferrier considered these two monkeys so important that he chose to describe them in detail at the 1881 International Medical Congress in London. These were the hemiplegic and deaf animals seen by the attendees who followed Ferrier and Goltz from the lecture site to King's College Hospital on that momentous summer afternoon.

Ferrier's Day in Court

The antivivisectionists were mobilized and clearly on a rampage in London during the summer of 1881, precisely when Ferrier was describing his monkeys.[47] Realizing that they were determined to put an end to experimental medicine, some prominent speakers at the International Medical Congress used their platform time to justify the need for animal research. Although the atmosphere remained charged after the congress ended, Ferrier did not expect the summons he received early that November.

For the scientific brotherhood, the unexpected summons of one of their most respected leaders exploded like a powerful bomb. The latest persecution served as yet another example of how outside forces could disrupt medical advancement. In the past, however, the opposition had been high clergy or conservative keepers of the old world order. Now the inquisitors were an odd assortment of activists who believed that scientists were butchering defenseless pets. Even the British fox hunters, once described by Oscar Wilde as "the unspeakable in full pursuit of the uneatable" were not as despicable, as reprehensible, or as amoral as the ghoulish physiologists.[48]

Both the animal-rights groups and the laws protecting animals in Britain were growing in number and in impact before Ferrier received his summons. In 1822 cruelty to horses and cattle was outlawed. Although the Martin's Act did not apply to cats and dogs, the matter of experimenting on any animal was soon raised in the House of Commons. By this time, irate Londoners had organized the Society for the Prevention of Cruelty to Animals, soon to have the word *Royal* fixed to its name (RSPCA).

At first the RSPCA was concerned with violations by the working class in the streets and on farms. As experimental medicine developed, however, increasing attention was given to the practices of both researchers and demonstrators in medicine. The medical community was sensitive to the growing resentment and in 1870 set up their own rules to reduce suffering by animals. By launching a preemptive move, they also wanted to assure the populace that only skilled, knowledgeable people would be conducting surgical experiments.

Many members of the large RSPCA criticized such self-regulation as nothing more than a smoke screen. They attracted more public attention by arguing that if the unthinkable could be done to animals, even more hideous experiments would soon be conducted on poor, unknowing people. They firmly believed that physicians were no more competent than good men and women from other walks of life to judge the morality of an act or, for that matter, its legality.

Led by Frances Power Cobbe (Fig. 11.9), an imposing woman not known for her patience, the militant Society for the Protection of Animals from Vivisection was assembled to make life hell for the vivisectors and to force the government to pass even stronger animal-protection legislation.[49] Because it was domiciled on Victoria Street, this organization soon became known as the Victoria Street Society. Its supporters included members of Parliament, clergymen, aristocrats, and even the poets Browning and Tennyson. Queen Victoria herself was sympathetic to the cause, as were the editors of several popular periodicals.

Cobbe's Victoria Street Society and other groups committed to total abolition of animal research now began to push tough new legislation through Parliament. Their efforts led to the Cruelty to Animals Act of 1876, under which experimenters had to be screened and licensed each year by the government. This act also stipulated that research could only be performed at registered places subject to inspection. Moreover, animal researchers would have to (1) justify their work in terms of saving or prolonging human life, (2) follow strict procedures to minimize pain, and (3) obtain special certificates if animals were allowed to emerge from anesthesia for additional study.

Yet for the antivivisectionists, this was not enough. The new law fell short of total abolition and did not spare cuddly cats and faithful dogs from the dreaded scalpel.

FIGURE 11.9.

Frances Power Cobbe (1822–1904), leader of the antivivisection movement in Victorian England.

Hence, when the participants at the International Medical Congress of 1881 unanimously endorsed a statement supporting experimentation on living animals, the militant antivivisectionists felt as if the sadistic animal experimenters had just poured salt on an open wound. At her wits' end and with her temper raging, Cobbe saw no choice but to counterattack. When one of her watchdogs discovered that David Ferrier did not have a license from the Home Office for the research he presented, she and her associates felt sure they had found the perfect target.

The events that followed were covered by reporters from many newspapers, including the *Times* of London, as well as by physicians who wrote for medical journals around the world. Many periodicals also carried letters and editorials. In the November 19 issue of the *British Medical Journal,* the editors even tried to spell out the importance for medicine of Ferrier's research.[50] They cited cases of children and adults whose neurological problems were correctly diagnosed and successfully treated because of his cortical localizations. Their message was that animal research is not a sadistic, cruel game. Rather, research of the type conducted by Ferrier was immensely valuable and in fact is already saving human lives.

At last the criminal trial began in the Bow Street police court. Present were Charcot, Michael Foster, Yeo, Lister, Jackson, and many other practicing physicians and laboratory scientists. Ferrier was informed that he had a choice of a trial by jury or one to be decided by the judge, Sir James Ingham. Mr. Guilly, Q.C. (Queen's Counsel), on behalf of the defense, responded that his client preferred a ruling by the judge.

The specific charge leveled against Ferrier in the summons was that he had performed surgical experiments on some monkeys without a license. The defense responded that he had not performed the surgeries—the antiseptic operations had been conducted by Gerald Yeo, who held the appropriate license. Yeo also housed and cared for the animals, in accordance with the Act of 1876.

The prosecutor was forced to change his strategy after the defense's embarrassing disclosure. He now stressed that the law required all grievously injured animals to be killed before the anesthesia wore off. He argued that even observing how mutilated animals respond months later was a violation of the law. Therefore, if for no other reason, Ferrier was still guilty for his role in the experiments.

The defense responded that the animals emerged from surgery healthy and out of pain. Killing them before the anesthetic wore off not only was uncalled for, but would have defeated the purpose of the experiments. Ferrier did nothing inhumane, painful, or cruel. Moreover, his coworker Yeo had the certification required to keep the animals alive for this part of the study.

After listening to the arguments from both warring factions and questioning a number of witnesses, Sir James Ingham had heard enough. He announced that the summons was dismissed and sent everybody home.[51] He could not see how Ferrier's role in the monkey experiments constituted a criminal act. To the relief of Ferrier and Yeo, not to mention supportive members of the scientific and medical communities who were also present, there were no real violations of the Act of Parliament.

From Monkeys to Human Neurosurgery

During the sensational trial of 1881, Ferrier's supporters spoke about several patients who had been saved by surgeries based on the "functional maps" of the cortex he had constructed. For the first time, they argued, surgeons could feel comfortable conducting brain surgery solely on the basis of clinical symptoms. Prior to this time, brain surgery was performed only if cranial discolorations or skull deformities could be observed. As a result, deadly tumors lodged below the surface of the cortex were routinely left untreated.

The first man to bring functional maps of the cortex into the surgical arena was William Macewen, a Scottish surgeon who was not only a follower of Broca and Ferrier, but a disciple of Joseph Lister. In 1876, the year in which Ferrier's *Functions of the Brain* came off the press, he came close to operating on a boy with a suspected brain abscess, but was overridden by the family physician.[52] Sadly, the boy died and an autopsy revealed an operable brain abscess exactly where Macewen had said one would be found.

Macewen began to operate using Lister's antiseptic principles and Ferrier's maps three years after this tragic event.[53] His first case was a teenage girl with a tumor of the dura mater encasing the brain. She exhibited seizures of the right arm and face, suggesting that the growth was affecting the motor cortex on the left side. Macewen, who was an extraordinarily dexterous and careful surgeon, successfully removed her meningial tumor. To his credit, the girl lived. Additional brain surgeries for clots, abscesses, tumors, and splinters of bone soon followed, and were described by Macewen in papers sent to leading British medical journals (Fig. 11.10).[54]

The most famous surgical case of the period, however, involved a physician by the name of Hughes Bennett and a young London surgeon named Rickman J. Godlee.[55] Their case drew more attention than Macewen's earlier cases for many reasons. Not only did Godlee operate in the heart of London, but he was assisted by a rather influential group of insiders, including Ferrier, Jackson, and a promising young sur-

Fig. 5.—Lesion in Ascending Convolutions diagnosed from Motor Symptoms
alone.

Fig. 6.—Syphilitic Tumour in Paracentral Lobule. Lighter shading indicates
effusion on surface of upper part of central convolution.

Fig. 7.—Focal Lesion in Ascending Convolutions diagnosed from Motor and
Sensory Symptoms.

FIGURE 11.10.

*Drawings showing
three brain
surgeries performed
by Macewen in the
1880s. In each of
these cases, the
diagnosis was
based on the motor
symptoms. (From
Macewen, 1888.)*

geon by the name of Victor Horsley. Macewen, in contrast, was a combative Scottish outsider, perceived as a bull in the china shop by the conservative medical aristocracy. Moreover, although Macewen applied Lister's important principles to brain surgery before Godlee, many people found it easier to associate Godlee with Lister because Godlee was, in fact, Lister's nephew.

Godlee's tumor case first became known on December 16, 1884, as the result of a one-page letter criticizing the antivivisectionists and the laws limiting animal experimentation in Great Britain. This letter, entitled "Brain Surgery," appeared in the *Times* three years after Ferrier's trial. Signed "F. R. S.," it was probably written by none other than Ferrier's good friend, James Crichton-Browne. It began:

> Sir,—While the Bishop of Oxford and Professor Ruskin were . . . denouncing vivisection at Oxford last Tuesday afternoon there sat at one of the windows of the Hospital for Epilepsy and Paralysis, in Regent's Park . . . pale and careworn, but with a hopeful smile on his face, a man who could have spoken a really pertinent word upon the subject, and told the right rev. prelate and great art critic that he owed his life, and his wife and children their rescue from bereavement and penury, to some of these experiments on living animals which they so roundly condemned. The case of this man has been watched with intense interest by the medical profession, for it is of an unique description, and inaugurates a new era in cerebral surgery.[56]

F. R. S. ended his letter by stating that this surgical case "will be a living monument of the value of vivisection" and that the patient "owes his life to Ferrier's experiments, without which it would have been impossible to localize his malady or attempt its removal."

Within a matter of days, Bennett and Godlee published a complete description of their case in the medical journal *Lancet*.[57] Their patient was an unnamed Scottish farmer in his twenties, later identified as a man by the name of Henderson.[58] He had been suffering from Jacksonian motor epilepsy for three years and now had a useless left arm, weakness in his left leg, and visual problems.

Henderson's neurological symptoms, combined with severe headaches and vomiting, led Bennett to suspect a brain tumor. He hypothesized that the tumor was not very large and was probably situated in the right motor cortex. Having watched his own father die a few years earlier from an intracranial tumor that could have been removed, Bennett recommended immediate surgery. At the time, he might have known little or nothing about Macewen's successes in this field; all he knew was that Henderson would surely die if not quickly treated.

Henderson was told that he would not live much longer if he did not submit to the operation. Once informed, he agreed to the surgery. But Bennett was not a surgeon. Thus he asked Rickman Godlee to operate.

On November 25, 1884, Henderson was prepared for surgery at the National Hospital for Epilepsy and Paralysis. Because the hospital did not have an operating room, his surgery took place in some quarters modified for the event. Instruments were brought over from King's College Hospital, and a carbolic spraying machine powered by steam was used to disinfect the room.

After Henderson was put under chloroform, Godlee opened his skull and searched for the suspected tumor. There must have been great disappointment when nothing was observed on the surface of his brain. Godlee then probed deeper and, to everyone's relief, found exactly what had been expected—a growth about the size of a walnut, hidden just below the surface of the motor cortex. The tumor was removed and Henderson was bandaged and prepared for recovery.

Henderson now showed signs of improvement. He stopped vomiting, his headache ceased, he began moving his left leg, and his seizures ended. The clearest remnant of what he had endured was his still-paralyzed arm, most likely due to irreversible destruction of a section of the motor cortex by the tumor.

Possibly because his head was not treated with carbolic acid as thoroughly as possible, meningitis set in a few weeks later. Bennett and Godlee could not have felt worse when Henderson died at the National Hospital on December 24, 1884, almost a month after his surgery.[59] Rather than realizing what they had accomplished, they felt as if they had failed.

Neurosurgical interventions based on functional maps of the cortex began to be performed fairly regularly after Bennett and Godlee described their case in greater detail at a medical conference in 1885.[60] Also attending the meetings that spring were Ferrier and Jackson, both of whom looked upon the operation as a landmark in neurosurgery, even though the patient later died, and William Macewen, who spoke about his own cases. Macewen's talk did not receive the warmest of receptions, but after hearing him speak, Bennett called the Glasgow surgeon's findings "encouraging," and Godlee personally congratulated him for his pioneering efforts.

Final Years

By the time of the Bennett and Godlee case, Ferrier was no longer the productive laboratory researcher he had been. Although he still went to work early in the morning and rarely headed home before midnight, he was now much more involved with administration, consulting, and building his clinical practice than he was with animal research. In this respect, his later life paralleled that of Eduard Hitzig. After Hitzig became famous in the 1870s for his discovery of the motor cortex in dogs, he tried to reform psychiatric clinics, first as director of the Burghölzli Asylum in Switzerland and then in his native Germany.

Ferrier, who had married in 1874 and had two children, was knighted in 1912, four years after he retired from King's College Hospital. He succumbed to pneumonia in 1928, one year after Fritsch passed away and twenty-one years after Hitzig had died. Sir Charles Sherrington, who chose to go into neurophysiology after listening to Ferrier speak in 1881, was one of the scientists who eulogized him for his dedication and achievements.[61] Indeed, thanks to Ferrier, as well as to Fritsch and Hitzig, the concept of specialized motor, sensory, and association areas of the brain would now serve as a foundation for many future developments in the brain sciences and medicine.

Jean-Martin Charcot
Clinical Neurology Comes of Age

Let someone say of a doctor that he really knows his physiology or anatomy, that he is dynamic—these are not real compliments; but if you say he is an observer, a man who knows how to see, this is perhaps the greatest compliment one can make.

—*Jean-Martin Charcot, 1888*

To remove from neurology all the discoveries made by Charcot would be to render it unrecognizable. Indeed, in a neurologic service, not a single day passes in which we do not use some of the notions he introduced; his thinking is always with us.

—*Joseph Babinski, 1925*

Introduction

During the first half of the nineteenth century, most neurological diseases were categorized on the basis of a single feature, such as a paralysis or a tremor. Because these broad categories were not based on specific anatomical changes in the nervous system, unrelated disorders were often lumped together. Adding to the confusion, different stages of the same disease occasionally wound up with decidedly different names.

Labeling and classifying neurological diseases took on a different, more modern form between 1862 and 1870, when observable clinical signs were tied to underlying pathological changes. The idea of learning more about diseases by conducting autopsies was to a large extent a French achievement. The individual most responsible for applying the *méthode anatomo-clinique* to the nervous system became one of the most honored physicians of all time. His name was Jean-Martin Charcot (Fig. 12.1).

Due largely to the efforts of Charcot, paralyses, tremors, and changes in sensitivity were finally looked upon as signs (what the observer sees) and symptoms (what the patient reports), rather than as diseases. At the same time, more attention began to be paid to syndromes (groups of signs and symptoms that go together) than at any time in the past. As succinctly put by the medical historian Fielding Henry Garrison, "Charcot found neurology a muddle and left it a highly organized science."[1]

Charcot's Rise to the Top

Jean-Martin Charcot was born in Paris in 1825, shortly after the emperor Napoleon had been laid to rest.[2] The boy's father, an artisan and carriage builder, was a man of average means. With his wife, he saw to it that Jean-Martin and his three brothers were provided with a comfortable, middle-class home and fitting educations.

FIGURE 12.1.

*Jean-Martin
Charcot
(1825–1893).*

Jean-Martin was a thin child of average height with black hair. From an early age, he displayed certain personality traits and skills that would remain with him throughout his life. Being rather shy, he was said to be somewhat cold and aloof. Not one for the great outdoors or to deal with crowds, he preferred to be left alone to read and think.

Although he entertained the idea of becoming an artist, Charcot made up his mind to head into medicine after he completed his secondary schooling, probably in 1843. His choice was based on more economic stability, upward mobility, and intellectual excitement in the healing arts than in the fine arts. At the time, wealth was not a prerequisite to enter medicine, rendering it an attractive option for bright students from working-class French homes who wished to make something of themselves.

From all indications, Charcot's skills were perceived as only average when he began to attend university classes in Paris. Nevertheless, the pensive young man impressed those around him with his excellent ability to draw the significant clinical details of the patients he was seeing. Later in life, Charcot would describe himself as a *visuel,* French for a person who relies on his eyes and is able to see things that others may overlook.

After completing his coursework and clinical rotations, Charcot passed a series of academic examinations. In 1846 he became a medical extern and two years later began his internship. The latter consisted of a series of one-year rotations at different hospitals, including the Salpêtrière, the famous Paris hospice for sick and elderly women.

Charcot completed his doctorate in 1853. In his thesis he showed how rheumatism can be distinguished from gout. His work was so impressive that he was made *chef de clinique* at the Faculté de Médecine. He then received *agrégé* appointments given to only a few select candidates, and eventually became *médecin de la*

Salpêtrière, or senior physician at the Salpêtrière Hospital, a coveted position that brought him back to the site of one of his clinical rotations.

In time, Charcot would rise even higher in the French academic and hospital systems. But to describe what happened to him in the 1880s would be jumping ahead too fast. To understand Charcot as a physician-scientist, and to appreciate what he did for neurology, we must first look at the history of his home away from home, the Salpêtrière.

The Salpêtrière: A Paris Landmark

The Salpêtrière takes its name from the French word for saltpeter, a key ingredient in gunpowder.[3] It was originally built to replace an old arsenal on the right bank of the River Seine, not far from the heart of Paris. Large quantities of saltpeter were stored there. After having been rocked by violent explosions a few too many times, the increasingly powerful merchants and residents in the rapidly growing neighborhood near the old arsenal sent a petition to King Louis XIII. They asked him to move his munitions to a safer location, one far from their homes and places of business. In 1634 the king granted their request and constructed a new saltpeter store outside the city limits on the left bank of the Seine, not far from the royal gardens.

By the middle of the seventeenth century, Paris was overrun with beggars and others in need of public help. Charitable organizations sprung up to provide medical supplies, food, and clothing. In this environment, the Salpêtrière was converted into a hospice and was integrated into the general hospital system of the city. Its new role was to serve sick and destitute women; the Bicêtre, a second large Paris hospital, was given the job of serving the men.

In order for the Salpêtrière to do this, new buildings had to be constructed. As construction took place, the number of inhabitants swelled from eight hundred to two thousand, and then to well over three thousand. The hospice continued its expansion during the eighteenth century, with many more buildings springing up to accommodate approximately eight thousand elderly, disabled, mentally retarded, psychotic, and chronically ill women. At its peak, the Salpêtrière had well over a hundred buildings. It was not only the largest asylum of its type in the world, it was a veritable city within a city.

After the French Revolution of 1789, new attention was drawn to the plight of the women in the massive asylum. The beggars and the destitute prostitutes were now sent to other institutions, and better care began to be given to the women who were indeed sick. A landmark event occurred in 1798, when reform-minded Philippe Pinel removed the shackles from forty-nine severely mentally disturbed women. For the first time they were permitted to see the beautiful gardens and walk in the courtyards of the massive institution.

Nevertheless, when Charcot returned to the Salpêtrière as senior physician in 1862, the state of the hospice still left much to be desired. Many of the buildings were in ruins and, just as important, he found the approximately five thousand residents thrown together in ways that made little sense. The situation was so poor that Charcot described what he saw as a badly organized museum of living pathology. Sweeping changes would have to be made at "this great asylum of human misery."

Organizing and Classifying

When Charcot walked back through the gates of the Salpêtrière in 1862, he met Edmé Félix Alfred Vulpian, with whom he had done a clinical rotation. Vulpian had accepted a position at the Salpêtrière at about the same time as Charcot, and he was every bit his match when it came to his intelligence, thirst for knowledge, and motivation to succeed.

Charcot and Vulpian quickly reestablished their friendship and then began to inventory the chronically ill women, more than half of whom had been classified as indigents and noninsane epileptics. They literally went from ward to ward, taking notes about the physical and mental status of each woman they saw, excluding the severely insane (the so-called *aliénés*), whom they left for their psychiatric colleagues.

It did not take long for the two physicians to realize that many of the women had been badly misclassified. In particular, the label "epilepsy" appeared almost indiscriminately. To learn more about the various diseases that plagued the patients, they concluded that autopsies would have to be performed when the women died. Hence, they expanded the facilities for conducting autopsies and built a laboratory to study pathological anatomy (Fig. 12.2).

Located in a tiny kitchen next to a cancer ward, their laboratory was a far cry from the well-equipped glass and stainless-steel laboratories of today. At first, it held little more than some microscopes and a stack of specimen jars. Nevertheless, Charcot and Vulpian felt that they could now begin to search for the anatomical changes that might account for the various signs and symptoms these patients were displaying.

FIGURE 12.2.

Charcot examining a brain while wearing an autopsy apron and a top hat. (From E. Brissaud, Leçons sur les maladies nerveuses. *Paris: Masson, 1895.)*

Charcot also became interested in medical photography after arriving at the Salpêtrière. His eyes were opened to the value of photographic work by Guillaume Duchenne de Boulogne, an unconventional French scientist who was respected for his research on movement disorders and electricity. The talented photographer joined Charcot's service in 1862 and went on to make hundreds of helpful slides, both of patients and of autopsy material.

Determined to have a knowledgeable, well-trained staff under him, Charcot began to give clinical lectures. His talks started without much fanfare in an improvised lecture room that had been a part of an old pharmacy. There, before relatively small audiences, he presented his thoughts on illnesses of the aged and chronic diseases.

Charcot did not seem particularly interested in neurology when he began his reorganization of the Salpêtrière. But, after beginning to work with patients suffering from chronic diseases, and guided by Duchenne de Boulogne, his interest in brain and spinal cord disorders grew rapidly. Within a few years, Charcot was providing medical personnel around the world with the best descriptions to date of many neurological disorders.

Georges Guillain, one of Charcot's biographers, noted the change when he explained:

> The classical medical texts of 1850 . . . devoted only a few pages to the vascular, infectious, and degenerative diseases of the brain and to brain tumors, which were then not adequately classified. The descriptions in these texts of spinal cord pathology and disease were spotty and vague, and their inadequate accounts of epilepsy, chorea, and tetanus were relegated to sections on the neuroses. By the time of Charcot's death, however, the . . . major categories of neurologic disease had been distinctly identified and exquisitely correlated with their anatomical and pathologic substrata; the clinicoanatomic approach to the nervous system, designed and developed by Charcot, is what created the foundations of neurology.[4]

A Typical Morning

Charcot could not have accomplished what he did had he not been adept at organizing the patients, his clinical rounds, the services at the Salpêtrière, and his own time. His routine was to awaken by eight A.M. and have breakfast while a newspaper was read to him. He then took a hired carriage to the hospital, to be met by his chiefs, interns, and the nursing supervisor. They would brief him, after which he would head to his office, where he would examine patients capable of being brought to him.

The neurological examination when Charcot later began to work with male as well as female patients was described as follows by two of Charcot's students:

> He would seat himself near a table and immediately call for the patient who was to be studied. The patient then was completely undressed. The intern would read a clinical summary of the case, while the master listened attentively. Then there was a long silence, during which Charcot looked, kept looking at the patient while tapping his hand on the table. After a while he would request the pa-

tient to make a movement; he would induce him to speak; he would ask that his reflexes be examined and that his sensory responses be tested.[5]

Charcot then examined a second and possibly a third patient in the same way, always showing an uncanny ability to see right to the root of the problem. He would then talk to his staff about what he had seen, what to expect, and how to manage these patients. Afterward, he would walk to his waiting carriage and head home for lunch and private consultations. Never one to waste time, he always read during the short trip home, often unaware that he had already arrived at his destination.

Home Life

Charcot married at age thirty-nine, two years after he began to work at the Salpêtrière. Madame Augustine Victoire Durvis, then a young widow, was a small but distinguished woman, ten years younger than himself. She was friendly, intelligent, wealthy, and a skilled artist with paints, enamels, and ceramics—all of which made her a perfect companion for her second husband.

Madame Charcot cared deeply for her spouse. She knew he had difficulty confiding in others and respected his privacy when he was reading or writing. Charcot's students occasionally asked for her help when they were apprehensive about approaching the master directly. If going through Madame Charcot failed, the case was usually considered lost.

The Charcots had two children. The oldest, Jeanne, was born in 1865; her brother, Jean-Baptiste, was born in 1867. Although authoritarian at home as well as in the office, Charcot loved both of his children and enjoyed being with them. They saw some of the patients who came to their home for private consultations. In addition, because Charcot liked to invite people for dinner on Tuesday nights, the children also met many of his interns, colleagues, and famous guests.[6]

Like his wife, Charcot drew and painted at home.[7] His works included caricatures of patients, sketches from trips, and elegant paintings. He even copied some of the works of the great masters. He was a traditionalist when it came to art, music, and literature. The French Impressionists, notably Manet, Monet, and Renoir, were too imprecise for his taste; he preferred the Flemish and Dutch masters. Similarly, his favorite composers were Gluck, Mozart, and Beethoven. For pleasure, he also read the Greek and Roman classics. But in his mind, Shakespeare had to be the undisputed master when it came to understanding how complex motives and emotions can underlie human behavior—a subject he forever found fascinating.[8]

Charcot also loved to play with the pet dogs and monkeys at his home. One of the monkeys, Rosalie, learned to sit in an infant's high chair and to eat wearing a bib. He enjoyed feeding her and was known to break into laughter when she snatched some of his own food. Charcot was an animal lover, a man who opposed bullfighting and all forms of hunting. He was also not one for animal research. At the entrance to his office in the Salpêtrière there was a sign that, when translated, read: "You will find no dog laboratory here." Indeed, all of his research involved human patients.

Multiple Sclerosis

Early in his career, Charcot became interested in multiple sclerosis (MS), a chronic neurological disorder that destroys the white myelin sheath encasing some long axons. MS usually manifests itself when a person is in the prime of life. The exact pattern of demyelination can vary greatly from case to case, although the disease typically affects the spinal cord, the brainstem, the cerebellum, the cerebrum, and the optic nerves. As the axons become demyelinated, sclerotic (hard) plaques appear. These plaques impair the ability of the surviving nerve fibers to conduct impulses. Because some remyelination can take place, the signs and symptoms of MS may wax and wane, especially early in the disease.

Depending on the locus and extent of the demyelination, MS sufferers can experience fatigue, tremor, paralyses, and sensory problems, including double vision (diplopia). There may also be cognitive deficits, such as memory problems and naming difficulties, as well as some unusual emotional changes. Some MS patients laugh and cry spontaneously, and they may also seem surprisingly unconcerned about the future, overly optimistic about what they can do, and even cheerful.[9]

Very little was known about MS before Charcot began to study it in the 1860s. In fact, there is reason to believe that MS was a relatively new disease at the time.[10] Augustus d'Esté, the grandson of King George III and a cousin of Queen Victoria, provided the first good written description of some of the changes experienced by a person with MS.[11] He did this in his diaries, which date from 1822 to 1847. But because his diaries were not made public until 1940, not even Charcot, who was an avid reader of medical literature in English, knew what this unfortunate man had written.

What Charcot did know was that some features of MS were described in two massive pathological-anatomy books from the 1830s and 1840s. One book was by Robert Carswell, who saw the yellowish-brown plaques in two individuals who had developed paralyses before they died.[12] The second was by Jean Cruveilhier, who had been the first professor of pathological anatomy at the Paris School of Medicine, which he had attended.[13]

One of Cruveilhier's cases was a woman named Dargès, a cook at the Salpêtrière whose extremities had become weak and uncooperative. As Dargès' disease progressed, she became paraplegic, experienced visual problems, and showed tremors. Although her intelligence seemed unaffected, when Cruveilhier spoke to her "she became seized with an emotion difficult to describe. She blushed, laughed, and cried."[14]

Several brief clinical and anatomical descriptions of MS appeared in the German literature after this time, but, as Charcot saw it, none of the early clinical descriptions of MS had depth. Further, no author even came close to bringing the signs, symptoms, and pathological anatomy together in a satisfactory way.

Charcot learned more about MS after he asked a charwoman who showed awkwardness and a peculiar form of tremor to serve as his personal housemaid. As a result, he was able to track her condition on a daily basis, even though she broke some of his costly plates in the process. When his sickly housemaid could no longer take care of herself, she was admitted to the Salpêtrière, where she later died. An autopsy revealed the hardened, discolored plaques of MS.

Charcot collaborated with Vulpian on most of his early research on MS. Between 1863 and 1865 the two physicians paid special attention to the anatomical features of the disease. At the same time, they saw how easy it could be to confuse the clinical features of MS with those of "the shaking palsy," soon to be called Parkinson's disease. Tremors, they noted, can characterize both disorders early on, and both MS and Parkinson's patients can lose their ability to move about as the years pass.

Charcot then employed a device called a sphygmograph to help him differentiate between the tremors of the two diseases. He also asked patients to hold very large plumes in front of a screen, which allowed him to see their tremors in more detail. Once he recognized that the MS patients exhibit tremors only when they are engaged in willful movements, and not when resting, whereas the Parkinson's patients continue to show tremors when at rest, he knew he had found the key to separating the two disorders. With this distinction to build on, he observed that only the MS patients also have visual and other sensory problems. In addition, he noted that the signs and symptoms of MS are much more likely to wax and wane over time.

In 1866 Vulpian and Charcot gave a report on both the clinical and anatomical features of MS to the Société Médicale des Hôpitaux.[15] Vulpian cited one case and handled most of the discussion, and Charcot presented two cases. Two years later Charcot gave two important lectures on *sclérose en plaques*, his term for MS, as a distinct disease entity.[16]

After Vulpian left the Salpêtrière in 1868 to assume a post at the nearby Charité Hospital, Charcot continued to work on MS. He now devoted increased attention to less severe cases than the ones described in his earlier presentations.[17] Indeed, his basic strategy for researching all diseases was first to define the full-blown clinical entity and then to turn to incomplete, slight, or partial forms of the disease. He referred to the latter as *formes frustes,* a French term meaning "blurred form."

Many of Charcot's insights about MS appeared in the first volume of his *Leçons sur les maladies du système nerveux* (Lectures on the Diseases of the Nervous System), published in 1872–1873.[18] In the first of the three MS lectures published in it, he described the disorder's main pathological sites. He also expressed his belief that plaques in different locations can be associated with different clinical features.

His second lecture was devoted to signs and symptoms. Here he described the tremor of intention, the condition of the legs, and the cephalic symptoms in sequence. The latter included visual defects, speech problems (difficulty enunciating syllables), and intellectual changes. He even alluded to some of the emotional changes shown by MS patients in this context. Although he concluded that the three most revealing clinical signs of MS are tremor of intention, unusual eye movements (nystagmus), and odd, scanning speech, he later realized that his "triad" was far from sacred. Indeed, he found that some of his patients never developed all three features of the disorder.

Once Charcot drew attention to the detailed clinical and anatomical features of MS, many more articles about it started to appear. Samuel Wilks, an Englishman, wrote that Charcot had successfully opened his eyes.[19] Similar sentiments were echoed by an American contemporary of Wilks, Meredith Clymer of the University of Pennsylvania.[20] When summarizing what was known about MS in 1870, he wrote: "To Dr. Charcot unquestionably belongs the credit of distinguishing this condition [MS] from other paralytic disorders and notably from *paralysis agitans* [Parkinson's disease], recognizing its pathological individuality, and tracing its clinical history."[21]

Parkinson's Disease

Parkinson's disease was studied just as intensively as MS by Charcot in the 1860s. This disorder was named after James Parkinson, a physician, paleontologist, and political activist who was born in a London suburb in 1755. In 1817 Parkinson published a pamphlet entitled *An Essay on the Shaking Palsy*.[22] Nevertheless, his insights remained unknown to the world at large before Charcot began to lecture on his so-called shaking palsy.[23] In fact, only after Charcot called the disorder *la maladie de Parkinson* did Parkinson achieve name recognition.

Parkinson's sixty-six-page pamphlet was based on just six cases. Of these, only three were examined in detail; he wrote that two others were simply met "within the street" and one "was only seen at a distance." Nevertheless, Parkinson correctly and succinctly described the disorder as one of "involuntary tremulous motion, with lessened muscular power, in parts not in action, and even when supported; with a propensity to bend the trunk forward, and to pass from a walking to a running pace: the senses and intellects being uninjured."[24]

He further noted that the disease progresses so slowly that an afflicted person may not accurately recall when his or her symptoms first appeared. Because Parkinson's patients were still alive when he composed his essay, he was unable to present autopsy material. He simply speculated that the affliction may have its basis in the lower brainstem or spinal cord.

In 1861 and 1862 Charcot and Vulpian gave several detailed accounts about "this strange affection" resembling both MS and chorea (rheumatic fever).[25] When the second volume of Charcot's *Lectures on the Diseases of the Nervous System* ap-

FIGURE 12.3.

A Moroccan man with Parkinson's disease sketched by Charcot. (From Meige, 1898.)

peared in 1877, the stages of the disorder were made clearer and its features were better defined.[26]

Charcot drew new attention to the blank stare, to the stolid and motionless nature of the face, and to a number of postural changes (Fig. 12.3) Witness the detail in which he wrote about hand movements in Parkinson's disease:

> At this stage of the disease . . . we occasionally find the rhythmical and involuntary oscillations of the different parts of the hand recalling the appearance of certain coordinated movements. Thus, in some patients, the thumb moves over the fingers, as when a pencil or paper-ball is rolled between them; in others, the movements are more complicated and resemble what takes place when crumbling a piece of bread.[27]

Charcot seemed more than a little surprised that Parkinson did not adequately describe the single most disabling feature of the disease—the rigidity that immobilizes these patients long after they first show tremors.[28] He emphasized how difficult it can be for even a moderately advanced Parkinson's patient to get up from a chair and how these patients will eventually become stiff and bedridden. The stooping or bent posture, the joints that appear "soldered together," the inability to turn the head, and the deformations of the hands, Charcot maintained, can all be related to the muscle stiffness.

Charcot now added some excellent descriptions of the forward running movements (propulsion) of the less advanced Parkinson patients. He then pointed out that when jerked backward by their clothes, they also exhibit an unavoidable tendency to begin running backward (retropulsion), which sometimes continues until they hit a solid object. Oddly, they remain stooped forward even while running backward.

Some of the more subtle changes were also described. For example, he commented on how the afflicted seem to speak "between their teeth." As for the mechanics of their handwriting, he contrasted their tremulous upward strokes with their firmer and smoother downward strokes (Fig. 12.4). Later, he would write about the shrunken size of the handwriting (micrographia).

Some of the Parkinson's patients described by Charcot traveled considerable distances to see him. After long train or carriage trips, these individuals did not seem to shake as much as the patients who lived in the neighborhood or were housed at the

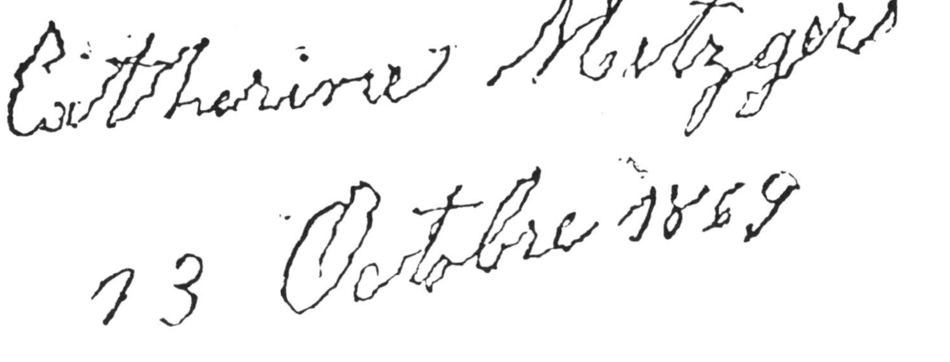

FIGURE 12.4.

An example of the handwriting of one of Charcot's patients with Parkinson's disease. Note the unsteady upward strokes and smoother downward strokes. (From J. M. Charcot, Leçons sur les maladies du système nerveux. *Paris: Delahaye, 1872–1873.)*

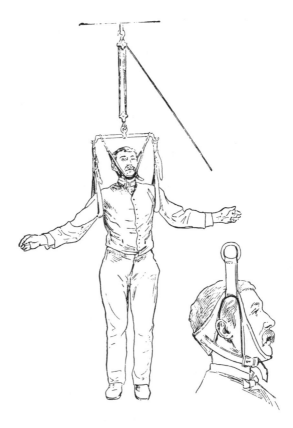

FIGURE 12.5.

The stretching apparatus tried by Charcot for treating Parkinson's disease. This device was the subject of his "Tuesday lecture" of January 15, 1889.

Salpêtrière. Thinking that bouncing or vibration may be therapeutic, he built a trembling armchair as well as a harness that bounced patients in the air to test his idea (Fig. 12.5).[29] Neither the vibrating, the stretching, nor the bouncing proved to have lasting effects.[30] After electrical stimulation also proved ineffective, Charcot became even more skeptical about slowing the course of Parkinson's disease.

When it came to the underlying pathology, Charcot agreed with Parkinson that the central nervous system must be involved. But, being unable to find a causal lesion, he was forced to label the disorder a *névrose*—a term that signified a disorder of the nervous system for which a causal lesion had yet to be discovered. Unfortunately, his word soon became *neurosis* in English, falsely conveying the impression that Charcot believed he was dealing with a mental condition.

Charcot died before Parkinson's disease was correctly associated with degenerative changes in the substantia nigra, an area of the brainstem so named because it is darkly pigmented.[31] But his expansion of the syndrome described by Parkinson has withstood the test of time. Even today, his astute observations about this disorder make worthwhile reading.

Amyotrophic Lateral Sclerosis (ALS)

Amyotrophic lateral sclerosis (ALS) is also a progressive disorder, but one that typically results in death within five years.[32] It is often called Lou Gehrig's disease in the

United States, in remembrance of the great New York Yankee's baseball player who was forced to give up the game in 1939, at the height of his career. ALS is also known among neurologists as Charcot's disease, a fitting eponym for the man who first recognized it as a distinct disease entity.

By the mid-1860s Charcot had become interested in the amyotrophies, a group of disorders characterized by muscle atrophy or wasting. But whether the shrinkage could be attributed to inadequate nutrition of the muscles, to a problem within the muscles themselves, or to some other factor was anything but clear.

To learn more about the amyotrophies, Charcot began to track patients at the Salpêtrière with muscular wasting. He also carefully examined the muscles, spinal cord, and brain when these women died. The autopsy material was revealing. Most patients with muscular atrophy showed a marked loss of cells in just one part of the spinal cord, a gray region called the anterior horn. Knowing that the nerves from this region control the skeletal muscles, Charcot correctly recognized that the muscles of the arms and legs of these patients had atrophied because they could no longer be stimulated by their nerves. That is, they shrank from disuse.

Charcot now began to divide his amyotrophic patients into subgroups. In this context, he saw that some patients contracted the disorder as adults, whereas others showed paralyses in childhood. The early-onset cases were, in fact, polio victims. Although polio had been studied previously by others, including his colleague Duchenne de Boulogne, Charcot now added to what was known about its neuroanatomical features.

While engaged in this work, Charcot also studied patients who developed severe paralyses and muscle contractures but did not have muscle wasting. Autopsies on these individuals revealed pathological changes in a different part of the spinal cord, a glossy white region called the lateral columns. These columns contain the axons of neurons located in the motor cortex, and they end on the anterior horn cells. In the contracture patients, the lateral columns showed scarring, hardening, and degeneration.

During his search for more cases of amyotrophy and for additional cases of paralyses without muscle atrophy, Charcot encountered a small number of patients who showed both debilitating conditions at once. That is, they showed paralyses with contractures and muscle wasting. Their autopsies revealed anterior horn cell losses as well as pathology of the lateral columns. In 1869 Charcot and his assistant Alex Joffroy gave the compound condition an incredibly cumbersome name, chronic deuteropathic spinal amyotrophy.[33]

Charcot's insights on what we now call ALS became much better known in 1877, when his book of lectures appeared.[34] He explained that the disorder typically begins in the arms, which quickly become paralyzed and assume abnormal postures. The paralysis and atrophy then extend to the legs and the face and mouth, affecting speech, chewing, and swallowing. In contrast, the senses and the intellect seem to remain normal. The prognosis, explained Charcot, is extremely grim: "There does not exist, as far as I am aware, a single example of a case where, the group of symptoms just described having existed, recovery followed."[35]

Charcot continued to study ALS during the 1880s.[36] He last lectured on it in 1888, when he evaluated a fifty-seven-year-old man in a fairly advanced stage of the disease.[37] As the debilitated man was taken from the room, he told him that he would soon instruct him what to do to get better. He then informed his audience that there really was no hope. He explained that the deception emanated from his compassion for the dying man.

A sustained interest in ALS and the amyotrophies led Charcot to an overlooked but related disorder shortly before he gave his last ALS lecture. The affliction he and his subordinate Pierre Marie discovered was a slowly developing, hereditary disease that leads to selective muscle wasting of the lower parts of the body.[38] Three months after their report appeared, the same disorder was described in a Cambridge University doctoral dissertation by Howard Henry Tooth, who acknowledged the contributions of Charcot and Marie.[39] This more limited amyotrophic illness now bears the name peroneal muscular atrophy. It also goes by another appropriate eponym: Charcot-Marie-Tooth disease.

After the Franco-Prussian War

The Franco-Prussian War of 1870–1871 had a dramatic effect on life in the French capital and on Charcot, who was forced to slow his neurological research in order to treat those now afflicted with cholera, smallpox, and other diseases. To make matters worse, the Salpêtrière, which temporarily served as a hospital for sick and wounded soldiers, was bombarded in January of 1871. Charcot never forgave the Germans for this "barbaric" act. Yet he took it in stride when a bullet flew through both windows of a carriage in which he was riding. According to the coachman, he never even closed the book he was reading and merely muttered the French equivalent of "Humbug!"[40]

Charcot's wife and children were sent to London during the war. Once an armistice was declared, Charcot escorted them back to Paris. Jean-Baptiste Charcot, who described the reunion, later mused that his father arrived "with a most extraordinary beard . . . that my mother caused to be shaved in a few minutes."[41]

A year after the war, life returned to normal for most Parisians. For Charcot, it was the beginning of the next phase of his career. First he was made professor of pathological anatomy in the Faculté de Médecine. This was the position that had been occupied by Cruveilhier and then by Vulpian. Second, he was elected to the French Académie de Médecine, the most elite medical organization in France. And third, he began to devote increased attention to cerebral localization, which had become a fashionable subject for clinical research after Paul Broca's seminal papers on language and the brain appeared in the 1860s (see Chapter 10).

Charcot on Cerebral Localization

Charcot approached cerebral localization with the same *méthode anatomo-clinique* that he had used in the past.[42] He recorded behavior after cortical damage and, if an interesting patient died, he performed a postmortem examination with the hope of correlating brain pathology with behavior. Once his hypothesis was formulated, he searched for more "textbook" cases to confirm the suspected relationship between pathological anatomy and clinical symptoms. His last step, as before, was to look at less characteristic forms of the disorder, his so-called *formes frustes*.

Charcot remained most interested in the motor system, the system he knew best from his earlier research on MS, Parkinson's disease, and ALS. As we saw in Chapter 11, the motor cortex was discovered by Fritsch and Hitzig in dogs in 1870 and described in other animals, including monkeys, by David Ferrier a few years later. By 1875 Charcot and his pupils were conducting their own large study on motor-cortex damage, but this time by examining human cases.[43]

Charcot and his associate Pitres showed that the human motor cortex is organized just like that of the other mammals studied, with the opposite leg represented on top, the arm in the middle, and the face on the bottom.[44] Thus, they confirmed the correctness of John Hughlings Jackson's earlier thoughts about the organization and location of the human motor cortex—ideas based on observing what Charcot now called Jacksonian seizures (see Chapter 11). In addition, Charcot and Pitres were the first to show conclusively that damage to the human motor cortex causes degeneration of the pyramidal tracts and lateral columns of the spinal cord.

When it came to cortical localization, Charcot was clearly in the mainstream. He even tried to help Paul Broca by sending his fellow Parisian detailed reports on a number of patients with well-documented speech problems. Nevertheless, Charcot was bothered by the fact that some of his cases seemed to violate Broca's frontal-lobe theory of speech. Most troubling were a few cases of severe language disturbances without apparent frontal-lobe involvement. As early as 1863, for example, he did an autopsy in Broca's presence on a severely aphasic woman with a perfectly intact Broca's frontal lobe area but degeneration throughout her temporal lobes.

Charcot became even more perplexed two years after seeing this case when he came across a second instance of aphasia without damage to Broca's critical frontal region. This case was reported by one of Charcot's assistants (Bouchard) to the Société de Biologie. It too left Broca and his followers scratching their heads.

To explain these "anomalies," Charcot reasoned that different individuals may process language in dissimilar ways, that is, with different parts of the brain. This being the case, was there any reason to assume that every patient must show a classic Broca's motor aphasia after damage to the third frontal convolution of the left hemisphere? For Charcot, the answer was clearly no. As he enjoyed reminding his students, theory has its place, but it does not stop things from existing.

The First Chair of Neurology, and Changes

Delighted by what was being accomplished on French soil, Charcot's powerful political friends, led by republican statesman Léon Gambetta, managed to establish the first Chair of Diseases of the Nervous System at the Paris School of Medicine. This neurological post was tailor-made for Charcot, who was offered the position and officially accepted the honor in 1882. The new position was hailed as a landmark; it represented the first time that neurology was officially given a chair and recognized as a separate discipline.

In 1883, the year in which obligatory church attendance was ended at the Salpêtrière and trained nurses were hired to replace the nuns, Charcot received more welcome news. He was nominated to the prestigious French Académie des Sciences, the leading French scientific organization.

A number of significant changes now took place at the Salpêtrière. First, a large *consultation externe* (outpatient clinic) was established near the entrance to the hospital. The outpatient facility served a rapidly growing number of working-class patients, both men and women, with problems that did not require hospitalization. Second, a better-equipped photography laboratory was set up, one that could even deal with a sequence of movements. And third, a new teaching amphitheater was erected to seat the large crowds of physicians, dignitaries, and inquisitive professionals from other walks of life who were now showing up to hear Charcot lecture.

Charcot used the new auditorium to turn his Friday lectures into lavish presentations. He used colored chalks to draw diagrams on blackboards, showed plaster casts of patients with deformities, had specimens prepared for display, and turned on a projector when he wanted to show lantern slides of patients or of diseased parts of the nervous system. Everything was arranged beforehand to make these presentations unforgettable for the audience.

The master was always punctual and enjoyed entering the hall with his associates and students following in a formal procession. He then took the stage, behind which there was displayed a huge painting of Pinel removing the irons from the insane at the Salpêtrière. Charcot's diction was slow and clear, and he had a remarkable ability to imitate his patients, some of whom appeared with him on the stage. The situation could not have been better for demonstrating how a disease may progress or for showing how the signs and symptoms of different diseases can be differentiated.

Charcot seemed all the more impressive because he never lectured from his notes. Nevertheless, his Friday lectures were not spontaneous presentations; they were meticulously prepared days in advance and written only after he had consulted his journals, books, and files. He then memorized, polished, and rehearsed what he wanted to say until he was satisfied. At the end of a lecture, he usually passed his handwritten material to an assistant, who had the task of reviewing it and submitting it for publication.

In addition to his Friday clinical lectures, Charcot started a Tuesday lecture series at the outpatient facility. The Tuesday lectures were unrehearsed and usually involved his outpatients. He focused on one or two interesting cases, assessing the problem, making a diagnosis, prescribing a treatment, and giving his prognosis. Explaining what he was thinking as he went along, these lectures showed Charcot at his clinical best. They delighted his students, who hoped to apply his logic and methods with equal success when summoned to the bedside.

Although Charcot fought for better medical treatment for his hospitalized patients, some of his critics, including satirical author Léon Daudet, accused him of being abusive and insensitive during his demonstrations. They portrayed him as a dictator and a despot—the so-called Napoleon or Caesar of the Salpêtrière—a man who acted as if the laws of civilized society did not pertain to him. In private, they even referred to his students and followers as the *Charcoterie*, a play on the French word *charcuterie*, meaning a butcher shop for pork products.

Charcot, however, had more than his share of loyal supporters. They appreciated what he was trying to do and understood why he was doing it. From their perspective, he was a brilliant clinician and an enlightened reformer who cared deeply about his patients and his students. He could appear cold, but he was the dedicated physician who transformed the old Salpêtrière into the world's greatest teaching hospital, and for this he deserved more respect.

Gilles de la Tourette's Syndrome

Charcot's interest in movement disorders never waned. During the middle of the 1880s, for example, he helped to establish Gilles de la Tourette's syndrome as a distinct disease entity. The road to this new movement disorder was anything but direct. It began when he asked the young physician who would become his most faithful disciple to help him classify movement disorders.

Georges Gilles de la Tourette was born in 1857.[45] He was a brilliant but restless man and a literary talent who had become an intern at the Salpêtrière in 1881. He was drawn to movement disorders by a paper entitled "Experiments with the 'Jumpers' or 'Jumping Frenchmen' of Maine," which he had read and translated from English into French in 1881.[46] In this paper, George Beard, a highly respected American psychiatrist, had described some Frenchmen from the northeastern United States and nearby Canada who exhibited intense startle responses, made unusual noises, and imitated movements and sounds.

Gilles de la Tourette reasoned that if "jumping Frenchmen" could be found in North America, they should also be present in Paris. Thus he began his own search for them. At about the same time Charcot drew his attention to two articles describing possibly similar conditions in other parts of the world. One was written by H. A. O'Brien, a man without medical training; it dealt with a disorder called *latah* by the Malaysians.[47] Like "jumping," *latah* was characterized by extreme startle reactions and imitative movements and sounds. Another striking feature of *latah* was the shouting of obscenities.

The other article was written by William Hammond, once surgeon general of the United States and the author of the first American textbook of neurology. Hammond described a condition called *miryachit,* and his report was based on accounts given by American naval officers stationed off Siberia.[48] His depiction of individuals with *miryachit* was much like that of the jumping Frenchmen of Maine, although the affected Siberians did not make strange sounds.

In 1884 Gilles de la Tourette published an article about these three strange movement disorders.[49] He wrote that all three were probably one and the same disorder. Further, he saw the condition as distinct from the real choreas, illnesses characterized by intense, involuntary movements. At the end of his paper, he mentioned that there was a teenage boy under Charcot's care who seemed to exhibit a similar condition. The boy was smart, but he suffered from extreme hyperexcitability and he repeated sounds. The boy also exhibited involuntary movements of his head and waist, and almost always screamed *"merde"* (shit) when he did so. In a footnote, Gilles de la Tourette mentioned that he had just come across several other cases in Paris, and promised to describe them in a follow-up paper.

Gilles de la Tourette's next important paper appeared in 1885.[50] He began by presenting some earlier descriptions of the movement disorder he was now seeing in Paris. He then pointed out that the earlier case studies were incomplete, poorly documented, and improperly classified. After making these points, he presented his French cases, beginning with the Marquise de Dampierre, who had first been described sixty years earlier by Jean Itard.[51]

Gilles de la Tourette did not change Itard's original wording when he described the marquise's cursing:

In the midst of a conversation that interests her deeply, all of a sudden, without being able to prevent it, she interrupts what she is saying or listening to with bizarre shouts and with words that are even more extraordinary and which make a deplorable contrast with her distinguished manners. These words are, for the most part, gross swear words and obscene epithets, and something which is no less embarrassing for her than for the listeners.[52]

Although he probably never met the cursing marquise, she served as his standard or anchor case. Other cases, many of which were less convincing, were presented in her shadow.[53] Gilles de la Tourette drew attention to the small, involuntary movements as the first indication of the disease. He maintained that the jerky movements usually started in the face and spread to other parts of the body. He also pointed to imitation and animal-like sounds as additional signs of this strange illness. In his opinion, however, the signature feature of the disorder had to be the involuntary cursing.

The major problem Gilles de la Tourette now faced was how to classify and label the individuals with this unusual condition that begin in childhood, is progressive, and leaves intelligence and sensation unaffected. After some thought, he took the stance that the Paris disorder is probably the same as "jumping," *latah*, and *miryachit*.

Charcot stepped to the fore after Gilles de la Tourette published his 1885 paper. In his lectures he made several new and salient points about the cases seen in Paris.[54] In contrast to what Gilles de la Tourette had written, Charcot emphasized that the small convulsive tics, and not the cursing, should be considered the real signature of this strange disorder. He also associated the disorder with obsessions and compulsions.

Charcot introduced the term *la maladie des tics de Gilles de la Tourette* to honor his assistant for first drawing attention to the tic disorder, even if he had made some mistakes. Unfortunately, many foreigners, not knowing better, immediately assumed that *Gilles* was a perfectly disposable part of his assistant's name. To the horror of the French, the condition soon became known as Tourette's syndrome in America and many other parts of the world.

Over the next few years, the idea that Gilles de la Tourette had described a unique disease generated considerable debate. Georges Guinon, one of Charcot's most popular residents, argued that the disorder is only an extreme form of hysteria.[55] Charcot, Gilles de la Tourette, and their supporters disagreed. Hysterics, they conceded, may show tics, but their tics are imitative, purposeful, and not spontaneous. Additionally, hysterics usually complain about sensory changes, whereas sensation remains normal in this tic disorder. A third difference is that the new disorder is incurable, whereas there are effective treatments for the hysterias.

In 1968 Arthur and Elaine Shapiro, a psychiatrist and a psychologist in New York, successfully treated a young woman with this disorder using a drug called haloperidol, which blocks the neurotransmitter dopamine.[56] Their work enhanced the view that Gilles de la Tourette's syndrome is, in fact, every bit as much a biochemical disorder as Parkinson's disease. They also showed that Charcot was wrong to have predicted that *la maladie des tics de Gilles de la Tourette* would resist all treatments.

Hysteria

During the final period of his productive life, Charcot was most concerned with the hysterias, a term used to refer to physical signs that are not accompanied by underlying anatomical changes that can account for them. Although the word *hysteria* can be traced to the Greek word *hystera,* meaning "uterus" (as in *hysterectomy*), some of the cases he saw at his new outpatient clinic—like some of the patients seen by Thomas Willis (see Chapter 7)—were men with imaginary physical complaints.[57]

During the 1870s Charcot tried to distinguish different types of hysteria (now called conversion reactions) from each other and from the epilepsies. His task was tricky because some of the hysterics at the Salpêtrière had become adept at mimicking the epileptics. Nevertheless, he soon had these women separated, in part by noting that hysterics show purposeful movements that only approximate the spasms of a real seizure. Once he recognized this, he debunked the reigning belief that "hystero-epilepsy" is a "hybrid" disease, one that can show itself as hysteria on one occasion and as epilepsy on another.

Charcot now worked diligently to make the hysteria diagnosis, considered by some the "wastepaper basket of medicine," a fitting subject for proper scientific study.[58] With his need to organize and classify, he lectured about the signs and symptoms associated with hysteria, including false epileptic seizures, tremors, and paralyses, strange losses of sensitivity, and trancelike states. He also showed that hysterics tend to respond fairly well to conventional therapies, including therapeutic baths and galvanism, although many relapse when again allowed to mingle with the sick and injured to model.

Charcot hypothesized that mental events can act as *agents provocateurs,* or triggers, for hysterical reactions, at least in an individuals with weak constitutions. He found provoking agents in the loss of a loved one, fears about a real illness, and work-related trauma.

These dynamic ideas had a major impact on Sigmund Freud, who traveled to Paris expecting to study neuroanatomy and neuropathology during the winter of 1885. When Freud began his trip back to Vienna after listening to Charcot lecture on the hysterias, the seeds for the development of psychoanalysis were planted in his fertile mind. Freud was so excited by what he had learned that he quickly translated and published some of Charcot's lectures on hysteria. His intents were to acquaint German-reading physicians with this material and, like other foreign visitors to the Salpêtrière, to enhance his own stature at home by associating his name with Charcot's.[59]

From his perspective, Charcot did not see the hysterias as anything new. He believed that individuals thought to be possessed by demons and cured by faith healing were, in fact, the hysterics of old (Fig. 12.6).[60] In this context, he published two notable art books with Paul Richer showing "hysterics" having fits and being cured across the ages.[61] The Church was not pleased, to say the least. But as a result of these efforts, more hysterics began to be treated as impressionable individuals with real medical problems, rather than as possessed souls, malingerers, or outright frauds.

Still, some of Charcot's loose speculations about hysteria, which stemmed from his failure to find a causal brain lesion, proved embarrassing.[62] Among other things,

FIGURE 12.6.

"Saint Ignace Délivrant un Jeune Possédé." In this facsimile of a 1610 illustration by Jean Collaert, a demon is shown leaving the body of a young boy. Charcot and Richer interpreted the drama as an earlier example of hysteria. (From Charcot and Richer, 1877.)

he believed that only hysterics and other *névroses* with susceptible nervous systems could be hypnotized. In accordance with his theory, his diagnoses began to rely more heavily on whether the patients could be hypnotized than on detailed clinical examinations.

Hippolyte Bernheim, a soft-spoken professor who worked in the northeastern French city of Nancy, emerged as one of the leading critics of Charcot's theories of hysteria. He argued that hysteria and hypnotizability are not twin traits of an abnormal neurological condition that runs in families. He found that almost anyone could be hypnotized and that the same therapeutic effects could be achieved by suggestion without a trance.

In addition, Joseph Delboeuf, a Belgian, visited the Salpêtrière and left convinced that Charcot's hysterics were actually primed by the staff to behave in particular ways before being presented to the master. The subjects in the hypnosis demonstrations—the most famous of whom was Blanche Wittman (the "Queen of the Hysterics")—not only relished the attention, but received rewards for convincing performances (Fig. 12.7).

The debate about the nature of hysteria made headline news in 1890, during a sensational murder trial. Members of Bernheim's Nancy school argued that the accused woman, whose name was Bompard, had committed a murder after being hypnotized by her lover. Hence she should be acquitted. In contrast, the men of the Salpêtrière claimed she was not a hysteric, and therefore could not have been hypnotized when she committed what was really a cold-blooded murder.

Although the forces of the Salpêtrière won this legal battle, they suffered a humiliating defeat in the larger war, meaning their stances on hysteria. Only after the damage had been done did Charcot realize that he had relied too much on his staff and that he had put unjustified faith in what was only a theory about the pathological basis of hysteria. He now recognized that he had deviated from the scientific methods that had served him so admirably in the past and he was embarrassed.

Charcot's Death

Jean-Martin Charcot succumbed to pulmonary complications from a lingering cardiac condition on August 15, 1893. He had been a compulsive worker who always believed that more could and must be done in the field of medicine. Not one for exercise, never knowing how to rest, and a heavy cigar smoker who loved rich foods, he, like his friend Paul Broca, was a perfect candidate for heart disease.

His funeral services were held at the Salpêtrière chapel. The ceremony was simple and there were no speeches or religious ceremonies for the man who appeared to others to be anticlerical. Eulogies for Charcot appeared in more than seventy newspapers and medical periodicals. Five years after his death, his students and friends even erected a bronze statue of him at the entrance to the Salpêtrière. Sadly, it was melted down by the Germans who needed metal after they occupied Paris during the Second World War.

The passing of Charcot freed his son Jean-Baptiste from a somewhat forced career in medicine and allowed him to become very famous on his own as a polar explorer. Although Jean-Baptiste chose another path to follow, neurology as a discipline would not only stay the course but flourish as a scientific-medical enterprise in the post-Charcot years. Largely as a result of Charcot's efforts and insights, the chaotic and nascent field of neurology had taken on the form of a modern discipline during the second half of the nineteenth century.

FIGURE 12.7.

"A Clinical Lesson at the Salpêtrière," painted by Louis Brouillet in 1887. Joseph Babinski is holding Blanche Wittman, the hypnotized "Queen of Hysterics," with Charcot to his right.

Santiago Ramón y Cajal
From Nerve Nets to Neuron Doctrine

Is it too much to say of him that he is the greatest anatomist the nervous system has ever known?

—*Charles Scott Sherrington, 1949*

Introduction

Scientists knew very little about neurons in the 1870s, the decade in which Eduard Hitzig and David Ferrier published their most important experiments on cortical localization and Jean-Martin Charcot gave neurology its sound footing in science. There was uncertainty about whether closely associated fibers grow from the cell body and about their respective functions. Scientists also wondered whether neural conduction is in only one direction and just how the impulse passes from one unit to another.

To get a better feel for how little was known about the elements comprising the nervous system during the 1870s, we need only look at the terminology of the day. *Dendrite* and *axon*, the two kinds of processes emanating from the nerve cell body, were not yet a part of the vocabulary; these words will be introduced by Wilhelm His in 1890 and Albrecht von Kölliker in 1896, respectively.[1] Even the word *neuron* will have to wait to make its debut, in this case by Wilhelm Waldeyer in 1891.[2] As for the term *synapse*, when Charles Sherrington introduced this new word in 1897, it was only to describe a purely hypothetical junction.[3]

Three ingredients were sorely needed to allow nineteenth-century investigators to gain a better understanding of the elements of the nervous system. First, there was a need for better microscopes to allow scientists to see their specimens at high magnification without distortion. Second, improvements in histological techniques were necessary to make the cell bodies and their processes stand out from the background. Finally, there had to be greater willingness to look with an open mind at slides containing pieces of stained brain. As we shall now see, all three factors gradually came together by the end of the nineteenth century.

The First Microscopes

Two seventeenth-century scientists stand out for constructing early microscopes and using them to look into the world of the infinitely small. One is Robert Hooke, an English physicist, astronomer, and paleontologist, and the other is Anton Van Leeuwenhoek, an inquisitive linen draper from Delft, Holland.

Hooke combined a lens and an eyepiece to construct the first compound microscope, a notable advance over the simple magnifying glass. He examined different materials with his instruments, such as strands of wool and thin slivers of cork, and published his beautiful drawings in his *Micrographia* of 1667.[4] Interestingly, when he first saw the honeycomb matrix of some shaved cork under one of his microscopes, the image of cells in a prison came to his mind (Fig. 13.1). Thanks to his vivid imagination, his word *cell* became a fixture in the life sciences. Hooke did not, however, apply his descriptive word to animal material, nor did he study the nerves or brain substance under his microscopes.

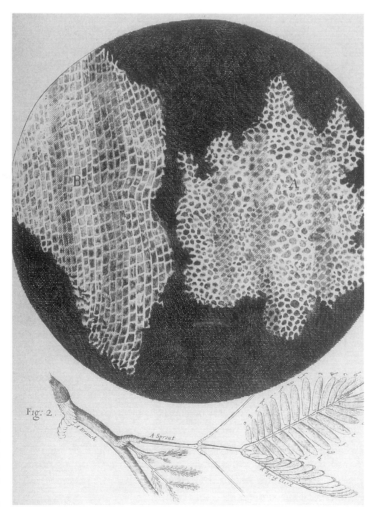

FIGURE 13.1.

A piece of cork with "cells" from Robert Hooke's Micrographia.

Anton Van Leeuwenhoek was equally mesmerized by what could be seen with the microscope. To satisfy his seemingly insatiable curiosity, he constructed hundreds of simple instruments and investigated innumerable plant and animal specimens.[5]

One of the first structures explored by the inquiring Dutchman was the optic nerve. He wanted to know whether the nerves from the back of the eye are really hollow, and therefore ideal for the transmission of tiny spirits (or pneuma), as suggested by Galen (Chapter 4). In 1674 and 1675 he sent short communications to the Royal Society of London in which he explained that he had examined the optic nerves of several cows but could not find any canals.[6] Perplexed, he suggested that perhaps vision has something to do with Newtonian vibrations in the nerves.

In 1675 and 1677 Van Leeuwenhoek examined nerve cells from animal brains and spinal cords under his microscopes. More than forty years later, he even noted that the nerves were associated with a fatty material.[7] This might have been the first reference to the white myelin sheath that covers long axons and makes them glisten in the light. He also made drawings of some nerves freshly sliced and "viewed sideways" (Fig. 13.2).

Although his powers of observation were outstanding, Van Leeuwenhoek did not always realize the significance of what he was seeing. Perhaps this was because he typically wrote about his specimens while the images were still fresh in his mind, not after taking the time to think things over. In addition, he had to endure two serious technical problems. First, his primitive microscopes tended to produce distorted images. Second, the cells he wanted to study usually did not stand out from the surrounding fluids or the background, making their finer features hard to discern.

When he was in his eighties, Van Leeuwenhoek attempted to solve the latter problem. He tried to stain some thin sections of muscle with saffron, a yellow material obtained from crocuses.[8] His decision to use saffron did not come out of thin air. It had been used for well over two thousand years to dye materials, and in the recent past it had been injected into blood vessels to stain them.[9]

Van Leeuwenhoek mixed his saffron with "burnt wine" (brandy) and wet his already-cut muscle specimens with it. To his delight, he found that the blurry muscle fibers could now be seen more clearly. There is no evidence, however, to suggest that

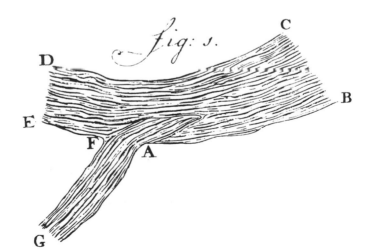

FIGURE 13.2.

A drawing published by Anton Van Leeuwenhoek in 1719 showing a piece of nerve from a cow or sheep sliced and viewed sideways. A branch of the main nerve is indicated by the letter G.

he also tried to stain pieces of nerve or brain with saffron or any other substance for better microscopy.

Van Leeuwenhoek's studies on the nervous system did not lead to an immediate flood of microscopic research on the retina, the peripheral nerves, or the central nervous system. More than a century would have to pass before the microscope would become more than a plaything for adventurous amateurs and eccentric dilettantes. During this lengthy period, many highly respected scientists stayed clear of the microscope, viewing it as an untrustworthy instrument that could quickly destroy a person's career.[10]

Better Instruments and Techniques

The road to a better understanding of the fine structure of the nervous system was paved in the 1820s when a new kind of microscope appeared, one that overcame some of the distortions encountered by Hooke and Van Leeuwenhoek by allowing scientists to focus on different colors simultaneously. The use of doublet lenses of lead and flint glass immediately opened the door for a wealth of new discoveries. This proved to be especially true in Germany, where there were many excellent instrument makers and scientists interested in microscopy to take advantage of the technical breakthrough.

German researchers also discovered better ways to prepare their specimens. This was important, since most investigators in the seventeenth and eighteenth centuries did not even harden or "fix" their specimens for thinner slicing and better preservation. Occasionally they boiled specimens in oil or soaked them in wine, but these procedures were not standardized and were often less than satisfactory.

Early in the nineteenth century Johann Christian Reil found that pieces of brain could be made more suitable for fine dissection by first soaking them in alcohol.[11] A few decades later Adolph Hannover introduced chromic acid as an even better hardening agent for nerve tissue.[12] Formaldehyde, the substance used by most scientists to harden and preserve tissue today, did not appear until the nineteenth century was drawing to a close.

The Birth of Cell Theory

Armed with their new microscopes and better methods for hardening their specimens, some nineteenth-century microscopists focused their attention on the nervous system. Jan Evangelista Purkyně was among the first of the new explorers of the brain.[13] Born and educated on Czech soil, Purkyně joined the faculty of the University of Breslau (now Wroclaw) in 1822 and, a few years later, requested a new achromatic microscope. After a seven-year wait, he finally received his coveted instrument. Because of inadequate space at the university, he took his microscope home, where he built a laboratory and invited his students to work with him.

In 1837 Purkyně described a cell from the cerebellum at a scientific congress in Prague. It had a cell body with a nucleus and some elongated fiberlike processes extending from it. A year later he published his talk with a drawing of some of his cerebellar cells (Fig. 13.3).[14] His description was considered so good for the time that these cells were called "Purkyně cells" (or "Purkinjě cells") in his honor.

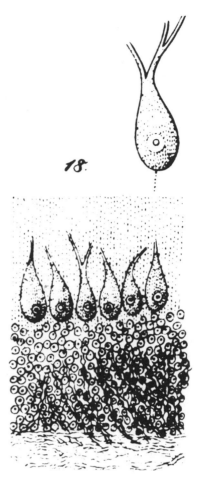

FIGURE 13.3.
*Purkyně's
illustration of the
cerebellar cells now
known as Purkyně
(or Purkinjě) cells.
This plate was
published in 1838.*

Shortly after Purkyně described his cell, cell theory—the idea that all living forms are composed of cells—came into its own. The two men usually recognized for their pivotal roles in developing the broader theory are Matthias Schleiden and Theodor Schwann. In 1838 Schleiden, a German botanist, concluded that all plant parts are constructed from cells.[15] A year later Schwann, who was working in Belgium, convinced his friend Schleiden that cell theory also holds for animal tissues.

Schwann went on to describe the fatty myelin covering that gives some axons a glossy white appearance.[16] He astutely recognized that myelin is a "secondary deposit" and not a part of the nerve cell itself. In the peripheral nervous system, the cells that wrap around some axons and make them look white are now called Schwann cells in recognition of his observations.

Schwann and Schleiden's cell theory was quickly accepted for all parts of the body except the nervous system. Some cautious scientists called for additional information because they did not know enough about the fiberlike structures found near the cell bodies. Another issue was whether the units maintain their independence like other cells, or fuse in some odd way.[17] Still more revealing histological techniques were needed to help solve the disputes.

The Carmine Stain

The secrets of the nerve cell continued to come to light during the 1860s, thanks in part to the advent of new dyes for staining tissue.[18] The first widely accepted stain to be used by brain scientists was carmine, a reddish substance extracted from the bodies of certain insects just before they lay their eggs. Alphonse Corti, an Italian anatomist, was one of the first scientists to try this stain on animal tissue. He used it to study the structure of the inner ear in 1851, and was the first person to described the part of the cochlea we now call the "organ of Corti."[19]

Nevertheless, another individual received greater recognition than Corti for using carmine to study animal tissues. His name was Joseph von Gerlach. During the 1850s he injected some carmine into the bloodstream and found it was readily picked up by nearby cells, much like a living plant might pick up dye dissolved in water. He then tried wetting some specimens from the cerebellum with a carmine solution, but the results were unimpressive and he went home dejected, forgetting that he had left a few pieces of cerebellum in the solution. The next morning he returned to his laboratory and, to his delight, discovered that some cerebellar nerve cells and their fibers were wonderfully stained.

Gerlach was a professor of anatomy and he recognized the importance of his accidental discovery. He then worked hard to perfect the stain for histologists and to promote it as his own.[20] More interested in advancing his own reputation than in being fair to Corti, he did not mention that the Italian had used the stain before him, even though he had read his paper.

The Theory of Nerve Nets

Armed with "Gerlach's" carmine method for staining nonmyelinated nerves, and using chromic acid as a hardening agent, another talented German microscopist, Otto Friedrich Karl Deiters, now observed that each nerve cell body (soma) could have multiple "protoplasmic extensions" (dendrites) but only one "axis cylinder." But before he could make his new findings public, he died. Fortunately, Max Schultze, his chief in Bonn, recognized the importance of Deiters' writings, edited them, and published them in 1865.[21]

How nerve cells communicate with one another was one of the issues that Deiters had tried to address. After giving this matter considerable thought, he considered the possibility that the axis cylinder endings of one nerve cell and the dendrites of the next might fuse. But he proceeded cautiously, only mentioning this idea in words. On his drawings, he showed no more than what he had been able to observe with his own eyes.

Many scientists now began to believe that nerve cell processes probably anastomose (fuse) with each other, just like physically joined pieces of glass tubing. Such a union, they thought, could best account for the rapid and faithful transmission of messages in the nervous system. Albrecht von Kölliker, a leading authority in the field of histology, postulated that just the dendrites of neighboring cells united with each other.[22] But there were other variations on the theme as well, some with axons fusing with dendrites, and others with axons fusing with other axons.

Joseph von Gerlach, of carmine stain fame, was enthusiastic about the fusing idea.[23] He maintained that the nerve impulse travels from cell to cell across fine fiber nets or trellises. In large part because of his authority, the nervous system began to be looked upon as a giant net, web, or reticulum, made up of a huge number of physically interconnected parts.

The "Black Reaction"

In retrospect, it is easy to understand the mistakes made by Gerlach and the reticularists. These pioneering microscopists did not have the tools they needed to see how axons and dendrites really ended, and their fusing theories could in fact account for neural transmission from one cell to the next. But once again, improved methods were needed to shed light on the fine anatomy of the nervous system.

The call for a more revealing stain was answered in 1873, when Camillo Golgi (Fig. 13.4) described a new procedure that turned nerve cells and their processes black against a yellowish background.[24] Golgi was about thirty years old when he introduced his silver stain, and he did not work at a university or at a major medical center at the time.

Golgi came from the Lombardy region of Italy and was educated at the University of Pavia.[25] He made his breakthrough while working as a physician in Northern Italy at a *casa degli incurabili,* the term for a small hospital for the chronically ill. Golgi was experienced in research when he took his post at the hospital; he had studied pellagra, brain tumors, and the structure of the cerebellum while in Pavia. He went to Abbiategrasso because the position offered more job security, not because he wanted to spend more time in clinical practice. He never intended to stop doing research.

Abbiategrasso did not serve as a magnet for talented young scientists. In fact, its *casa degli incurabili* was relatively unknown to people outside the region. Never-

FIGURE 13.4.

Camillo Golgi (1843–1926), the Italian anatomist who developed a new silver stain for nerve cells in 1873.

theless, Golgi was determined to make a name for himself. He arrived at the hospital immensely proud of his heritage, the quality of his education, and his prior achievements.

His breakthrough took place in a kitchen at the hospital, which he had converted into a simple laboratory. It contained little instrumentation other than a microscope, and much of his work took place in the evening by candlelight. Although he never explained how the idea of developing a silver stain came to him, historians note that several other scientists had also tried to impregnate cells with silver, an agent then starting to be used for photography.[26]

Golgi first hardened his specimens for a few days. He next immersed the tissue in a solution of silver nitrate for another day or two. His specimens were then treated with alcohol baths and oils, washed, and mounted on slides. Recognizing that his stained specimens tended to fade in the light, Golgi stored his slides in the dark until needed.

The new silver stain did not darken all nerve cells in a piece of brain or spinal cord. If it did, the result would have been a jet-black smudge with little or no detail. Instead, it stained only about 3 percent of the nerve cells, which allowed these cells to stand out vividly from the background. This was the beauty of *la reazione nera*, or "black reaction." It produced sharp images that looked like fine silver and black etchings on a light yellow matrix—images much better than any seen before (Fig. 13.5).

Golgi published his first paper on the remarkable new stain in the *Gazetta Medica Italiana, Lombardia.* His brief article bore the title "Sulla struttura della grigia

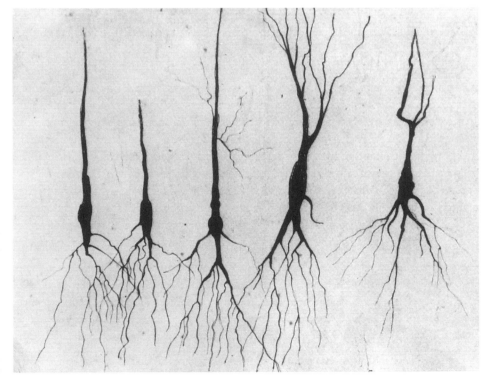

FIGURE 13.5.

An early drawing by Camillo Golgi showing five nerve cells impregnated with his silver stain. (From B. Zanobio, Golgian memorabilia. . . . *In M. Santini,* Golgi Centennial Symposium Proceedings. *New York: Raven Press, 1973.)*

del cervello" (On the Structure of the Gray Matter of the Brain). In it he described the nerve cell body and its attached processes (a single axon and several branching dendrites) with wonderful clarity. He also presented the first good description of axon collaterals, side branches that can split off from the main axon stem. But he did not describe his methods in detail, nor did he illustrate what his stained specimens looked like.

Golgi followed up on his 1873 paper with a number of other publications, many with better descriptions of his procedures and detailed drawings. He turned to the cells of the cerebellum and also wrote about the fine structure of the olfactory bulb. He distinguished between nerve cells with long axons and those with short axons that ended in complex, bushlike structures. These two different types of nerve cells are now called Golgi I and Golgi II cells in his honor.

In 1875 Golgi accepted a faculty position at his alma mater, the University of Pavia, where he continued to use his silver stain to study nerve cells.[27] He now wrote about the fine structure of the cerebral cortex, the corpus callosum, and the spinal cord. He also studied glial cells and correctly recognized that they can be distinguished from nerve cells by the absence of an axon. Another of his contributions was the discovery of special organs in the muscles to provide feedback about muscle tension. Today, anatomists call these structures Golgi tendon organs.

Golgi learned much about axons, dendrites, and glia with his new stain. But because his stain was so selective, he was unable to see how the axons and dendrites really ended as they approached other cells. Still, he embraced the popular belief that nerve cell processes physically fuse with one another. His ideas about axons fusing to form nets led him to question the concept of distinct, highly specialized cortical areas. Functional localization with sharp boundaries between areas, he maintained, is inconsistent with the reality of nerves fusing into nets and with the potential of neurological patients to recover from specific brain injuries.[28]

Must Nerve Cells Fuse?

In the fall of 1886 Wilhelm His broke with the past when he concluded that transmission of nerve impulses may be possible without cells fusing.[29] He, like many others, knew that the endings of motor neurons do not fuse with their muscle fibers. Anatomical studies of growing neurons convinced him that the same is true for the nerves originating in the eye, ear, nose, and throat. If fusing is not a prerequisite for transmission in the periphery, why, he asked, should the central nervous system be different?

August Forel independently reached the same conclusion at about the same time as His.[30] While working in Munich years earlier, he had removed sensory organs and nerve cell bodies from animals. When these animals were later sacrificed, he saw only sharply defined areas of degeneration. The failure to find broader areas of degeneration, and now his newer studies with the Golgi stain, were not supportive of the theory of nerve nets. Forel therefore postulated that simple contact could better explain neural communication.

Although both His and Forel wrote their papers independently in 1886, Forel's article did not appear until January 1887. At first, the publication delay that cost Forel priority by three months upset him. But he later cooled down and concluded that it probably made no difference at all. As he put it: "Well, our two papers suffered the fate of the majority of new ideas; they were simply ignored."[31] Indeed, the findings of His and Forel seemed tentative, inconclusive, and little more than vague suggestions that did not merit much attention, at least not yet.

Santiago Ramón y Cajal

The man who did the most to topple the idea of nerve nets and promote a more realistic neuroanatomy was Santiago Ramón y Cajal (Fig. 13.6), a Spaniard.[32] In many of his own books and papers, Ramón y Cajal dropped the surname of his father (Ramón) and went by Cajal, the surname of his mother. For this reason, we too shall call him Cajal.

Between 1901 and 1917, Cajal wrote about his early life in a work entitled *Recuerdos de mi vida*.[33] This wonderful autobiography appeared in English in 1937 as *Recollections of My Life*. With the pen of a poet, he vividly portrayed his formative years and told how he discovered the microscope and heard about the stain with the potential to unlock the mysteries of the nervous system.

Cajal was born in 1852 in the impoverished country town of Petilla de Aragon, in the northeastern province of Zaragoza, near the French border. He described himself as a rebellious, headstrong child who was constantly engaged in battles with his father, Don Justo, a village surgeon and later a physician of modest means. Don Justo

FIGURE 13.6.

Santiago Ramón y Cajal (1852–1934), whose many contributions to the fine structure of the nervous system provided modern neuroanatomy with a solid foundation.

was disciplined, uncompromising, and determined to keep a respectable family image. For his part, young Santiago was headstrong and not one for endless rules.

Santiago's nonconformist personality caused problems for him at the different schools he attended. Truancy, lame excuses, and poor grades led to constant whippings and other physical punishments. Stone throwing, fruit stealing, and mindless acts of vandalism (such as blowing up a neighbor's gate with a homemade cannon) also did not endear him to people in the several small communities in which the family lived.

Yet as rebellious as he tended to be, he was never aggressive, lazy, or cruel. And he did have interests, including nature, athletics, novels, and art. In fact, he fervently hoped to become an artist. His father, however, had no intention of allowing his son to pursue such a worthless career. At his wits' end, Don Justo even apprenticed Santiago to an ill-tempered barber and then to some cobblers, hoping he might learn a trade.

During the summer of 1868 Don Justo launched a campaign to try to interest his teenage son in anatomy, his own area of expertise. The two visited graveyards to find human remains exhumed from burial plots that had been leased but not renewed. The opportunity to sketch bones proved to be the turning point in Santiago's life. He listened to what his father had to say about the skeletal system and soon found himself answering even his most challenging questions. He slowly began to conclude that studying medicine and proceeding in his father's footsteps might not be such a bad idea after all.

He now settled down to work toward a bachelor's degree, and began preparatory work for his medical training. He then breezed through medical school in Zaragoza, where he "saw in the cadaver, not death, with its train of gloomy suggestions, but the marvelous workmanship of life."[34] Along the way, he even made some money assisting in dissections and tutoring in anatomy.

Although he now possessed a medical degree, his training was such that he had never seen a sick patient, been in a clinic, or witnessed a birth. But before he could do these things on his own, he had to deal with compulsory military service. Hence he enlisted in the army and took an examination to serve as a medical officer.

At the time, the Spanish army was fighting to suppress the independence movement in Cuba. In 1874, after a quiet period at home—no bullets, no wounded—Santiago's name was drawn by lot and he was sent off to a desolate, mosquito-infested outpost in Cuba. Like many of the soldiers he treated, he soon caught malaria, the dreaded disease then thought to be caused by bad (*mal*) air (*aria*), not infected mosquitoes.

Once transferred to a healthier location in Cuba, Cajal managed to regain some strength, after which he was reassigned to yet another jungle outpost, where his malaria flared up again. Now fearing the worst, he requested, and finally was given, an honorable discharge. He boarded a steamer, praying for the endurance to make it back to Spain alive.

Spain and the Microscope

After leaving Cuba, Cajal's health improved and he was able to accept a temporary position as a professor in Zaragoza. Because he did not have a secure appointment, he began studying for the examinations for open professorships in Spain. While in Madrid to take one of these competitive tests, he had the opportunity to look through a microscope. Excited by what he saw, he returned home determined to go into microscopy. The fact that most Spanish professors still despised the instrument mattered little to him.

Cajal used his hard-earned savings to purchase a microscope he had spotted in a Madrid shop, some basic accessories, and a few manuals. Soon afterward he bought a microtome for cutting very thin slices of tissue. Then he received some welcome news: He was offered the directorship of the Anatomical Museums of the Faculty of Medicine in Zaragoza.

With the security of the directorship and some additional income from giving lessons to doctoral students, Cajal concluded that he finally could support a wife. In his autobiography, he compared the young woman he asked to marry him to the Madonna of the Renaissance artist Raphael. Throughout his long life, Doña Silvería remained cheerful, modest, and completely dedicated to him and the happiness of their many children. Fortunately, marriage and a large family did not signal the end of Cajal's career as a scientist, as some thought it would. As he later expressed it in his wonderfully poetic way: "It was thus demonstrated that, contrary to the expectations of my friends, the children of the flesh did not smother the children of the mind."[35]

In 1883 Cajal won the coveted chair in anatomy at Valencia. Soon after his arrival, a deadly cholera epidemic swept through the region. By boiling water and adequately cooking their food, he and his family escaped its wrath. Nevertheless, the death rate shot up so rapidly that government officials begged him to put his own studies aside in order to learn more about cholera and claims of a new vaccine.

Cajal now dedicated himself to the study of cholera. He wrote a monograph on the nature of the disease, proper hygiene to prevent it, and inoculation. The thankful government sent him a magnificent Zeiss microscope as a token of its appreciation. There could not have been a better way to thank him for his efforts.

The Spaniard and the Silver Stain

While in Madrid in 1887 to administer examinations in anatomy, Cajal made the rounds to see some of the better-equipped Spanish laboratories. This activity brought him to the door of Luis Simarro, a neuropsychiatrist. Simarro explained that he had learned about Golgi's silver stain while on a trip to Paris, and was using it to look for degenerative changes in the brains of psychiatric patients.

Cajal looked at some of Simarro's slides and immediately saw the superiority of Golgi's stain over other stains for studying the finer features of nerve cells. Thus, at about the same time that His and Forel were beginning to question reticular theory, Cajal headed back to Valencia to begin his own research with the stain developed by "the savant of Pavia."

Once in his laboratory, he began to stain brain sections from animals. Nevertheless, it did not take him long to understand that the uncertain character of the Golgi stain represented a significant obstacle that had to be overcome. He also realized that the city of Valencia, although better than Zaragoza, was not the ideal place to devote himself to this sort of work, and so he moved to Barcelona.

Cajal now tried to improve Golgi's silver staining technique. He found that he could produce better slides by staining more intensely and cutting thicker sections. Even more important, he observed that the Golgi stain worked best when he stained axons not covered with a fatty myelin sheath. The search for brains containing non-myelinated axons led him primarily to birds but also to very young mammals.

After making these and other adjustments, Cajal had a more reliable staining procedure than before. Thus, unlike many others (including Simarro himself) who were quick to give up on the Golgi stain because it was so capricious, he was convinced that he could finally study the elements of the nervous system in a systematic way. Indeed, now that most of the bugs were worked out, new findings quickly came to light, one rapidly following another.

But where should he publish his observations and accompanying thoughts? Cajal realized that he needed a journal that would publish a large amount of text and accompanying illustrations quite rapidly. Not satisfied with existing Spanish publications, he decided to found his own journal. He called it the *Revista Trimestral de Histología Normal y Patológica* (Trimonthly Review of Normal and Pathological Histology).

Starting and operating a periodical was expensive. To make matters worse, Cajal now had six children. But he was not about to stop in his quest to show that world-class science could come from neglected Spanish soil. To save money, he walked away from his beloved chess club (in his mind, his only vice), where he was able to take on four skilled opponents simultaneously. He then dug deep into his pockets and came up with the money to publish his journal and mail sixty copies of each issue to leading anatomists in other countries.

Cajal's first paper based on his work with the modified Golgi stain appeared in the inaugural issue of the *Revista*, in May 1888.[36] It dealt with the bird cerebellum and described how certain axons ended in "nests" or "baskets." Nowhere could he find evidence for either axons or dendrites fusing and forming nets like those described by Gerlach or Golgi. For this reason, he wrote that either the modified Golgi procedure is insufficient to show the fibers physically joining or there must be some other mode of transmission. At the time, and unaware of what His and Forel had to say on the issue, he thought that transmission by simple contact was by far the most likely possibility.

In another paper in the same issue of the *Revista*, he examined the retina and came to the same negative conclusion about nerve nets. He then turned to the olfactory bulb, the cerebral cortex, the spinal cord, and the brainstem. No matter where he looked and no matter how hard he tried, he could find nothing suggestive of a web or reticulum. Reviewing his material in 1889, he argued that nerve cells should be regarded as independent elements, just like all other cells comprising the body.[37]

The Trip to Berlin

Cajal knew all too well that his country, even more than Golgi's Italy, stood in relative isolation from the scientific centers of Europe. It also did not take him long to recognize that sending copies of his journal to England, France, and Germany was a waste of money, because few recipients in central or northern Europe read Spanish. This seemed painfully clear from the limited mail he received and from the absence of citations of his work in the scientific literature, even though he had published fourteen articles in 1887 and 1888 alone.

Feeling desperate about communicating with the rest of the world and wanting more recognition for his efforts, Cajal decided to pursue two different strategies. First, he would have some of his papers translated into French. Second, he would join the German Anatomical Society in order to present his material at its upcoming meetings in Berlin.

Working without the help of university students or professional colleagues, Cajal prepared more slides of the cerebellum, the retina, and the spinal cord. Then, as the summer of 1889 gave way to fall, he carefully packed his best slides and his precious Zeiss microscope for the journey to Berlin. He planned to make stops along the way, hoping to learn more by visiting other laboratories.

The meetings of the German Anatomical Society took place on the campus of the University of Berlin in October 1889. The conference opened with members of the group taking turns speaking about their research. Cajal found it difficult to follow the oral presentations. He was not skilled in German and he was obsessed about how he would do in the practical demonstrations scheduled to follow.

Once the platform sessions were over, he made his way to his assigned table in the demonstration area. His own Zeiss microscope, joined by a few others requisitioned for the occasion, was on the table, and his slides of Golgi-stained cerebellum, spinal cord, and retina were on display. His magnificent silver images, he hoped, would speak for themselves, but if necessary he would try to communicate in his halting French.

The initial response to his exhibit left much to be desired. Because many histologists had their own displays, they only slowly wandered over to see what he had to offer. These scientists tended to be skeptical at first, thinking that only second-rate work could come out of a backward country like Spain. But when they looked at Cajal's beautiful slides and observed nerve cells with a clarity not seen before, their frowns gave way to smiles and warm words of congratulations.

The viewers, many of whom had been frustrated with the Golgi method, now asked about his modified silver stain. Among other things, he told them of the importance of using birds and young animals, and of cutting thicker sections. They also wanted to know what the observant visitor thought about nerve nets. Cajal explained as best he could that his slides provided absolutely no evidence for nerve cell processes fusing with one another. The axons reached out to the dendrites of the next cell, but they did not physically fuse. Nor did axons appear to fuse with other axons, or dendrites appear to join with other dendrites. Each nerve cell seemed to be an absolutely independent unit.

Albrecht von Kölliker was among the respected scientists who attended these demonstrations, and he did not hesitate to congratulate Cajal for his accomplishments. After examining his slides and listening to the dark-eyed Spaniard speak, the acknowledged patriarch of German histology found himself so impressed that he invited Cajal to a fine dinner and introduced him to several other German histologists. For the first time, Cajal felt really appreciated on the world stage.

When the meetings ended, Cajal thanked Kölliker for what he had done. He then started home, again with several stops along the way. One was in Pavia, where he hoped to meet Camillo Golgi. The Italian anatomist, however, was in Rome, where he (like Broca in France) was serving in the Senate. In nineteenth-century Italy, a position in the Senate was one way in which scientists were honored for significant contributions. Years later, Cajal would lament that had Golgi been in Pavia when he passed through, the two probably would have enjoyed getting together and might even have become good friends.

Kölliker now did several things that must be considered exceptional, especially for a man already in his seventies. First, he tried to confirm many of Cajal's observations with the modified Golgi procedure. Within a few months, he was satisfied that his Spanish friend was correct. As a result, he publicly abandoned the nerve net theory that he had endorsed before the Berlin meetings. In addition, he began to learn Spanish in order to translate Cajal's manuscripts into German.

The important role that Kölliker played during and after the Berlin meetings was not lost on Cajal. In his autobiography, he expressed "profound gratitude" to Kölliker for his friendship, support, and efforts to disseminate his new findings to the scientific world. He had indeed found a worthy friend in Kölliker, as well as a respected and powerful ally. Now energized as never before, he worked from nine in the morning to midnight day after day on his neuroanatomy. He turned out an amazing twenty-seven books and articles on the fine structure of the nervous system in 1890 and 1891 alone.

The Term *Neuron* Is Born

Wilhelm von Waldeyer, the director of the Anatomical Institute of the University of Berlin, had a reputation for synthesizing information from different fields, ranging from anthropology to anatomy. Impressed by what Cajal showed him in Berlin and the discussions after the demonstrations, Waldeyer decided to write a review about nerve cells being independent units.

Waldeyer's paper appeared in six parts during the winter of 1891.[38] His highly influential article covered earlier work and the recent material presented by Cajal. He stated that the terminal arborizations of nerve cells do not fuse, and he resolutely concluded that the "neuron" is the anatomical and physiological unit of the nervous system. With his penchant for new words (*chromosome* being one), Waldeyer now introduced the term *neuron* into the vocabulary.

Waldeyer's neuron served as a rallying cry for the supporters of the theory of independent nerve units, soon to be known as neuron theory or neuron doctrine.[39] Although he did not contribute a single personal observation to support the new theory, his exposition, terminology, and prestige figured prominently in its acceptance and the downfall of reticular theory.

The article written by Waldeyer was so influential that many people erroneously assumed that some of Cajal's contributions were really Waldeyer's. This misperception disturbed Cajal, who tried to set the record straight over the years.[40] Cajal, however, never presented the idea of the neuron as an independent unit as his own. Without question, no one had conducted as many good anatomical studies as he did to show that nerves do not fuse, and no one had fought harder for neuron doctrine. But in his own mind, there were many worthy contributors to the theory, including Forel and His. Cajal was also generous in his praise of Golgi for providing the stain to see the nervous system in a fresh new way. But most of all, he acknowledged Kölliker, who confirmed his own findings and was now helping him to promote neuron doctrine.

Neural Transmission: A One-Way Street

One of the next problems to be faced was whether dendrites and axons are both involved in the transmission of information, and if so, how. Golgi, for one, theorized that dendrites function in nutrition, not in conveying information. To Cajal, however, it seemed obvious that both dendrites and axons must play roles in conveying nerve impulses.

But in what direction do impulses travel under natural conditions? This issue was even more of a challenge because some scientists who applied strong shocks to the middle of long axon cables found that an impulse can spread in both directions. Does transmission naturally occur in either direction, or just in one direction?

The idea that the nervous impulse may flow in only one direction under natural conditions made the most sense to Cajal, who had traced the nerves from the sensory organs, such as the eye or nose, to the brain. These sensory nerves have their dendrites in the periphery and axons that project inward toward the brain. Precisely the opposite is true for the motor nerves. Their dendrites are in the brain or spinal cord and their axons go toward the muscles.

To Cajal, these facts suggested three things: (1) the dendrites are receptive, (2) the cell body is executive, and (3) the axon is responsible for transmitting information to other cells in the series. In 1891 he called this physiological one-way street from dendrites to cell body to axon the "law of dynamic polarization."[41] He gave due credit to Arthur Van Gehuchten, a Belgian with whom he had been corresponding, for reaching the same conclusion and helping him to see the situation clearly.

At the same time that Cajal was correctly deducing the directional flow of the nerve impulse, he was also wondering how these impulses pass from one neuron to another. If neurons do not fuse, what then? Do they just touch? Could they influence each other by producing tiny electrical sparks that can jump small gaps?

When he first began to write about this issue, he suggested that the axons and dendrites of different cells may communicate by touching one another. But even early on he wrote that physical contact was not an absolute necessity for one cell to excite another. Thus although Cajal was strong in his conviction that axons and dendrites do not fuse together, he remained open-minded.

[It is important to remember that the microscopic gap between two neurons, or between a neuron and a muscle, could not be seen at this time. There were no electron microscopes in the 1880s, and even the most powerful instruments then available did not come close to having the power to show a minuscule junction. As noted at the start of this chapter, the tiny gap did not even have a name until 1897, when Charles Sherrington saw fit to coin the term *synapse* for the suspected but still unseen junction (see Chapter 14).]

Some "Speculative Cavorts"

In 1890, two years prior to his move to the University of Madrid, Cajal described how an embryonic neuron sends forth an axon with a clublike ending.[42] Its battering ram pushes other neurons and glia aside as it makes its way to its target (Fig. 13.7). He called the ending, which seemed to move forward like the pseudopod (false foot) of the lowly ameba, a growth cone. Its existence confirmed afterward by the tissue culture studies of American biologist Ross Harrison, the mobile growth cone had the effect of making the units of the nervous system seem even more dynamic than before.[43]

With images of growing axons now in their minds, some of Cajal's contemporaries began to speculate about how neural extension and retraction, even in adulthood, could relate to behavior.[44] One particularly alluring idea was that axons and dendrites of different cells could grow closer to each other at frequently used junctions to enhance communication. This sort of action, some scientists thought, could account for simple forms of learning (conditioning) and perhaps even the association of higher ideas. A corollary of this theory held that forgetting involves a "spreading of the gap." Disuse, diseases of the nervous system, substance abuse, fatigue, fevers, and aging were listed among the factors that could cause processes to pull away from each other at the synapse.

The idea that synaptic growth could account for memory was anticipated in 1872 by Alexander Bain, David Ferrier's teacher in Scotland and a leading mental associa-

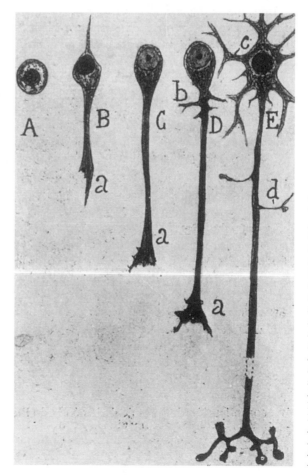

FIGURE 13.7.

A drawing by Cajal (1937) showing stages of a developing neuron (A–F). The growth cone at the end of the growing axon is signified by the small letter a. The small letter d is used to show an axon collateral on the most mature neuron. (From S. Ramón y Cajal, Recollection of My Life. *Philadelphia: American Philosophical Society, 1937.)*

tionist.[45] Bain, however, proposed his theory before scientists had the mind-set to give it serious consideration. In the 1890s, when the same theory surfaced again in the hands of the anatomists, more people were willing to listen. In its favor, the idea of neural "ameboidism" could account for many things, not just learning; the list included sleep and wakefulness, the effects of anesthetics, and even hypnotic states.

Critics of the idea snapped back that there was no direct evidence for neural movements during learning. Too often, the naysayers maintained, the experiments cited as support for neural ameboidism lacked proper controls, were plagued by artifacts, and involved pathological brain states. Further, the changes needed to account for learning or even sleep and wakefulness, are too fast to be due to neural growth and retraction.

Cajal saw problems with the theory as applied to such things as states of consciousness, but initially chose to remain agnostic. In 1895 he wrote that he could not determine whether such ideas are correct or incorrect, but then went on to suggest that a theory based on glial cell movements makes more sense.[46] Glial cells, he pointed out, are capable of growing appendages between neurons and then pulling them back fairly quickly. Expansion of a glial appendage could break a functional connection between two neurons, whereas its retraction could reopen the communication channel.

Both the neural and glial versions of the growth and retraction theories slowly languished from lack of good experimental support. Among the men who questioned these ideas was Cajal's good friend Albrecht von Kölliker.[47] Many scientists agreed with the fair-minded Kölliker, who thought that even the Spaniard's glial theory was long on imagination but woefully short in substance. Thinking things over, Cajal was compelled to admit that Kölliker was right—too much was being made of the kinds of growth seen only during early development.

A second theory that Cajal entertained at this time was that the proliferation of neural connections in the cerebral cortex may correlate with intelligence. In a lecture given in England in 1894, he surmised that even learning a musical instrument may increase the growth of certain dendrites and the branching of axon collaterals.[48] Genius, he maintained, can be achieved economically—without adding more neurons and without demanding more space—by increasing dendrite and axon branching.

During the 1960s and 1970s many studies were conducted on animals brought up in different environments. They showed that animals raised in "enriched" or stimulating environments have more synapses in the roof brain than their brothers or sisters not so raised.[49] They also showed that the greater number of connections correlated positively with how fast these animals learned mazes and solved other problems. Largely overlooked at the time was that these findings were very much in accord with the neural theory of intelligence proposed by Cajal many years earlier.

Cajal called his theories of the mid-1890s his "speculative cavorts." He wrote in his autobiography that, thankfully, his lapses into "unfounded imagination" were few in number. He enjoyed describing himself as "an unconquerable fanatic in the religion of facts." Theories may be intellectually challenging, but he, for one, felt more comfortable returning to the firmer terrain of things he could actually touch and see.

The Nobel Prize of 1906

Cajal and Golgi shared the Nobel Prize in Physiology or Medicine for their contributions to neuroanatomy. The letters announcing the prize were sent out in the autumn of 1906. Both men were instructed to prepare speeches for the award ceremony to be held in Stockholm that December. They were further informed that the pageantry would be attended by the Swedish royal family, Nobel's descendants, diplomats, other scientists, and famous people from various walks of life. King Oscar II himself would personally award the medals.[50]

The expectation was that Golgi would talk about the stain that allowed scientists to see neurons better than ever before, axon collaterals, his discovery of the Golgi type I and II cells and Golgi tendon organs, and his other pioneering anatomical observations. Cajal was expected to describe the studies that led him to neuron doctrine, his subsequent research on the fine structure of the nervous system, his theory of directional conduction, and neural growth.

The idea that the two men would stick to their important contributions and present them in the light of new advances proved to be only half true. To the surprise of those assembled, including Cajal, the speech given by Golgi was largely devoted to resurrecting the defunct nerve net theory.[51] To make matters worse, Golgi came off as less than diplomatic in his presentation. As soon as he began to talk, he launched an attack on the theory championed by the man with whom he was sharing the honor. He stated that neuron doctrine was a fad already going out of favor, argued for axons fusing with each other, and made it very clear that he did not see any hope for the reductionistic philosophy of cortical localization.

Tremendous advances had been made since 1873, but Golgi did not budge one iota from where he had stood when he first described his silver stain in 1873. Cajal and the other knowledgeable scientists in the audience were beside themselves. How could this man, of all men, be so blind to new findings? How could the individual who had provided the scientific community with the means to see the nervous system in a new way and who had made so many important discoveries be so opposed to neuron doctrine? How could he now be so disparaging to the leading proponent of an obviously better theory?

Golgi's words and demeanor clearly hurt Cajal, who had not met the Italian anatomist previously. He now viewed him as arrogant, discourteous, and inexcusably egotistical. As Cajal later put it: "What a cruel irony of fate to pair, like Siamese twins united by the shoulders, scientific adversaries of such contrasting character!"[52]

When Cajal followed with his own Nobel address, he was not caustic or abrasive to Golgi.[53] Instead, he remained gracious as he spoke on "The Structure and Connections of Neurons." After he mentioned the importance of the silver stain and what he was able to do with it, he even referred to Golgi as "my illustrious colleague."

Once back in Spain, Cajal found himself besieged by letters and telegrams congratulating him on his Nobel Prize. The response was wonderful, but in private he bemoaned the fact it would take him months to reply to all those who were kind enough to write. He also complained about the endless banquets and the upset stomachs caused by the award. To Cajal, who preferred to be in the laboratory, the aftermath of the award was rather like being tortured by the forces of the Inquisition.

Later Achievements

Cajal was only fifty-four years old when he received his Nobel Prize (Golgi was sixty-three). Recollecting what Paul Broca had to say about science being for the young, we might think that Cajal would now turn to administration, go on the lecture circuit, or spend more time with his family. One can only guess what Broca might have thought if he could have watched how Cajal energetically turned his attention to regeneration in the damaged nervous system and to other overlooked phenomena during the remaining years of his life.

Cajal wrote over a hundred articles and more than a dozen books after he won the Nobel Prize. Four of the books that still figure prominently on the shelves of many neuroscientists are *Histology of the Nervous System, Degeneration and Regeneration of the Nervous System, Studies on the Cerebral Cortex,* and his inspirational autobiography, *Recollections of My Life.*

In his autobiography, he reflected on his own success and quoted from a speech he had given to his Spanish audience at the turn of the century. The oratory was classic Cajal:

> I am really not a savant, but a patriot; I am a tireless worker rather than a calculating architect. The history of my merits is simple; it is the quite ordinary history of an indomitable will determined to succeed at any cost.[54]

In 1922 Cajal retired from his university position. Although his health was declining, he did not stop writing at home, where he maintained a library of over eight thousand books. In 1934, eight years after Golgi had died in Pavia, Cajal passed away in Madrid. When word spread of Cajal's death at age eighty-three, family members, scientists, and friends around the world mourned the passing of a giant.[55]

Santiago Ramón y Cajal had risen from a rebellious childhood in a scientifically backward country to receive almost every honor that could be bestowed upon a scientist. He was a Nobel Prize winner and the recipient of numerous other awards. He received doctorates from many schools, including England's Oxford and Cambridge Universities, had been a member of many esteemed foreign societies, and even saw his image on postage stamps and currency. But probably giving him more delight than any of these distinctions was the fact that he had managed to make a difference in terms of how people viewed the fine structure and organization of the nervous system, and that he accomplished what he did on previously neglected Spanish soil.

Charles Scott Sherrington
The Integrated Nervous System

The uniqueness and significance of Sherrington's concept [nervous integration] lies in the fact that it provided the first comprehensive, experimentally documented explanation of *how* the nervous system, through the unit mechanism of reflex action, produces an "integrated" or "co-ordinated" motor organism.

—*Judith Swazey, 1969*

He was a man who knew how to ask the right question and who knew when he had got the right answer.

—*E. G. T. Liddell, 1960*

Introduction

Charles Scott Sherrington (Fig. 14.1), the scientist best known for his work on how the elements of the nervous system join together functionally, was a wiry man who stood only five feet six inches tall.[1] He was polite, shy, gentle, and self-effacing. As one of his students put it: "He was one of the mildest men I have ever known, rarely being vexed and at most saying, 'Dear me' or 'That is most annoying.'"[2] His personality was such that he was not one to make unkind remarks about professional colleagues, and he seemed oblivious to his ever-increasing reputation as a world-class scientist.

As we shall see, Sherrington was more than just a shy but dedicated neurophysiologist with an insatiable appetite to learn more about the functional organization of the nervous system. He defined his science extremely broadly, had first established himself as a pathologist, and was skilled when it came to writing history, poetry, and philosophy. Among his other interests, he also loved collecting old books and manuscripts, reading in foreign languages, and studying art.

Sherrington was born in 1857 in London. Because his father died when he was young, he and his two younger brothers were raised in Ipswich by his mother and her second husband, Caleb Rose. More than a well-read physician, Rose was also a classical scholar, an amateur geologist, and an archeologist. He covered the walls of his attractive home with fine paintings, invited creative people to tea, and loved to talk about new discoveries and achievements. As much as anyone, Sherrington's stepfather was responsible for instilling in him his boyish sense of wonder.

Formal education began at the Ipswich Grammar School, where he immersed himself in literature, history, and languages. He was especially inspired by the poetry of Keats. But like the other boys, young Sherrington also found time for sports.

FIGURE 14.1.

*Charles Scott
Sherrington
(1857–1952).*

After talking with his stepfather and prominent London physician Samuel Wilks, he made up his mind to study medicine. Although he was only eighteen and did not possess a college degree, he still passed the preliminary examinations given by the Royal College of Surgeons. One year later, in 1876, he started his clinical training at St. Thomas's Hospital, an institution located near the Houses of Parliament in London.

Cambridge and Europe

In 1879, after studying medicine for three years in London and Edinburgh, Sherrington made his way to the university town of Cambridge. There he began his academic studies as a "noncollegiate student." A year later he was officially admitted to one of the Cambridge colleges, Gonville and Caius. Learning about David Ferrier's research on the cerebral cortex led him to physiology, although he still managed to balance his interests in science and medicine with his love of poetry and his insatiable thirst for the humanities.

At the time, the physiology department at Cambridge was run by Michael Foster, who had already founded the *Journal of Physiology* and published a popular multivolume textbook in the field. Although Foster was a poor lecturer, he was an outstanding organizer with a good eye for finding and inspiring talent. His department boasted both Walter Gaskell and John Newport Langley, both of whom would become world-famous for their research on the autonomic nervous system. All three men would play important roles in Sherrington's life—Foster because of his textbook; Gaskell because he was levelheaded and always a source of good advice; and Langley because he opened the doors of his laboratory to the young man.

Sherrington's first publication appeared while he was still an undergraduate. Langley was given a part of the brain and spinal cord of the dog Friedrich Goltz had brought to the 1881 International Medical Congress in London (see Chapter 11). The dog had been sacrificed along with one of Ferrier's monkeys to determine whether the two rivals in the localization debate had properly described the lesions in their animals. Langley asked Sherrington to help him with the tedious assignment of assessing the neuronal degeneration. The two researchers published their findings in 1884.[3]

During the winter of 1884 Sherrington left Cambridge to return to St. Thomas's Hospital to continue his clinical studies in medicine. His residence at the hospital, however, was short-lived. He received a "studentship" in physiology that allowed him to travel to Bonn to work with noted physiologist Eduard Pflüger, a scientist who had published several so-called laws of spinal reflex action,[4] and then to Strassburg to study neuronal degeneration with Goltz.

Under Goltz's supervision, Sherrington did more research on the anatomical and behavioral changes that follow motor cortex lesions in animals.[5] One phenomenon that caught his attention at this time was "spinal shock," a term first used in 1850 by a pioneer in the study of reflexes, Marshall Hall.[6] Sherrington noted that his animals exhibited a temporary depression of all limb reflexes after they sustained large lesions of even a single cerebral hemisphere. With the passage of time, however, they came out of this state and some limb reflexes even "rebounded" to become brisker than before.

The Pathology of Cholera

In 1885, the year in which he completed his medical degree, Sherrington was asked by two British societies to investigate an epidemic of Asiatic cholera in southern Europe. This was the same epidemic that Santiago Ramón y Cajal helped with soon after he moved to Valencia (see Chapter 13). Sherrington complied with the request and with two colleagues traveled to Spain, where the team found cholera bacteria in the nasal mucous membranes and the feces of the men and women who had just died. They also investigated claims about the effectiveness of a new cholera vaccine, only to conclude that they did not see much good in it.

Soon afterward Sherrington went to Italy to help with another outbreak of cholera. After doing what he could, he took some cholera-infected tissue with him to Rudolf Virchow's laboratory in Berlin for further study. After a short stay, Virchow sent Sherrington off to study with another talented Berliner, Robert Koch, the founder of medical bacteriology.

Sherrington wound up spending a year with Koch, who had discovered the bacillus responsible for tuberculosis. While learning all he could about pathology from Koch, he also attended lectures given by several German leaders in the brain sciences, including Hermann Helmholtz, Emil du Bois-Reymond, and Wilhelm Waldeyer.

The Brown Institution

When Sherrington returned to England in 1887, he was thirty years old. He was welcomed with two appointments, one as a fellow of Gonville and Caius College and the other as a lecturer in physiology at St. Thomas's Hospital Medical School.

Although his interest in pathology was strong at this time, he also continued to work on anatomical changes after damage to the motor cortex. Now, however, he made smaller, more focal lesions and turned from dogs to monkeys for his studies.[7]

In 1891 he accepted the position of professor-superintendent of the University of London's Brown Institution. The new appointment gave him the financial security he needed to marry Ethel Wright, to whom he had been engaged since 1888. His bride was a skilled linguist who enjoyed reading, travel, the arts, and architecture as much as he did. She also possessed good organizational skills and helped him with his manuscripts, correspondence, and accounting.[8]

The Brown Institute had excellent facilities for doing a wide range of animal research. Within its confines, Sherrington and a young colleague by the name of Armand Ruffer worked on inoculating horses with small doses of diphtheria toxin. They hoped that the animals would produce an antitoxin that could be of practical use to humans. At the time, many English physicians were highly skeptical of this sort of work, but Ruffer had visited the Institut Pasteur in Paris and returned inspired.

One night in 1894 Sherrington got word that his eight-year-old nephew had come down with diphtheria. By the time he arrived in Sussex with his flasks containing antiserum, he was told by the family physician that there was no hope, the child would surely be dead within hours. With nothing to lose, Sherrington went ahead and treated the boy with the antiserum. To his delight and the surprise of the family physician, his nephew recovered within a day.

Upon returning to London, he told Ruffer what had happened. The two ecstatic physicians then ran off to tell Lord Lister, a strong supporter of such work. At Lister's residence they were stopped by a butler, who informed them that his master was entertaining some surgeons from the Continent. They then scribbled a note that was delivered to the famous promoter of antisepsis. After Lister read the note, the doors flew open and Sherrington and Ruffer were invited to join the guests for dinner. Lister told those present that, to the best of his knowledge, this was the first time diphtheria antiserum had been used to save a life in Great Britain. The evening was a joyous affair and nobody seemed particularly bothered by the fact that the two unexpected guests were not properly dressed for the occasion.

A Change of Focus

Sherrington could have continued to work in the field of immunology and pathology, but he was already finding the study of the working nervous system even more alluring. As a result, he began to devote increasingly more time to nervous system research. Now, however, he approached the nervous system with a new philosophy and in a different way.

The stimulus for these changes came from his old teacher, Walter Gaskell, who managed to convince him that he had begun his career "at the wrong end." Gaskell maintained that it made more sense to study the nervous system from its base up, rather than from the top down. Sherrington agreed with Gaskell that the simpler spinal cord would be a better starting place than the highly complex cerebral cortex. In addition, he had approached the nervous system mostly as a neuroanatomist in the past. Now he was inspired to learn more about its function. His new mission would be to try to understand how the nervous system works, and this required combining anatomical, physiological, and behavioral measures.

Sherrington's move to the spinal end of the nervous system began with the knee jerk. But before turning to this simple reflex, it is necessary to look at how scientists viewed reflexes before 1891. In addition, we must also bring neuron doctrine up again, since this theory guided him like a beacon from one experiment on reflexes to the next.

The Reflex Before Sherrington

Well before Sherrington even began his work on the knee jerk, the distinction between involuntary and deliberate muscle actions had been established. More than anyone else, it was the seventeenth-century philosopher René Descartes who put the concept of reflex action into play (see Chapter 6).[9] Thanks largely to Thomas Willis (see Chapter 7), terms such as *reflexion* and *reflected* made their way into the vocabularies of medical scientists later in the same century.[10] These words, from which the noun *reflex* would be derived, were used by Willis to signify how spirits in the nerves to the central nervous system could be "reflected" back to the muscles. The rapid, automatic input-output action was akin to light bouncing off a mirror.

During the eighteenth century Robert Whytt became the leading authority on reflexes.[11] Working in Edinburgh, he studied frogs, tortoises, and other animals, and attempted to list the different reflexes and explain their protective functions. Whytt was intrigued by the fact that many involuntary actions, such as birds flying when startled and frogs hopping when touched, could still take place after decapitation.

By the middle of the nineteenth century, Marshall Hall, an Englishman, was being recognized for his research on reflexes. In the introduction to his last book, he stated that he had logged more than twenty-five thousand hours studying reflex actions.[12] Hall looked upon the spinal cord as having two distinct components: one specifically for reflexes and the other for conveying impulses to and from the higher centers of consciousness. The use of the term *arc* to describe the reflex circuit was another of his lasting contributions.

In 1863 the Russian physiologist Ivan M. Sechenov left his own enduring mark on the field.[13] He demonstrated that reflexes can be affected by cutting higher brain parts off from lower parts or by stimulating certain brainstem structures. A salt crystal placed on the frog's optic lobe, for example, inhibited its leg withdrawal reflex.

When Sherrington began his own work on reflexes, inadequate knowledge about the underlying circuitry and how reflexes could be inhibited were two problems that stood out as worthy of further study. A third seemed to be the general lack of awareness of just how complex some reflexes could be. Sherrington had the genius to see a real need for someone to tie reflex anatomy, physiology, and behavior together into a coherent scheme—one with the potential to shed light on the functional organization of the whole nervous system, not just the spinal cord.

Sherrington's Synapse

Well before Cajal's neuron theory became widely accepted, Sherrington was convinced that neurons must be independent entities. His bias in favor of neuron doctrine over nerve nets stemmed from his nerve degeneration studies. Repeatedly he found that the pattern of neuronal degeneration after limited cortical lesions was restricted, not diffuse. Thus when Cajal and other anatomists began to publish articles supporting neuron doctrine in the 1880s and 1890s, he was already strongly biased in favor of what these men had to say.

In 1897, three years after Ramón y Cajal stayed with him while lecturing in London, Sherrington introduced a term for the functional junction that he felt had to exist between neurons. His term was *synapsis,* soon to become *synapse.*[14]

Sherrington looked upon the synapse as a physiological construct, since neither he nor his contemporaries could see the gap in the pre-electron-microscope era. Nevertheless, only a functional junction between neurons could explain the focal findings of his own degeneration studies or, for that matter, the one-way nature of neural transmission. Moreover, thanks to Hermann Helmholtz in Germany, he now knew that reflexes were many times slower than the speed of nerve conduction. Central reflex time, or "lost time," was a phenomenon of considerable interest to physiologists and psychologists, and it too could be accounted for by postulating gaps in the cables or circuitry that would take more time to cross.

Many years after the term *synapse* was accepted, Sherrington was asked how he had come up with his new word. In a letter to John Fulton, who had been one of his later students, he explained that he felt a need to give a special name to the postulated junction while editing a chapter for a new edition of Michael Foster's textbook.[15] At first he intended to use the term *syndesm* to describe the gap, but a friend of his suggested *synapsis* instead. The latter term, which came from the Greek word meaning "to clasp," was deemed better, and so, without ever really seeing a synapse, he introduced the word in the chapter.

The Knee Jerk

Now let us return to the knee jerk, the reflex familiar to everyone who has undergone a routine physical examination. This "stretch reflex" is easily elicited by gently hitting the tendon just below the patella (kneecap) with a small hammer. It was first described in 1875 by two Germans working independently, Wilhelm Erb and Carl Westphal.[16]

Although Erb and Wesphal were both astute observers, they interpreted the knee jerk differently. Erb contended that the quick extension of the knee is a true reflex, because it seemed to involve a nerve to the spinal cord and one back to an extensor muscle. Wesphal, however, saw the action as nothing more than a mechanical twitching of the tightened muscle. After their reports came out, scientific opinion was split over the nature of the knee jerk, mainly because it took place so rapidly that it did not appear to involve nerves to and from the central nervous system, but also because it appeared to lack purpose and seemed unaffected by narcotics.

In 1891 and 1892 Sherrington published two important papers on the knee jerk.[17] Using monkeys, rabbits, cats, and dogs, he found strong support for Erb's contention that the knee jerk is a true reflex. He demonstrated that its arc is made up of sensory and motor nerves, which he proceeded to identify. Further, he not only traced the motor neurons to the quadriceps muscle, but showed that the sensory neurons originate in the musculature. Not stopping with this pertinent information, he proved that the knee jerk could, in fact, be blocked.

With the advent of Sherrington's papers, the remarkable speed of the knee jerk reflex no longer seemed so mysterious. It could still be explained by neuron theory. One had to assume only that there are fewer synaptic gaps in its circuit than in slower reflex circuits. Today we know that there is just one synapse in the arc for the knee jerk.

The Muscle Sense

Sherrington's work on the sensory nerves underlying the knee jerk led him to pursue the idea that there are specialized sensory end organs in the muscles. He therefore began to investigate what he at different times referred to as the "muscular sense," "kinesthesia," or "proprioception."

When he began this line of research in the mid-1890s, most scientists did not accept the idea of a special sense for muscle feedback.[18] The doubting Thomases looked upon the postulated sixth sense as something not proven. They pointed to the paucity of evidence for sensory nerves leaving the muscles and to minimal evidence for specialized muscle receptors. In addition, they held that sensory organs near the surface of the skin already conveyed all of the needed information about the state of the muscles.

This negativism began to give way when Sherrington turned to cats and monkeys to learn more. To his astonishment, he discovered that many nerves associated with the limb muscles do not degenerate when the motor nerves from the spinal cord to the muscles are severed. Equally telling, he traced a large number of nerves from the muscles to the sensory ganglia, where their cell bodies are located. Both new anatomical findings strongly suggested that between one-quarter and one-half of the nerves associated with the muscles actually convey sensory information to the spinal cord, providing feedback for better-controlled motor actions.[19]

But what about the receptors themselves? Were they the same as those found under the skin surface? Sherrington was convinced that the muscle receptors are quite different from the skin receptors. Specifically, he pointed to the muscle spindles, which run parallel to the muscle fibers, as the most likely end organs to signal muscle stretching.[20] He also pointed to the Golgi tendon organs, found in the bands of inelastic fibrous tissue that connect muscles to bone. Based on their location and structure, he conjectured that these tendon organs probably serve as tension recorders.

Modern neuroscientists tell us that Sherrington was correct when he postulated that the rodlike muscle spindles serve as muscle "length recorders" and that the Golgi tendon organs provide feedback about tension. To his credit, it is now impossible for a scientist to think about movement without bringing these receptors and his broader concept of a proprioceptive system into play. As Sherrington liked to remind others, the muscles really do have a voice in their own management.

Spinal Roots and Dermatomes

Sherrington's work on the knee jerk took him in two directions. One, as we have seen, was to the so-called muscular sense and the concept of a proprioceptive system. The other was to map the sensory nerves from the skin to the spinal cord and the motor nerves going from the spinal cord to the muscles. He knew he could not move on to study other reflexes without first knowing more about the sensory receptive fields and muscle connections of each spinal nerve.

This huge endeavor, much of which he looked upon as boring and pedestrian, occupied him for about a decade. Fortunately, he was still able to accumulate a wealth of new information about reflexes while mapping the spinal nerves. This was because he used reflexive movements as a behavioral assay in many of his mapping studies.

Sherrington's research on the spinal root distributions was conducted primarily on monkeys, an animal not used for research of this nature in the past. The more usual frogs, rats, rabbits, cats, and dogs were also employed, but they played secondary roles. His strategy was to sever one or more spinal roots above and below the one he wanted to isolate for study. Because the endings from nearby spinal nerves overlap each other, cutting a single nerve and looking for areas of lost sensitivity or motor losses would not show the true borders of the skin region or the specific muscles served by that nerve.

Once a motor root was isolated, Sherrington stimulated it electrically. He made note of the muscles still affected by the stimulation and the features of the response. In contrast, when his main objective was to learn more about the sensory roots leav-

Fig. 18.

Fig. 19.

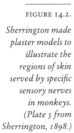

FIGURE 14.2.

Sherrington made plaster models to illustrate the regions of skin served by specific sensory nerves in monkeys. (Plate 5 from Sherrington, 1898.)

Fig. 20.

Fig. 21.

ing the skin, he isolated a sensory root and stimulated the skin with electrical or natural stimuli. He then made records of the regions of the skin from which he could still trigger reflexive movements.

Sherrington's first important paper on the anatomical distributions of the spinal roots appeared in 1892, and it helped him to become a Fellow of the Royal Society a year later.[21] Although he called his publication a "note," it was approximately 150 pages in length. Other "notes" and papers followed, with information on the motor nerves and drawings, photographs, and even plaster casts illustrating the regions of skin, or dermatomes, served by the different sensory nerves (Fig. 14.2).[22]

Sherrington was not, however, the only individual to take on the problem of mapping the peripheral nerves at the turn of the century. Another talented physician-scientist, and a man destined to become one of his friends, also began to map the dermatomes in the 1890s. Henry Head's strategy for learning about the regions of skin served by the different nerves, however, was very different from Sherrington's.

Head examined humans with shingles, a localized skin rash caused by a herpes virus. Infections of different sensory nerves, he noted, can produce well-defined rashes over different parts of the body. Head made a composite diagram of the sensory root distributions from the individual rash patterns. He did this with the help of Alfred Campbell, and the two men published their findings in 1900 (Fig. 14.3).[23]

The Sherrington and Head maps of the spinal roots not only were important for laboratory scientists, but were greatly appreciated by practicing physicians. The clinicians now realized that an alteration of sensation neatly confined to a specific spinal root territory is likely to reflect a problem in the peripheral nervous system. In contrast, abnormalities spreading across the dermatomes, especially in unusual patterns, are likely to signal central nervous system involvement. This is because the sensory

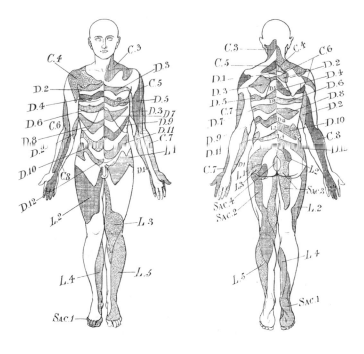

FIGURE 14.3.

A composite map of the human dermatomes. (From Head and Campbell, 1900.)

nerves from the different body regions come together in the spinal cord and stay close together or even overlap in the brain.

The Liverpool Years

In 1895 Sherrington moved to University College in Liverpool as Holt Professor of Physiology. The years he spent in Liverpool were probably the happiest of his life. He knew what he wanted to accomplish in the laboratory, his health was good, and he and his wife were able to enjoy many outdoor and cultural activities.

Sherrington's Liverpool era was marked by the further growth of his reputation as an exceptional scientist. He published approximately 125 articles during his eighteen years at Liverpool, including sixteen in 1897 alone. He also wrote his most important book during this period.

Today it is hard for us to imagine how Sherrington did so much without secretarial assistance at the university. The only real help he had came from his wife, who also handled the paperwork from his two long-standing positions: one as secretary of the International Physiological Congresses and the other as secretary of the Physiological Society, another of Michael Foster's creations.

Sherrington's first laboratory at Liverpool was nothing to brag about; it was housed in a small, dilapidated building. Fortunately, a new home for physiology and pathology was completed in 1898. To Sherrington's delight, one of the attractive female figures gracing the emblem of the new building was none other than Ethel. His wife's image symbolized Physiologia, whereas another professor's wife served as the model for Pathologia.[24]

Among the students who went to Liverpool to study with Sherrington were the physiologists T. Graham Brown, F. W. Mott, and E. Schuster. The famous American neurosurgeon Harvey Cushing also spent some time with him, as did Robert S. Woodworth, a pioneer in the new field of experimental psychology. Other prominent guests included Alexander Forbes from Harvard (see Chapter 15), and Rudolf Magnus, H. E. Hering, and Alfred Fröhlich, all from Continental Europe. With so many talented students, congenial colleagues, and foreign visitors to serve as coworkers, and a myriad of wonderful new ideas filling his head, it is no wonder that Sherrington enjoyed his Liverpool years as much as he did.

Reciprocal Innervation

Sherrington discovered or further described so many different things as he continued his research on reflexes at Liverpool that it is hard to single out one as being most important. He gave some indications, however, that he personally looked upon a better understanding of the principle of reciprocal innervation as his most significant contribution.[25] Nevertheless, he never claimed to be the discoverer of this principle, which has roots in the work of René Descartes, Charles Bell, and Marshall Hall.[26]

Sherrington defined reciprocal innervation as a form of coordination in which "inhibito-motor spinal reflexes occur quite habitually and concurrently with many of the excitato-motor."[27] In plain English, he meant that what one muscle does will also influence what other muscles do. By looking around us, we can see numerous examples of reciprocal innervation. Take the body movements we make when we are walking. Because some leg muscles extend when others are flexed, we do not fall flat on our faces. Moreover, as we move one leg forward, we also make certain head and arm movements to maintain balance, even though we do not think about these important, supportive actions.

Between 1893 and 1909 Sherrington published fourteen notes on reciprocal innervation in the journal *Proceedings of the Royal Society*. He also gave simple demonstrations in which he showed how anyone could feel the lessening of tension in one muscle when another linked to it contracts.

A better understanding of reciprocal innervation as a unifying factor emerged when Sherrington turned to the decerebrate preparation in 1896. Decerebration is achieved by cutting through the midbrain of an anesthetized animal. The operation effectively cuts off the lower nervous system from its command centers higher in the brain. His main intent was to produce a lasting state of unconsciousness in order to avoid the continuous use of anesthetics that could interfere with the phenomena he wished to study. But he was also drawn to decerebration because it made things simpler. The operation removed consciousness and voluntary control over the muscles from the picture.

Sherrington's decerebrate animals became so stiff that he referred to them as being in a state of "decerebrate rigidity" or "exaggerated standing." In a paper from 1898 he even included a set of drawings to show a cat's body in this condition and the effects of stimulating a front or hind limb while the animal was in this state (Fig. 14.4).[28] He showed that if the left foreleg were stimulated, it moved forward while the hind leg on the same side moved backward. Because of reciprocal innervation, the two legs on the opposite side of the body exhibited opposing movements as these responses took place. The result looked like a normal walking pattern, but for the remarkable fact that these animals had no forebrains—no "will" or conscious control over their actions.

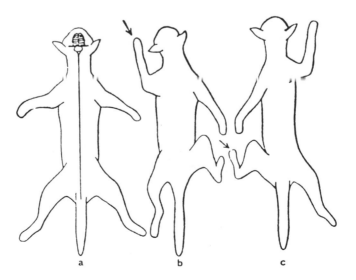

a   b   c

FIGURE 14.4.

Three postures of a cat in a state of decerebrate rigidity: (a) no external stimulation; (b) after stimulation of the left foreleg; and (c) after stimulation of the left hindleg. (From Sherrington, 1898.)

The Scratch Reflex and Chaining

Once his appetite was whetted by a better understanding of how reciprocal innervation operated at the spinal level, Sherrington conducted many experiments on the mechanisms responsible for coordinated stepping, shaking, and running.[29] His work on these higher reflexes is noteworthy because it further showed that seemingly complex movement patterns can be programmed into the seemingly primitive spinal cord.

He now drew attention to the scratch reflex in mammals, a reflex that had caught the attention of some mid-nineteenth-century researchers, most of whom studied frogs, toads, and salamanders.[30] To elicit reflexive scratching from a hind limb, he and his assistants first brushed or tickled a cat near its ear, or a dog in the region of the shoulder flank.

Hoping to get even better control over the stimulus, however, Sherrington turned to the "artificial flea." His so-called flea was really nothing more than a thin pin that could be inserted into the skin. A wire from a stimulator was attached to the pin, and it allowed the experimenter to control the "flea bites" at will. By using multiple pins, he could even approximate the jumping movements of a real flea across the skin.

Careful examination of the scratch reflex revealed the area of skin from which scratching could be elicited, the nerves and muscles involved, and the highly complex nature of the response.[31] When triggered, the animal first extends the digits of the appropriate leg to form a claw for scratching. It then goes through an alternating series of knee, ankle, and hip movements to make repeated scratching movements. At the same time as it is vigorously scratching, it is also executing a series of other reflexes that enable it to stand without falling. A host of potentially competing reactions, such as continued locomotion, are inhibited while the scratching takes place.

Remarkably, the full-blown scratch reflex, like the stepping reflex, was found not to require a forebrain; both could be elicited in a spinal animal. A "brainless" cat or dog can even follow the "flea" as it ventures over the body surface, although not with normal pinpoint accuracy.

The working model Sherrington envisioned from the study of compound reflexes is that relatively simple automatic reactions can be tied together by excitatory and inhibitory processes at synapses in the central nervous system to generate more complicated behavioral acts. This knitting together of reflexes, so that the end product of one reflex can serve as the stimulus for another in a linked sequence, is called "chaining."[32]

By emphasizing the basic biology of chained reflexes, Sherrington moved in a more biological direction than had either Alexander Bain or Ivan Sechenov, both of whom had studied complex reflexes.[33] These two men had envisioned the chaining idea several decades before Sherrington. Both, however, became more interested in applying it to learned responses or conditioning. In contrast, Sherrington never looked upon learned habits as real reflexes and was decidedly more concerned with underlying mechanisms, especially excitatory and inhibitory interactions at the synapse.

Mapping the Ape Motor Cortex

Although Sherrington was more interested in the spinal cord than the brain, he did map the ape motor cortex while he was at Liverpool. He gave several reasons for this work, which he did not view as going against Gaskell's kindly advice to start lower in the system. First, having worked for years on the spinal cord, he was ready to learn more about the central machinery that could affect spinal cord activity. Second, he wanted to study the ape brain to enhance his understanding of the evolution of the nervous system. And third, he knew that maps of the ape brain would be of use to neurosurgeons and neurologists. Indeed, there was only one published study on the great-ape cortex prior to his own publications, and it was based on a single, young orangutan.[34]

Between 1901 and 1903 Sherrington and A. S. F. Grünbaum published several short papers in which they mapped the motor cortex of the chimpanzee, gorilla, and orangutan.[35] In 1906 their studies were extended to the baboon, and between 1911 and 1918 seven more Sherrington papers on apes appeared with various coauthors.

The most complete paper in the cortical mapping series was voluntarily withheld from publication until 1917.[36] The lengthy delay had to do with a dispute started by neurosurgeon Victor Horsley over who did what first.[37] When the article finally appeared after Horsley's death, it was authored by Leyton and Sherrington—Leyton being none other than Grünbaum, who had changed his Germanic last name to one having a distinctly British sound as anti-German feelings grew in Great Britain.

The new maps of the motor cortex, all made under light anesthesia, represented a significant step forward from the maps made by Fritsch and Hitzig on dogs and Ferrier and monkeys (see Chapter 11). To begin with, Leyton and Sherrington found that they could elicit considerably finer movements from the ape motor cortex than those reported with other species. Rather than moving all its fingers at once, a chimpanzee or gorilla may move individual digits when gently stimulated.

They also added to what was known about the boundaries of the excitable cortex in higher primates. They found that the motor cortex runs the length of the precen-

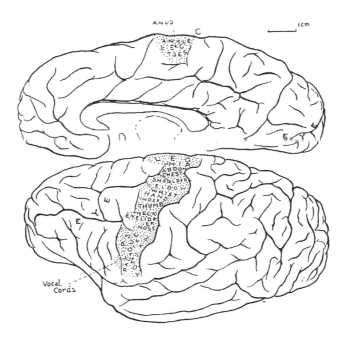

FIGURE 14.5.

The gorilla motor cortex as revealed by electrical stimulation. The top figure shows the medial surface of the hemisphere and the bottom drawing shows its lateral surface. (From Leyton and Sherrington, 1917.)

tral gyrus, but that about one-third of it is hidden in the Rolandic fissure separating the frontal and parietal lobes (Fig. 14.5). In no case did the motor cortex extend into the parietal lobe, still thought by many investigators to be a part of the motor system.

A third point emphasized by the two researchers was that some of the smaller grooves in the cortex (the sulci) may not be reliable markers for where the individual body parts are represented. Not only do the surface markers differ across the great apes, but they can also vary among individuals of the same species and even from side to side in the same animal.

Fourth, Leyton and Sherrington discussed how the various parts of the body are connected to each other at the cortical level. As they saw it, the very same facial muscle that contracts when a tiny cortical area is stimulated must be tied into wider circuits for chewing, making sounds, and grimacing. The number of possible interactive combinations increases tremendously as one goes up the phylogenetic scale.

Long after these mapping studies were conducted, Wilder Penfield, a Sherrington student from a later period, succeeded in electrically mapping motor, sensory, and association areas of the human cortex.[38] Sherrington was over ninety when Penfield, who stopped by on a visit from Montreal, gleefully told him what he was doing with conscious human patients who were being treated neurosurgically for severe epilepsy. With a twinkle in his eyes, Sherrington responded: "It must have been great fun to put a question to 'the preparation' and have it answer."[39]

Brain Damage and Recovery

Leyton and Sherrington wrote about more than just mapping the motor cortex in their lengthy article of 1917. They also described the behavioral effects of making small lesions in the motor cortex. They reported that they surgically removed parts of the ape cortex and found the use of corresponding body parts significantly impaired. Although the most severe problems occurred when the finger areas were ablated, they were surprised by how good the recovery was within just a few weeks.

What part of the brain could be taking over for a structure seemingly as unique as the hand area of the left motor cortex? To answer this question, the experimenters remapped the motor cortex in some of their recovered animals. They observed that the motor cortex was not reorganized after brain damage; that is, there was not a new hand area in a nearby region to replace the one that had been ablated. They also studied the effects of enlarging the first lesion. In accord with the negative data from their electrical recording studies, removing additional cortex in the neighborhood of the previously damaged hand area did not reinstate the deficit.

Could the other hemisphere have taken over the lost function? This idea was very popular at the time, and it was deemed worthy of study. Hence the experimenters set about making additional lesions in the cortical hand area of the opposite hemisphere. The second surgery temporarily affected movements of the opposite hand, but it too failed to reinstate the earlier deficit. Lesions in the nearby parietal cortex also only minimally affected movements of the recovered hand.

Thus, when it came to discovering what part of the brain may take over for a damaged area, Sherrington and his associates went away scratching their heads. Their experiments seemed only to suggest where functional reorganization may *not* be taking place following cortical injuries. Today, researchers with finer assays are still asking how behavioral recovery can take place after lesions in cortical areas known to be associated with specific functions.[40]

The Integrative Action of the Nervous System

In 1903 Sherrington celebrated his forty-sixth birthday. By this time he had published more than a hundred papers that were so highly regarded that he was invited to Yale University to give the Silliman Lectures. He did this in 1904. In his talks he not only described many of his experimental findings, but also laid out the logic and theories that were still guiding his work.

He explained to his American audience that reflexes are purposeful acts and great evolutionary advances. He then clarified why a "simple" reflex, such as the knee jerk, should be looked upon as the lowest level of integrative action. From this basis, he worked into how simple reflexes may be combined to account for more complex actions, as exemplified by walking or scratching reflexively.

The underlying theme of Sherrington's Silliman Lectures was that we and other animals function as unified wholes because of central integration. The integration takes place at the synapses, which have unique functional characteristics. He further explained that although integrative processes may begin with synapses at the spinal level, there is also longitudinal integration. As a result, even the cerebral cortex has a place in the schema.

Sherrington, in fact, referred to the cerebral cortex as our newest and most complicated switchboard, a great telephone exchange, and the abode of the psyche. He viewed it as the place where our distance receptors (ears, eyes) can trigger motor acts prior to actual physical contact. He even ended his Silliman Lectures by stating: "It is then around the cerebrum, its physiological and psychological attributes, that the main interest of biology must turn."[41]

Not brought up at this time was the age-old question of how the conscious mind is able to control the physical machinery of the nervous system and hence the body. But Sherrington knew that at some point he would have to come back to mind-body integration, so critical for completing his picture of the unified higher organism.

Sherrington's talks were well attended and greatly appreciated by the knowledgeable scientists in the audience, some of whom hosted a special dinner with a whimsical menu for his enjoyment (Fig. 14.6). Nevertheless, his "thinking-along-the-way" style of lecturing, with endless hesitations, digressions, and qualifications, drew less than favorable reviews from others in the audience.

In 1906 Sherrington's ten lectures appeared as *The Integrative Action of the Nervous System*.[42] This work of almost four hundred pages was immediately recognized as a landmark in neurophysiology. It had three reprintings in its first decade alone. In 1947 it was reprinted again and, remarkably, corrections were not needed.[43] The only change to appear in the 1947 edition was a new introductory section by Sherrington, who had reluctantly agreed to reflect on what he had accomplished.

Oxford

Sherrington came close to leaving Liverpool in 1908, when he was offered a chair in physiology at Toronto with financial support for his research. He loved Canada and

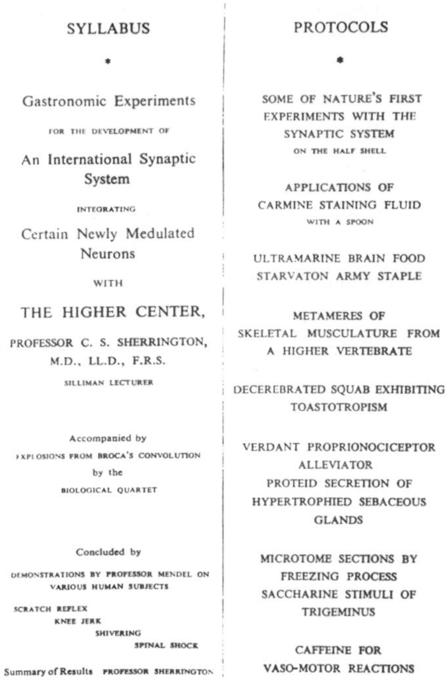

SYLLABUS	PROTOCOLS
•	•
Gastronomic Experiments	SOME OF NATURE'S FIRST EXPERIMENTS WITH THE SYNAPTIC SYSTEM
FOR THE DEVELOPMENT OF	ON THE HALF SHELL
An International Synaptic System	APPLICATIONS OF CARMINE STAINING FLUID
INTEGRATING	WITH A SPOON
Certain Newly Medulated Neurons	ULTRAMARINE BRAIN FOOD STARVATON ARMY STAPLE
WITH	
THE HIGHER CENTER,	METAMERES OF SKELETAL MUSCULATURE FROM A HIGHER VERTEBRATE
PROFESSOR C. S. SHERRINGTON, M.D., LL.D., F.R.S.	
SILLIMAN LECTURER	DECEREBRATED SQUAB EXHIBITING TOASTOTROPISM
Accompanied by	VERDANT PROPRIONOCICEPTOR ALLEVIATOR
EXPLOSIONS FROM BROCA'S CONVOLUTION by the	PROTEID SECRETION OF HYPERTROPHIED SEBACEOUS GLANDS
BIOLOGICAL QUARTET	
Concluded by	MICROTOME SECTIONS BY FREEZING PROCESS
DEMONSTRATIONS BY PROFESSOR MENDEL ON VARIOUS HUMAN SUBJECTS	SACCHARINE STIMULI OF TRIGEMINUS
SCRATCH REFLEX KNEE JERK	
SHIVERING	
SPINAL SHOCK	CAFFEINE FOR VASO-MOTOR REACTIONS
Summary of Results PROFESSOR SHERRINGTON	

FIGURE 14.6.

The comical menu for one of the dinners celebrating Sherrington's lectures at Yale University in 1904.

even mailed an acceptance letter. But after thinking it over, he changed his mind and spent a day at the Liverpool post office searching for his letter among the thousands being sorted for Atlantic crossings.[44]

Almost the same scenario was repeated a year later when a generous offer came from Columbia University in New York City. This time it rained so hard that he put off posting his acceptance letter until the next morning. By the time he finished his breakfast, he again had a change of heart and wrote a polite rejection letter instead.

The offer that Sherrington could not refuse came from Oxford in 1913. Repeating what had happened in Liverpool, where he took over the position then vacated by Francis Gotch, he was again asked to succeed Gotch, but this time because the elder physiologist had died. Sherrington did not hesitate when it came to accepting the Oxford professorship. He was fifty-six years old and as energetic as ever when he moved to the more prestigious institution.

The Sherringtons continued to entertain colleagues, visitors, and students after they settled in Oxford. Their home in the historic university town was happy, warm, and genial, and those who knew them looked forward to invitations to Sunday teas at "Chadland."[45] Sherrington often told stories and read poetry to his guests, all of whom enjoyed his fellowship, his culture, and the friendly, relaxed atmosphere he and his wife created.

Unfortunately, the First World War broke out about a year after his move, disrupting what promised to be an idyllic life. Sherrington's own laboratory work slowed to a trickle as eligible British men went into the army and foreign visitors remained home. The physiology department, which had boasted more than ninety students when he arrived, was so depleted by 1917 that he was left with only one student. Henry Viets, a fourth-year Harvard Medical School student, was so determined to work with Sherrington that he alone chanced crossing the submarine-infested North Atlantic.

Sherrington was anxious to aid in the war effort and, although almost sixty, he worked incognito seven days a week as an unskilled laborer at the Vickers-Maxim shell factory in Birmingham. He even lived in a hut with the men who were working twelve-hour shifts in the munitions plant, and he never told friends in Oxford where he was or that he was studying industrial fatigue for the government.

Sherrington stopped gauging shells and fitting them with fuse sockets three months later in order to write his report to the War Office. He recommended shorter hours and improved living conditions for the workers, personally convinced that such changes would increase productivity. Afterward he served on government committees to study tetanus and alcoholism.

When the "war to end all wars" finally ended, Sherrington returned to his teaching and research at Oxford. He continued to work on the same sorts of things that had interested him at Liverpool. John Carew Eccles and William Gibson, two of his students at Oxford, relate how he entered the laboratory one day, energized after watching a neighborhood cat successfully leap across a wide gap:

> One morning Sir Charles arrived with an inspired look on his face. He recounted vividly how he had seen a cat walking solemnly on a stone wall that was interrupted by an open gate. The cat paused, inspected the gap, then leaped exactly to the right distance, landed with ease and grace and resumed its solemn progression. A very ordinary happening, yet to Sherrington on that morning it was replete with problems for future research. . . . How had the strength of the muscle contractions been calculated so that the leap was exactly right for the gap? How had the motor machinery been organized so that there was this elegant landing on the far side of the gap? How after the landing was it arranged that the stately walk was resumed?[46]

In addition to Eccles and Gibson, Sherrington's other notable Oxford students were John Fulton, later to achieve fame for his work at Yale University on the roles of different parts of the frontal cortex in movement, learning, and emotion; Ragnar Granit, the Nobel Prize–winning visual neurophysiologist from Sweden; Thomas Graham Brown, another well-known neurophysiologist; Derek Denny-Brown, a leading Boston neurologist; and Wilder Penfield, the neurosurgeon who would have his "preparations" speak to him in Montreal.

With Sherrington at the helm, one of the group's primary objectives was to achieve a better understanding of synaptic excitation and inhibition. By hypothesizing two opposing active processes that could summate algebraically—a "central excitatory state" and a "central inhibitory state"—the Oxford team felt confident that even their most complex findings could be explained.

Sherrington's last physiological experiment was completed in 1931, four years before he gave up his Wayneflete Chair at Oxford. At the time, he was also finishing his last purely scientific book, one written with four of his students. It was called *Reflex Activity of the Spinal Cord.*[47]

Mind-Body Philosophy

In his "retirement," Sherrington engaged in a variety of activities, some related to his science and others not. In the latter domain, he published an expanded collection of his poems in 1940, some of which had appeared in 1925.[48] He also served as a trustee of the British Museum, helping the institution acquire rare books and manuscripts. He gave freely from his own extensive collection and was recognized as the most important donor of medical books in the museum's history.

Sherrington now thought more deeply about the relationship between mind and brain. Philosophy, of course, was not a new interest for Sherrington. In fact, one of his followers introduced him as the "supreme philosopher of the nervous system" in 1931.[49] This "knighting" took place at a neurological congress in Switzerland and led to a standing ovation from two thousand listeners.

In 1941, when he was eighty-four years old, Sherrington published *Man on His Nature,* one of his most philosophical and creative enterprises.[50] This work was based on the Gifford Lectures he had delivered at Edinburgh University in 1937 and 1938. In these lectures he discussed how the motor act evolved as a product of natural forces acting on the organism. In the beginning, reflexes moved the whole organism. But as life evolved, two changes took place. First, some reflexive movements were confined to individual parts of the body. And second, more complex reflexes appeared. The advent of the primitive brain permitted more regulation of the motor act. With larger brains, and finally consciousness as we know it, there was even more control. As Sherrington put it in his own special way:

> Lloyd Morgan, the biologist, urged that "the primary aim, object and purpose of consciousness is control." Dame Nature seems to have taken the like view.[51]

But just how does the thinking and perceiving mind interact with the basic reflex machinery of the body to achieve this highest level of integration? This was the question Sherrington had avoided in 1904 but brought up in his Rede Lecture of 1933 when he pondered: "What right have we to conjoin mental experience with the physiological?"[52]

He now answered by explaining that the mind is tied to the brain and both are still evolving. But, as a pragmatic scientist, he found he could say little more. How the mental world can affect the physical machinery of the brain, and how brain activity can affect the mind, were questions he could not begin to answer. All he could say was that we are who we are because we have achieved this higher level of integration.

The Historian of Neurophysiology

Throughout *Man on his Nature,* Sherrington contrasted what he now knew about the nervous system with the state of knowledge in the sixteenth century. His hero of the past was Jean Fernel, who was born in 1485, taught at the University of Paris, and died in 1558 (Fig. 14.7). Fernel was a deeply religious man, yet he was also a Renaissance scientist who argued against the idea that the ancients had already acquired all the knowledge humans needed to know. Our task, Fernel explained, is no longer just to gaze back at the wisdom of our forefathers, however wise they might have

FIGURE 14.7.

Jean Fernel (1485–1558), the French Renaissance physician about whom Sherrington wrote a part of one book and all of another.

been. We must now look ahead and develop new views based on careful observations of sick patients, skillful autopsies, and a better understanding of nature.

Fernel's fertile mind enabled him to discharge magic, superstition, the occult, and astrology from medicine. These notable breaks with the past can be found in his most important treatise, *De naturali parte medicinae* (On the Natural Part of Medicine), which appeared in 1542.[53] He even coined several new medical terms. It was Fernel who introduced the words that described the two main phases of Sherrington's own professional life, *pathology* and *physiology*. But although Fernel was a pioneer in many ways, he was unable to break completely from the past. He still looked upon the body as a puppet activated by the reigning soul and was not ready to abandon the theory of four humors.

Sherrington returned to Fernel in greater detail in his last major work, one completed in 1946. *The Endeavour of Jean Fernel* was the result of twenty years of study.[54] Sherrington had obviously found his kindred spirit in Jean Fernel. Both men were physicians and both were passionate about their calling. Although they lived four hundred years apart, both also thought about life and nature in a broad way. Each was a tireless searcher for the truth, determined to see, understand, and make a difference.

Legacy and Assessment

Sherrington, the neurophysiologist and student of animal behavior who showed neither pomp nor arrogance, died from heart failure in 1952. He had suffered from arthritis during the last few years of his long life, and his hearing had become poor. His wife had passed away almost twenty years earlier.

Shortly before his death at age ninety-five, he was visited by John Fulton, who had worked with him at Oxford. In his nursing home near the coast in Eastbourne, he confided to Fulton that he thought he had lived too long. Still, he never lost his engaging smile or his wry sense of humor. Catching Fulton off guard, he remarked: "At least I have outlived George Bernard Shaw."[55]

Another statement made late in life, however, better reflects the humility and insight of the man. Recognizing that his own explorations were merely a starting point for understanding behavior, and that reflex action is only one part of human behavior, he told Lord Russell Brain:

> The reflex was a very useful idea, but it has served its purpose. What the reflex does is so banal. You don't think that what we are doing now is reflex, do you? No, no, no.[56]

If Sherrington sounded modest or self-deprecating in his remarks, he did not have to be. In his lifetime, he had published over 320 books and journal articles that led people to look at neurophysiology and behavior in a new way. He received honorary doctorates from twenty-two universities and belonged to more than forty academies and scientific societies. His work was deemed so valuable that he shared the 1932 Nobel Prize with Edgar Adrian (see Chapter 15), was elected president of the Royal Society, and was knighted in his homeland.

As for his students, writing about their contributions would easily fill many large volumes in science and medicine. Especially notable, however, is the work conducted by John Carew Eccles, an Australian who spent ten years with him at Oxford. When Sherrington died, Eccles was already studying tiny postsynaptic changes with glass microelectrodes inserted into nerve cells.

During the 1950s Eccles and his colleagues in New Zealand and Australia discovered that nerve cell bodies and dendrites can show two different types of electrochemical changes.[57] They called the small opposing changes excitatory postsynaptic potentials, or EPSPs, and inhibitory postsynaptic potentials, or IPSPs. They then went on to elucidate the physiology and biochemistry of these "graded" synaptic changes, which could lead to all-or-none action potentials or, conversely, prevent them from occurring.

Eccles, who won the 1963 Nobel Prize for his extensive work in neurophysiology, is mentioned here because he, like many other scientists who fell under Sherrington's spell, looked upon his own discoveries as confirming his mentor's earlier thoughts. After all, it was Sherrington who developed the concepts of central excitatory and central inhibitory states, as well as the new idea of computation at the synapse.

In retrospect, Sherrington achieved greatness because he had a gift for designing experiments that allowed him to understand the functional organization of the nervous system—studies that served as guides for important new work and broad conceptual advances. He built an imposing edifice with many rock-solid bricks mortared together with insight and logic to provide structure where speculation, if not chaos, had previously reigned. He had a wonderful ability to see details but was even better when it came to grasping fundamental principles that could solve complex puzzles.

Henry Viets, the American student who crossed the Atlantic to work with Sherrington during the First World War, was saddened when he heard that Sherrington had died. In his eulogy to the inquisitive but modest man who had inspired so many other scientists to achieve greatness, he lamented:

> A great and good man has died—a man whose influence spread to every man and woman in this room and far beyond. We stand on mighty shoulders.[58]

15

Edgar D. Adrian

Coding in the Nervous System

> If these records give a true measure of the activity in the sensory nerve fibres it is clear that they transmit their messages to the central nervous system in a very simple way. The message consists merely of a series of brief impulses. . . . In any one fibre the waves are all of the same form. . . . In fact, the sensory messages are scarcely more complex than a succession of dots in the Morse Code.
>
> —*Edgar D. Adrian, 1932*

Instruments Guide Discoveries

Edgar Douglas Adrian once remarked: "The history of electrophysiology has been decided by the history of electrical recording instruments."[1] Adrian made this statement in 1932, not as an outsider but as one of the greatest neurophysiologists of all time. Others who have studied the history of electrophysiology are in agreement with his assessment.[2]

Little was known about how the nerves coded their messages when Adrian began to work on the nature of the nerve impulse early in the twentieth century. The problem was that researchers did not have the tools needed to amplify and record small, rapid electrical changes with sensitivity or fidelity.[3] The scientists of the early 1900s, however, were not without galvanometers and other measurement tools. One piece of apparatus they used was known as the capillary electrometer. A later instrument was called the string electrometer.

The capillary electrometer emerged in the 1870s, after Gabriel Lippmann recognized that a drop of mercury on some acid would change shape when he passed even small electrical currents through it. Étienne-Jules Marey then had the brilliant idea of putting some mercury and acid into a thin tube. By shining a bright light on the tube and positioning a strip of film behind it, he was able to see the small movements at the surface of the mercury even better, and could even capture them with his camera. In 1876 Marey proudly announced to the membership of the prestigious Paris Académie des Sciences that he and Lippmann had recorded frog and tortoise heartbeats on film.[4]

The ability of the capillary electrometer to follow cardiac events led other investigators to try the device on nerve impulses. In 1888 Francis Gotch and Victor Horsley, two prominent English scientists, discovered that the capillary electrometer could detect electrical changes in the peripheral nerves and in the spinal cord.[5] Eleven years later, Gotch even discovered a phenomenon called the refractory period with his instrument. He noticed that a nerve cannot discharge again immediately after firing; a small interval of time has to elapse between discharges.[6] This finding suggested that nerves must convey information with patterns of discrete impulses, not a single lingering impulse.

Although the discovery of the refractory period was very important, the capillary electrometers used by Gotch were inadequate for showing the fine features of the nerve event. The nerve impulses appeared as featureless blips, even when mirrors were aligned to enlarge the waves on moving strips of film. As a result, if action potentials (a term coined by Emil Du Bois-Reymond) varied in size, duration, and spacing, this was not something that could be discerned by those physiologists who wanted to learn more about coding in the nervous system.

Another drawback of the capillary electrometer was that its measurements were not particularly faithful. They were distorted because the mercury continued to move by inertia after the stimulus was withdrawn. In the 1890s scientists began to apply mathematical formulas to overcome this troublesome distortion, but their efforts were never really satisfying. For this reason, the demand continued to grow for more sensitive recording instruments.

The string galvanometer was introduced early in the twentieth century to overcome the distortions of the capillary electrometer. This device was the brainchild of Willem Einthoven, who held the chair of physiology at the University of Leiden in Holland. Einthoven had the mind of a physicist, one of the best laboratory setups in the world, a generous budget, and the patience of Job. He was interested in electrocardiography and had first tried using capillary electrometers in his work. But he became frustrated by the inability of the instrument to follow faster cardiac events. Like others before him, he first attempted to correct for the sluggishness of his instruments by applying mathematical equations to his data. Before long, he realized that he had to invent a new instrument.[7]

Einthoven was guided by the discovery that small oscillating currents can make a very thin wire ("string") vibrate if the wire were stretched between the poles of a strong electromagnet. His string galvanometer, which made its debut in 1901, took a few years to complete. It weighed several tons, filled an entire room, and required a cold-water source for cooling its electromagnet. But it worked so well, at least for cardiology, that Einthoven won a Nobel Prize for his instrument and discoveries about cardiac waves.

When neurophysiologists learned about Einthoven's device (which was soon made smaller, hardier, and easier to use), they jumped at the opportunity to try it. These researchers, however, were dismayed to find that the string galvanometer still lacked the sensitivity or amplification needed to show the features of the nerve impulse. To record even the tiniest blips, the neurophysiologists had to apply intense electrical shocks to thick nerve cables made of hundreds or thousands of fibers. Able to see only that some rapid changes were taking place, they were left scratching their heads once again.

The recording and amplifying instruments that would finally allow neurophysiologists to understand the nature of the nerve impulse would not emerge in Lippmann and Marey's France, Gotch's England, or Einthoven's Holland. The major technological breakthroughs would come from the United States, and not until after the First World War. To appreciate how the newer instruments came into being, and how they were used by Edgar Adrian to unlock more secrets of the working nervous system, we have to begin with Adrian's teacher, a brilliant and creative English physiologist by the name of Keith Lucas.

Keith Lucas

Lucas, born in 1879,[8] was the son of the managing director of Britain's Telegraph Construction and Maintenance Company. Like his father, he enjoyed working with mechanical and electrical instruments. Bright and energetic, he attended Trinity College in Cambridge, where he was attracted to physiology with its heavy reliance on instrumentation.

Even before he became a lecturer at Cambridge, Lucas was interested in muscle physiology. The question he really wanted to answer was why can a muscle flex less than completely? Said somewhat differently, why is it that we can make an arm muscle bulge either a little or a lot when we flex it? There seemed to be two ways of explaining the existence of both partial and complete arm muscle contractions. The first is that the muscle fibers can contract only part way. The second is that only some muscle fibers in a bundle contract, but that these select fibers do so fully.

Lucas hoped to determine which theory was correct by conducting experiments on frogs. His logic was straightforward: If the amount of muscle contraction depends upon how many individual fibers are contracting maximally, he should be able to see discrete steps in the contractile response as increasingly stronger shocks are applied to the muscle. In contrast, this function should be smooth if all muscle fibers contract only partially.

Lucas chose to work on a small back muscle (the dorsocutaneous) in frogs, one with less than two hundred fibers. His strategy was creative. He would divide the muscle into small strips, each having between twelve and thirty muscle fibers, and would stimulate just one strip with a few muscle fibers. At the same time, he would record the responses on film.

Lucas began by applying very mild electrical current to his muscle strip, and then slowly increased it. A staircase function emerged as individual fibers in the muscle joined in the response. Moreover, the number of steps in the staircase never exceeded the number of muscle fibers remaining in the cut strip. Hence, his results strongly supported the theory that partial muscle flexion must be due to some muscle fibers contracting fully, while the remainder do not respond at all. As he expressed it, individual skeletal muscle fibers behave in an "all-or-nothing" manner.

Lucas published his results in 1905.[9] By 1909 he was able to confirm the steplike function by stimulating the nerves to the same muscle.[10] In his second paper, he wrote: "In each muscle-fibre the contraction is always maximal regardless of the strength of the stimulus which excites the nerve-fibre."[11] Lucas was even more convinced than before that submaximal contractions of a skeletal muscle are really maximal contractions of only some of its fibers.

The latter experiment lured Lucas into nerve physiology and the question of whether nerve fibers also respond maximally or not at all. Although work already completed gave him reason to believe that axons behave just like muscle fibers, he ended his 1909 paper by writing: "We must therefore regard the question whether the response of a nerve-fibre is capable of gradation as being at present undecided."[12]

To continue his research in the neural domain and to find out if nerve fibers also obey the all-or-nothing principle, Lucas knew he had to have more sensitive recording instruments. Although the idea of using a string galvanometer was suggested to him, he thought he would be better off redesigning his capillary electrometer to improve its sensitivity. His technical skills were such that he not only came forth with a better instrument, but was able to combine it with an apparatus for delivering measurable electrical shocks. He added an analyzer to correct some of the distortions caused by the mechanics of the equipment, and also improved the system for making records of the nerve events. With his instruments in place, he was ready to bring a new assistant into his laboratory—a bright young scientist eager to help with his nerve experiments.

Adrian and the All-or-Nothing Principle

Edgar Douglas Adrian (Fig. 15.1) became Lucas' laboratory assistant. Born in 1889, Adrian came from a well-to-do London family.[13] Although he was first drawn to Greek, Latin, and the classics, by 1906 Adrian had become interested in the life sciences. Two years later he found himself attending classes at Trinity College, Cambridge, where he quickly gained a reputation for being very bright and extremely hardworking. In 1911 Adrian was awarded a first-class degree in physiology, enabling him to do postgraduate work in the field.

Adrian had taken tutorials with Lucas and was particularly interested in working in his laboratory. As for Lucas, he knew from the outset that Adrian had all of the ingredients he could wish for in an assistant. He was energetic, immensely talented,

FIGURE 15.1

Edgar Douglas Adrian (1889–1977).

emotionally stable, and anxious to take on a challenging problem. Hence Lucas welcomed Adrian into his cluttered basement laboratory and began to teach him the tricks of the trade.

After completing an experiment showing how stimuli close together in time can summate (have additive effects), Lucas asked Adrian to try to determine whether the all-or-nothing principle holds for nerves. Although Lucas now had better equipment than just about anyone else, his setup in 1911 still left something to be desired when it came to doing this sort of nerve work. As a result, Adrian was forced to rely on an indirect method to answer the question.

Adrian's strategy was to isolate a long nerve from a frog and to place a segment of it in a chamber with narcotizing alcohol vapors. He would expose the nerve to just enough alcohol to weaken the impulse, but not enough to stop it from making it through the block. He reasoned that the weakened impulse should immediately jump back to full strength once it comes out of the alcohol block if the all-or-nothing conduction principle holds. To test his hypothesis, he would measure how strong a second alcohol block would have to be to block the impulse further down the axon.

Adrian's findings from 1912 showed that the nerve impulse quickly returned to full strength after it passes through a weak alcohol block.[14] This was suggested by the fact that the second alcohol block had to be just as strong to stop the narcotized impulse as it had to be to stop a nerve impulse that was never weakened by the alcohol. He also found that it did not matter if the nerve impulse had been triggered by a barely adequate stimulus or a very strong one. So long as any stimulus could trigger an impulse, the action potential shot down the axon at full strength. These findings suggested that the nerves obey the same all-or-nothing law as the skeletal muscles.

In 1913 Adrian was awarded a medal for his experiments on all-or-nothing nerve conduction. Over the next two years, but now as a fellow of Trinity College, he continued to promote and defend the idea that axons convey information with discrete, all-or-nothing blips. At the same time he suggested that, under natural conditions, the yes-no axonal event is probably triggered by smaller, local changes at the synapse. This idea would be confirmed by others many years later.[15]

Unfortunately, the First World War broke out at this time, completely disrupting the new experiments that Adrian and Lucas had planned. In order to contribute to the war effort, Adrian left tranquil Cambridge to go to London to complete his medical training. Lucas also left when the war erupted; he joined the research staff of the Royal Aircraft Factory in Farnborough, where he used his engineering skills to design new bombsights and aircraft compasses.

Adrian obtained his medical degree in 1915. He then worked on shell shock, hysterical paralysis, and nerve injuries. Sadly, Lucas perished in a midair collision over Salisbury Plain in 1916. He had been testing new instruments for the Royal Air Force at the time of the crash, which robbed physiology of one of its most creative minds.

Following the loss of his mentor, Adrian took it upon himself to edit and publish the unfinished book Lucas had been writing. Lucas' *The Conduction of the Nervous Impulse* appeared in 1917.[16] It covered the period of quiet between successive nerve impulses, summation effects, and all-or-nothing conduction in axons. These were exciting advances, to be sure, but there was still the feeling that much more could be accomplished if only neurophysiologists had a better way to magnify nerve

responses without sacrificing clarity. Fortunately, as a result of the Americans joining the British, the French, and their allies in the war effort, existing technologies
were about to change for the better.

The Radio Tube Amplifier

Alexander Forbes hailed from a large, well-to-do Boston family and was blessed
with impressive genetic material.[17] His mother was the daughter of the poet Ralph
Waldo Emerson, and his father was a Civil War hero who worked his way to the
presidency of the huge Bell Telephone Company. With the brains, financial resources, and physical stamina to do whatever he wanted, Forbes chose to go to Harvard, where he starred on the football team and eventually became interested in
physiology. In 1905 he received his master's degree; five years later he had an M.D.
degree. He then opted to stay at Harvard as an instructor in Walter Cannon's highly
regarded physiology department.

The more Forbes read, the more he became impressed with the work of Charles
Sherrington (see Chapter 14). Hence, soon after receiving his appointment in physi-
ology, he asked for a leave of absence to study reflex physiology with Sherrington.
This request was granted, and Forbes and his wife headed to the docks for a boat to
Liverpool.

During his stay in England, Forbes took the time to see what was going on in the
physiology laboratories at Cambridge. Had Forbes decided not to stop off to see
Lucas and Adrian, neurophysiology would likely have had a very different history.
Once he began exchanging ideas with the Cambridge physiologists, he made up his
mind to conduct some new experiments with them before returning to the United
States. Because the studies were expected to take a few weeks to complete, he can-
celed the original bookings for himself and his wife for the trip back across the Atlantic. The tickets he exchanged had been for a lovely room on the crown jewel of
luxury liners—the new but ill-fated *Titanic.*

After arriving home on another famous ship, the *Lusitania,* Forbes filled his laboratory with equipment like that used by Lucas and Adrian. Among other things, he
added an improved capillary electrometer and an analyzer to the Einthoven string
galvanometer that his department had previously purchased. Capitalizing on what he
had learned from Sherrington, he then began to use the equipment to study reflexes
in cats.[18]

Forbes was an accomplished sailor who always loved the sea. Thus when the First
World War broke out, he joined the navy. With engineering skills like those of Lucas,
he soon found himself developing wireless radio receivers and "radio detection
finders" (radar) for the military. While in the service, he learned about some new
vacuum tubes (then called audions) that had the ability to enhance radio signals with
minimal distortion.

When the war finally ended, Forbes asked for and obtained some of the new vacuum tubes. The tubes enabled him to construct a new amplifier for physiological research. In 1919, when he put his thermionic amplifier into a circuit with a nerve and
an Einthoven string galvanometer, he found that he was able to multiply a tiny action
potential an unprecedented fifty times in size.

Over the next few years, Forbes described his amplifier in detail and published some studies using it.[19] Most of his nerve experiments on frogs and cats just confirmed earlier findings. To most historians, Forbes' major accomplishment was technological; he was able to show scientists how they could magnify a nerve responses far better than ever before.[20] The importance of this breakthrough was immediately recognized by other neurophysiologists, some of whom also wanted to delve deeper into the electrophysiological waters that had seemed so murky in the past.

The St. Louis Connection

The two axonologists (Forbes' term) who took the next big step were Herbert Gasser (Fig. 15.2) and Joseph Erlanger (Fig. 15.3).[21] Gasser had been Erlanger's student when both were at the University of Wisconsin. Although Gasser now served

FIGURE 15.2.

Herbert Gasser (1888–1963), a pioneer in the use of amplifiers and oscilloscopes to study peripheral nerve physiology. (Courtesy of the Washington University Archives, St. Louis, MO.)

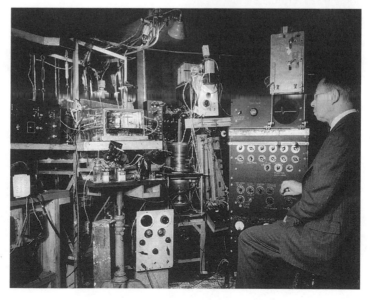

FIGURE 15.3.

Joseph Erlanger in his well-equipped Washington University laboratory. (Courtesy of the Washington University Archives, St. Louis, MO.)

under Erlanger in the physiology department at Washington University in St. Louis, he was the one who first began working on the instruments, ideas, and projects that would make both men world-famous. Prior to the 1920s, Erlanger's specialty had been the same as Einthoven's—namely, cardiology. But to his credit, the man occasionally called "the chief" quickly recognized the importance of what Gasser was doing and joined him, adding his own expertise to the projects.

Gasser knew he would have to find a way to increase the amplification of nerve currents even more than Forbes had if further progress were to be made in neurophysiology. For assistance, he turned to H. S. Newcomer, a man with a talent for building machines. With the additional help of some physicists at the Western Electric Company, Newcomer and Gasser built a multistage amplifier. Their device had switches that allowed them to feed the output from one vacuum tube amplifier into another amplifier, and then into a third amplifier if they so desired. The greatly magnified message was then sent to another instrument for display.

Gasser and Newcomer first used their multistage amplifier to record the messages from the phrenic nerve of the dog, which controls the muscles of the diaphragm and is important for breathing. At the time, they could use only two-stage amplification because the thin wires on their sensitive string galvanometer blew when they put the third amplifier in the series. But even with two-stage amplification, the action potentials were magnified so much that they were able to see electrical impulses shooting down the nerve with new clarity. Gasser and Newcomer described their new amplifier and published their initial findings in 1921.[22]

Just before their paper appeared, Gasser attended a scientific meeting in which he learned about another technological development that could be useful for this work—an improved cathode-ray tube for displaying electrical events. The cathode-ray tube had been invented by German physicist Karl Braun in 1897. Braun's idea was to shoot a threadlike electronic beam through a vacuum, where it could be deflected by external currents and displayed on a fluorescent screen. Braun's idea worked, but the early Braun tubes lacked the sensitivity to be of any use for physiological research.

The situation changed for the better in the 1920s. Gasser realized that when hooked to a circuit with a nerve and his three-stage amplifier, one of the newer-generation Braun tubes had the potential to display small and rapid electrical deflections, events that could not be picked up in the past. The cathode-ray tube also had another very attractive feature. Unlike earlier recording instruments, it was not subject to troublesome distortions produced by inertia or moving mechanical parts. It could produce faithful records.

Gasser tried to purchase the cathode-ray tube he needed from the Western Electric Company but, because the tube he wanted was still in the developmental stage, he met with no success. Consequently, he asked Newcomer to build a cathode-ray oscillograph from materials available at the university. The device that emerged was incredibly primitive; its central element was made out of a distillation flask (Fig. 15.4). To make matters worse, it exploded the first time the experimenters tested it, because they had forgotten to add a resistor to the circuit.

Erlanger now used his considerable pull to convince the cathode-ray tube manufacturer to give them the badly needed tube "strictly for experimental use." He also saw to it that additional equipment was built for connecting the cathode-ray tube to

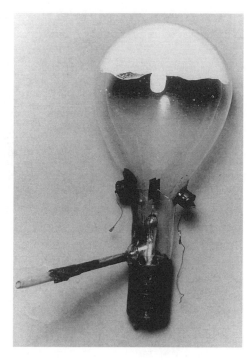

FIGURE 15.4.
An early cathode-ray tube used by Herbert Gasser and Joseph Erlanger. (Courtesy of the Washington University Archives, St. Louis, MO.)

the multistage amplifier. Before year's end, Gasser and Erlanger were able to publish an abstract describing how nerve potentials could be displayed on a low-voltage, inertialess oscillograph.[23] One year later, in 1922, they published a full account of what they were able to achieve.[24]

One of the St. Louis axonologists, who knew just how much trouble Gasser and Erlanger had with their first instruments, later described what life in this laboratory had been like in the "good old days":

> I can only compare their progress to the trek of the pioneers in oxcarts across the plains and mountains of the West. They were pioneers of electrophysiology and they encountered every kind of obstacle but the Indians. Far more time was spent on reconstruction than on recording of nerve, when one good record occasionally made a successful day.[25]

Although their early instruments had drawbacks, the Washington University scientists managed to amplify nerve events up to seven thousand times. The enabled them to see even more details in the impulses shooting through the cables to and from the switchboards of the central nervous system. Using frogs, rabbits, cats, and dogs, they discovered that the waves in a peripheral nerve cable are not smooth up-and-down events (Fig. 15.5). The farther the experimenters moved the recording electrode away from the place of stimulation, the bumpier the waves became.

The St. Louis axonologists now theorized that the nerve cables from which they were recording contained several different nerve fiber types, each of which conducted its action potentials at a different rate.[26] Based on what they knew about wire cables, they assumed that thick nerves must conduct more rapidly than thin nerves. From this premise, they argued that partially overlapping functions for different

FIGURE 15.5.

The "bumpy" compound action potential of the saphenous nerve. The first bump is due to the fastest fibers in the nerve, and other rises are due to the activity of more slowly conducting nerve fibers. (From Gasser, 1937; with permission of Academic Press.)

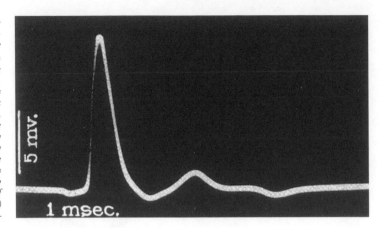

kinds of nerves can account for the bumpiness of the "compound" action potential in a peripheral nerve cable.

Subsequent studies showed that Gasser and Erlanger were correct—the different kinds of nerves in a cable can be identified by using anesthetics, cold, and other kinds of nerve blocks.[27] The St. Louis researchers also associated some nerve fiber types with different functions, such as light touch, dull pain, sharp pain, and temperature detection.[28] Because of their technological achievements and many discoveries about peripheral nerve physiology, Gasser and Erlanger were awarded the Nobel Prize in 1944.

Adrian's "Failed" Experiment

In 1919, while Alexander Forbes was still working on his single-stage amplifier, and two years before the axonologists at Washington University had their three-stage amplifier and oscilloscope up and running, Adrian returned to Cambridge. He was glad to leave the military behind him and come back to the college town with its historic old buildings, winding streets, small shops, and scenic bridges. He soon settled into quarters over the majestic gateway to Trinity College, where Sir Isaac Newton once lived. When he inquired about work space, he was handed the keys to the basement laboratory that had belonged to his mentor, Keith Lucas.

Although burdened with a heavy teaching load, Adrian immediately planned new experiments. He also began to communicate with Forbes, who agreed to spend part of 1921 back in Cambridge. When Forbes arrived, he unpacked some of his radio tubes and proceeded to build an amplifier for the two men to use.

The collaboration with Forbes allowed Adrian to conduct some new studies. Yet Adrian was not entirely satisfied with what he was able to accomplish with his friend's amplifier. Soon after Forbes left, Adrian read about the improved multistage amplifier that had been built at Washington University. Realizing what he could do with even better equipment, he contacted Gasser and asked him how to build one.

Gasser obliged with instructions, and Adrian soon had his coveted instrument. Stretched for money and afraid of burning out the expensive wire of a string galvanometer, Adrian now dusted off Keith Lucas' capillary tube electrometer and placed an order for some live frogs. He would test the sensitivity of his new amplification system as cheaply as possible.

Adrian was distressed to find that the baseline he obtained when he tried to record from a nerve in a suspended frog leg was highly irregular. His immediate thought was that his circuit was bad. If this proved to be the case, he might have to spend weeks or months rebuilding his apparatus. With thoughts of such a tragedy unfolding in his mind, he tried moving the suspended frog leg around, jiggling it, and then gently laying it down. Oddly, the oscillations that appeared when the muscle was hanging freely disappeared as soon as it was rested on a glass plate. Adrian suspended the muscle once more and saw the oscillations again, only to watch them disappear a second time when the muscle was laid flat. Obviously, the strange effect could be replicated. But what did it mean?

The next part of the story is best told by Adrian himself. His narrative appeared in a short paper appropriately entitled "Memorable Experiences in Research."[29]

> The explanation suddenly dawned on me, and that was a time when I was very pleased indeed. A stretched muscle, a muscle hanging under its own weight, ought . . . to be sending sensory impulses up the nerves coming from the muscle spindles, signaling stretch on the muscle. When you relax the stretched muscle, when you support it, those impulses ought to cease.
>
> I don't think it took more than an hour or so to show that was what the little oscillations were. I was able to make photographic records of them, and within about a week, I was nearly certain that many of these oscillations were action potentials coming up sensory fibres in the nerve, and what was more, many of them came from single nerve fibres.[30]

Adrian first presented his new findings at a scientific meeting in 1925 and published them a year later.[31] He knew he had confirmed a theory about proprioceptive feedback from the muscles that had been suggested years earlier by Charles Sherrington (see Chapter 14). The spindles were sending information about muscle stretch to the central nervous system, allowing movements like those needed for catching a ball to be properly regulated. Adrian could not have been more enthusiastic about his discovery, which came serendipitously from a preliminary test that at first seemed to go bad.

Recording from Single Neurons

While recording from muscle stretch receptors, it became clear to Adrian that even more could be learned if he could find a way to record messages from single neurons. He knew this had to be his next objective. As he put it in the last sentence of his 1926 paper on proprioceptive feedback: "More detailed analysis of these results is postponed until experiments have been made on preparations containing a known number of sensory endings, if possible only one."[32]

With this goal in mind, Adrian now set forth to try to record from single nerve fibers. Assisting him in this phase of his work was Yngve Zotterman, a young Swedish medical student who had studied in Cambridge six years earlier. Zotterman now had a Rockefeller Foundation fellowship, allowing him to spend part of his time in Adrian's laboratory doing neurophysiology.

Adrian and Zotterman began by looking for a nerve with very few fibers. They tried the small nerve to the dorsocutaneous muscle that Keith Lucas had studied before the war, but could not detect any changes in it when the muscle was stretched. They then turned to the nerve associated with another small muscle, but again could not isolate any nerve fibers. Zotterman next decided to try a Keith Lucas trick. He sliced off pieces of this small muscle to reduce the number of sensory end organs. With fewer end organs, fewer axons in the cable would be activated by muscle stimulation—just one if all went according to plan.

Zotterman recalled what happened on November 3, 1925:

> When we recorded from the nerve while stimulating the spindles by stretching the muscle our records showed that the electrical response derived from quite a few sensory nerve fibres. It was as if we were tapping a telegraph cable with many lines simultaneously in transmission. It did not permit any reading of the code. Before abandoning this preparation, however, I one day made cuts from the medial side of the muscle, successively cutting away one muscle spindle after the other from its connection with the nerve. Finally we were left with a tiny strip of muscle which obviously contained only one functioning spindle, signalling in one single sensory fibre.

He went on:

> Under strong emotional stress we hurried on, recording the response to different degrees of stimulation. Adrian ran in and out, controlling the recording apparatus in the dark-room and developing the photographic plates. We were excited, both of us quite aware that what we now saw had never been observed before and that we were discovering a great secret of life, how the sensory nerves transmit their information to the brain.[33]

Zotterman and Adrian successfully recorded from the only remaining active axon in the bundle. The action potentials were about the same height and traveled at the same speed. Adrian knew that his all-or-none law, which had been based solely on indirect evidence in the past, now had the confirmation it needed at the single-cell level.

A second finding of great interest emerged when they varied the weight on the muscle. The firing rate of the isolated neuron increased as the weight of the stimulus increased; a quarter gram of weight triggered twenty-one action potentials per second, whereas one gram of weight triggered thirty-three action potentials. From all indications, at least one part of the physiological code for conveying the strength of the stimulus is the rate of firing. The nerve cells do use something like a Morse code after all, but their code is restricted to the patterns of dots.

A third intriguing finding had to do with cells adapting to a steady stimulus. Adrian and Zotterman noticed that the burst of action potentials that follows the presentation of a new stimulus is not sustained if the stimulus remains in place. The decrease in the firing rate after the initial burst of activity suggested that the sensory nerves are programmed to respond to change, not to steady-state conditions.

This concept of adaptation so fascinated Adrian that he made plans to study it in other sensory systems. With nerve adaptation, he was convinced that he had a mechanism capable of explaining many intriguing everyday phenomena. Two common examples are why we have so much difficulty using the sense of smell to trace a gas leak, and why we are so sensitive to loud, monotonous sounds when we first hear them, yet less so after continuous exposure.

Adrian and Zotterman published their landmark findings in the *Journal of Physiology* early in 1926.[34] Scientists who read their report immediately recognized that another milestone had been passed—neurophysiologists finally were able to see how single nerve cells coded sensory information.

Many people who did not read their journal article had a hard time believing what Adrian and Zotterman had done when told of their accomplishments. Zotterman experienced this himself when he spoke with French scientists in 1926.[35] Another doubter was none other than Alexander Forbes. In a letter to Adrian, Forbes explained that he, of all people, had been skeptical of the accomplishment before the latest issue of the *Journal of Physiology* arrived in Boston from England.[36]

After conducting these experiments on muscle spindle receptors, Zotterman spent the rest of his limited time with Adrian trying to record from single units in other sensory systems. Although he wanted to do more, he had to leave Cambridge for additional training in cardiology, as required by his fellowship agreement. Adrian felt sorry to see the Swedish visitor leave, but other scientists now stepped into his laboratory to continue the work these two men had begun.

Isolating Single Motor Neurons

The next milestone was reached in 1928, when Adrian and another visitor to Cambridge, Detlev Bronk, an American physiologist, successfully recorded from single motor neurons.[37] To do this, Adrian and Bronk turned to the phrenic nerve controlling the diaphragm, the nerve that Gasser and Newcomer had recorded from in St. Louis. But unlike the Washington University axonologists, they took some rabbits, anesthetized them, and carefully dissected the small nerve until only two or three axons remained.

Adrian and Bronk now placed an electrode on the axons left in the fragment. With their powerful three-stage amplifier, they observed a burst of impulses each time the rabbit took a breath. There were no electric shocks or weights, just the impulses naturally triggered from the central nervous system. Yet the investigators still saw the separate actions of the individual motor nerve units and, as expected, the features of the action potentials were again constant.

Adrian and Bronk also did something else not done before. They decided to attach a loudspeaker to one of their amplifiers. This enabled them to hear a clicking noise every time a neuron under the electrode fired. Listening to nerves clicking away, as well as visualizing them, became routine in Adrian's laboratory, and the new technique was quickly picked up by electrophysiologists around the world.

The Nobel Prize

Twelve years before Gasser and Erlanger won their Nobel Prizes, Adrian was awarded the Nobel Prize in Physiology or Medicine for "discoveries regarding the functions of neurons." He shared his 1932 prize with Sir Charles S. Sherrington, who had spent a lifetime studying reflexes and the integrative actions of the nervous system (see Chapter 14). For people familiar with the contributions of the two English neurophysiologists, there could not have been more deserving winners.

In his introductory speech in Stockholm, Professor G. Liljestrand of the Royal Caroline Institute reviewed some of Adrian's major contributions.[38] They included his elucidation of the all-or-nothing principle, his work on single nerve cells, his insights about sensory coding, and his research on sensory adaptation.

Adrian then gave his Nobel Address.[39] He began by telling his audience how much his work had depended on new instrumentation, especially the three-stage amplifier developed in St. Louis:

> The nerves do their work economically, without visible change and with the smallest expenditure of energy. The signals which they transmit can only be detected as changes of electrical potential, and these changes are very small and of very brief duration. It is little wonder therefore that progress in this branch of physiology has always been governed by the progress of physical technique and that the advent of the triode valve [three-stage] amplifier has opened up new lines in this, as in so many other fields of research.[40]

Adrian then honored Keith Lucas by stating that "I cannot let this occasion pass without recording how much I owe to his inspiration. . . . In my own work I have tried to follow the lines which Keith Lucas would have developed if he had lived."[41] Adrian also paid homage to Forbes and Gasser for their technological achievements, and to Zotterman for coming up with the preparation that allowed both men to tap messages from single neurons for the first time. He then proceeded to review the experiments that had made him famous.

Adrian was far from ready to retire from laboratory life when he arrived home from Stockholm. For one thing, he was only in his early forties when he received the Nobel Prize. In addition, slowing down in any way was simply foreign to his constitution. Adrian was a intense man who thrived on adventure. For the most part, his need was fulfilled in the laboratory, although he could play just as hard as he could work. Outside the laboratory, his favorite "relaxations" were fencing, skiing, sailing, and mountaineering. Not one to spend time sitting in the park feeding the pigeons, he also had a fondness for zooming around normally quiet Cambridge at breakneck speed on a motorcycle with a sidecar.

Adrian's fast approach to life led others to characterize him as a lean, driven, and moody genius. Often happy and charming, he had a reputation for becoming sharp and difficult to approach when preoccupied. At such times, even his closest friends knew better than to bother him. His wife's personality was decidedly different. Hester was also bright and enterprising, but considerably easier to approach than the dedicated scientist who married her in 1923. Yet their marriage was a good one. She provided her husband not only with a fine home and three children, but with the support and encouragement he needed to remain a step ahead of the many bright, young scientists who were now heading into neurophysiology.

Adventures with the EEG

At the time of his Nobel Prize, Adrian knew that he next wanted to study the central nervous system. He said even so in the final sentences of his Nobel Address. He recognized, however, that the central nervous system had to be approached differently than the peripheral nerves. As expressed in his Nobel speech in Stockholm:

> In the latter [the nerves] our chief concern is to find out what is happening in the units, and this turns out to be a fairly simple series of events. Within the central nervous system the events in each unit are not so important. We are more concerned with the interactions of large numbers, and our problem is to find the way in which such interactions can take place.[42]

Adrian was obviously stimulated by what Sherrington was doing and saying about integration. In addition, a series of articles coming out of Germany had caught his attention. In 1924 Hans Berger, a professor of psychiatry and neurology in Jena, had succeeded in recording electrical waves from the human brain.[43] Berger wanted to learn more about the physiological correlates of mental functions, and he found that he could record brain waves through the skull with metal electrodes.

Berger spent years recording human brain waves. But he worked in complete isolation and remained silent about his achievements. Not once before 1929 did the secretive psychiatrist invite people to see what he was doing in the small building on the grounds of his clinic. Not once did he mention his EEG work in his lectures or present his findings at a scientific congress.

In 1929 Berger finally made up his mind to begin publishing his findings.[44] In Germany, however, his articles and talks on recording brain waves attracted little attention. If anything, many of his countrymen were skeptical about the whole thing. They doubted that Berger's oscillating waves even arose from the brain.

The reaction to what Berger had done was quite different in Cambridge, where Adrian immediately recognized that he had found a way to probe how the human cerebral cortex functions. To Adrian, the German psychiatrist went well beyond what Richard Caton had accomplished in the 1870s, when the all-but-forgotten English physiologist first used electrical recording techniques with animals to confirm cortical localization ideas put forth by David Ferrier.[45] In part because Berger was now studying human beings and had a technique that did not require any kind of surgery or anesthetic, his work seemed much more exciting.

Adrian read how Berger used several different kinds of electrodes in his studies. In some experiments, he inserted needles made of platinum or zinc-plated steel into the scalp. Later, he turned to large plate electrodes, which were strapped to the front and back of the head. Once his subjects were comfortable, he asked them to open and close their eyes, to endure painful stimuli or loud noises, and to attend to stimuli.

Berger observed brain waves of about ten cycles per second when a subject lay quietly with his eyes closed. When the same subject opened his eyes, the rhythmic waves were replaced by faster, low-voltage activity. Berger hypothesized that the ten-cycle-per-second pattern represented rhythmic discharges from cells spread over the entire cortex, or at least a huge expanse of the gray mantle. This conclusion, he thought, was justified by the fact that he could record the rhythm from any part of the scalp. Berger looked upon this discharge as the fundamental activity of the cortex, and the loss of the rhythm as the result of a blocking effect. In Berger's mind, activation of the visual cortex had the ability to inhibit the synchronous wave pattern from the rest of the cortex.

Adrian, however, was not convinced that all parts of the cortex contributed equally to the ten-cycle-per-second Berger rhythm. Nevertheless, he was so intrigued by Berger's neglected papers on the electroencephalogram that he made up his mind to do some recordings of his own.

In his first EEG experiment he was aided by Brian Matthews, a younger Cambridge man with exceptional instrumentation skills.[46] Among his accomplishments, Matthews built the oscillograph that they used for recording brain waves. With anesthetized rabbits and cats, Adrian and Matthews observed how the pulsating discharges of individual neurons synchronized to form rhythmic brain waves. After making some additional observations, Adrian was ready to turn to humans.

Using scalp electrodes, Adrian, Matthews, and others who joined in these experiments had no trouble confirming many of Berger's major findings with people.[47] In addition, the Cambridge physiologists made many discoveries of their own. For example, although the Berger rhythm can be abolished by having a subject open his eyes (Fig. 15.6), the rhythmic discharge will continue if the subject stares at an empty visual field. The investigators also found that the Berger rhythm disappeared when a subject closes his eyes and tries to solve a problem using visual imagery.

Adrian interpreted his findings very differently than had Berger. He argued that the Berger rhythm is not the fundamental rhythm of the brain. Instead, it probably reflects the negative side of cerebral activity, which manifest itself only when consciousness or attention is diminished. Moreover, the Berger waves probably come from just one part of the cerebral cortex, the occipital lobes, which are involved with vision. As he and Matthews wrote in 1934:

> We regard the effect as due to a spontaneous rhythmic activity of a group of cortical cells in some part of the occipital lobe. These tend to beat synchronously when they are undisturbed, but visual activity or widespread nonvisual activity in the brain breaks up the rhythm by exposing the cells to a mosaic of excitations which makes synchronous action impossible. Berger, if

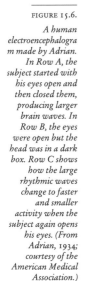

FIGURE 15.6.

A human electroencephalogram made by Adrian. In Row A, the subject started with his eyes open and then closed them, producing larger brain waves. In Row B, the eyes were open but the head was in a dark box. Row C shows how the large rhythmic waves change to faster and smaller activity when the subject again opens his eyes. (From Adrian, 1934; courtesy of the American Medical Association.)

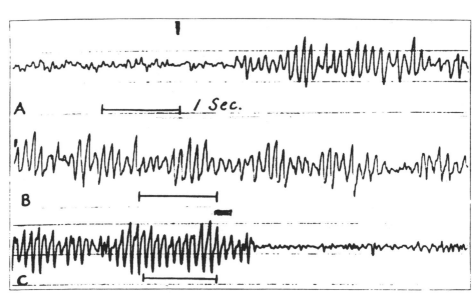

we interpret him correctly, regards the waves as having a much wider and less specific origin.[48]

Adrian's work played a major role in drawing attention to the EEG, especially in English-speaking countries.[49] During his lifetime, he saw the EEG rise to become an important tool for diagnosing seizure disorders. In 1936, long before modern brain scans entered the picture, he correctly predicted that with improved technologies and a better knowledge of cortical activity, scientists someday would be able to record and understand some of the brain changes accompanying thinking.[50]

Unlike Adrian, whose energy level was always at the upper end of the range, Berger became depressed and stopped publishing in 1938, at a time when he was finally being recognized for his pioneering work. Although his melancholia could have had a biological basis, it is also clear that the changes taking place in Germany affected him greatly. When he was nominated for the Nobel Prize just before the Second World War, powerful Nazi officials who looked upon the prize as political prevented him from accepting the greatest honor that could be bestowed on a scientist. Soon afterward, the Hitler regime forced him to retire from the university and give up his research. Sadly, when the Nobel Prize Committee tried to contact Berger after the armistice, they were told that the pioneer behind the human EEG had died in 1941. Severely depressed, suffering from cardiac problems and a viral infection, and with no lust for life in Nazi Germany, Berger took his own life by hanging himself.[51]

Multiple Representations on the Cortical Surface

After his foray into the world of the EEG, Adrian became interested in mapping the representation of the body on the cerebellar cortex. In 1943 he demonstrated that different cerebellar regions are associated with the forelimbs, hind limbs, and head.[52] He recorded from collections of cells as well as single cerebellar units in these experiments. His subjects were mostly cats and monkeys, but he also showed that his findings could generalize across mammals by including some dogs and even a goat.

Although Adrian found the cerebellar mapping work interesting, he became even more excited about mapping the somatosensory (skin sense) areas of the cerebral cortex. The first of his important discoveries occurred in 1940, after a large representation of one side of the body was mapped on the opposite parietal lobe.[53] Using cats, Adrian now found a smaller, second representation of its paws below the primary forepaw area.[54]

Adrian's second projection zone was quickly confirmed with dogs, monkeys, pigs, sheep, and rabbits, and important details about the new, second representation were now added by others elsewhere.[55] Scientists were soon talking about a large parietal lobe region depicting the body upside-down and facing forward, and a smaller area below it showing the animal lying flat on its back with its head toward the front. These two distinct regions became known as the first and second somatosensory areas. Within a few years of Adrian's discovery, descriptions of multiple projection areas for vision and hearing also appeared in the literature.

Although Adrian did not study the effects of damaging somatosensory areas I and II separately or together, he believed that each area must have a different function.

When he first discovered the second area, he thought it could have something to do with the claws. His theory lost currency after other scientists found a second facial region nearby, and when research on a variety of nonclawed animals revealed that they too have a second somatosensory area. Another theory now to draw attention suggested that the larger area may be involved with finer aspects of tactile discrimination than the second area. A third was that the first area is concerned with active sensorimotor exploration (haptics, palpation), whereas the second area processes tactile stimuli received passively by the organism, such as being brushed on the skin.

Even after years of study with a variety of techniques, the functions of the multiple representations are still being debated. The more that is learned, however, the more scientists are convinced that Adrian was right when he theorized that each cortical representation probably does a different thing.

Differences Across Species

After having discovered the second somatosensory area in the cat, Adrian carefully mapped the brain surface in a wide variety of animals, some of which had not been studied before. In addition to the usual cats, dogs, monkeys, and rabbits, he presented maps of the body projection areas in pigs, goats, sheep, and even a Shetland pony.[56] Adrian continued to use natural stimuli for much of this work. He tried to activate units by rubbing the skin, stroking with a feather or a glass rod, and pressing the skin with his own fingers.

The cortical maps showed certain common features. One is that each hemisphere receives most of its projections from the opposite side of the body; another is that the body surface always appears upside-down and facing foreward in the primary sensory cortex. But more interesting to Adrian were some of the differences that appeared across species.

He found that the size of the cortical representation for a given body part depends on how important that part is for each species, as revealed by how it behaves its natural environment. As a result, the cortical representation of the entire body surface can seem wildly out of proportion, but in different ways for different organisms.

For example, when he mapped the somatosensory cortex in monkeys, he found that the hands and lips occupied almost the entire sensory region. In contrast, the pig's snout occupies almost all of its somatic cortex (Fig. 15.7). As for the Shetland pony, the nose region has a cortical territory equal to all of the other body parts combined, even though the nose accounts for only a small fraction of the pony's real body surface.

Adrian came to believe he could predict certain features of an animal's brain physiology just by knowing how it behaves in the wild. Alternatively, by studying an animal's physiology, he could better understand why it behaves the way it does. The brain-behavior correlations were fun to think about and even more exciting to confirm.

Olfaction

Adrian's last experiments were conducted late in the 1950s and were devoted to the sense of smell.[57] He gave two reasons for studying olfaction. First, it was a relatively

FIGURE 15.7.

The cortical territory occupied by nerves from the pig's snout. (From Adrian, 1943; courtesy of Oxford University Press.)

neglected area. Second, he thought that smell must be a very important feature of animal life, because the olfactory organs are huge in many species.

He initially approached olfaction by trying to learn something about the olfactory world of organisms as varied as the hedgehog, the rabbit, and the cat. He then presented different odorants to his animals and recorded their olfactory activity. As expected, marked differences appeared across species. His rabbits, for example, were found to be very sensitive to grassy and fruity smells, whereas the cats did not respond to flowers but had olfactory systems that were activated by foul smells and the odors of other animals. Adrian commented that stimuli that may seem immaterial to us can act like blinding lights or loud alarms to other animals.

He next asked whether the olfactory system is made up of specific olfactory cells, each programmed to respond to just one odor, or whether the situation is decidedly more complex. When he recorded from the cells leaving the olfactory bulb to the brain, he found evidence for some specificity, but only with minimal stimulation. As he steadily increased the strength of an odorant above the threshold level, more and more neurons responded to it, although not with equal vigor.

These findings strongly suggested that olfaction depends on specificity within the system *and* on patterns of activity across neural populations. Adrian later speculated that if smells could be separated into fifty categories by the specific receptors involved, five more categories by the part of the olfactory epithelium affected, and another five by the pattern of discharge, the number of different smells that an animal could distinguish would be absolutely enormous.[58]

Administrative Demands and Final Years

Adrian was not one to shun administrative duties. He served as vice chancellor of Cambridge University, Master of Trinity College, and even as a Baron in the House of Lords. His additional responsibilities included acting as Chancellor of Leicester University, as President of the Royal Society of Medicine, and as a trustee for Rockefeller University. These responsibilities increasingly pulled him away from the laboratory, as did his steadily declining stamina for the lengthy recording experiments he loved to conduct.

If Adrian had ideas of doing more laboratory work, they were dashed when his laboratory was flooded early in the 1960s. With his equipment damaged beyond repair, he continued to channel his remaining energies in other ways. In particular, he devoted even more time to administration, national science policy, and the lecture circuit. As a spokesman or informal lobbyist for science, he had impeccable credentials. Moreover, few could match the great enthusiasm he showed for the future of the brain sciences. When Edgar Adrian died in 1977, the scientific community knew that it had lost not just a gifted spokesman but a bright and creative guiding light.

16

Otto Loewi and Henry Dale
The Discovery of Neurotransmitters

Who would have thought, years ago, that nervous stimulation influences the organs by releasing chemical substances, and that by such means the propagation of impulses from one neurone to another is effected?

—*Otto Loewi, 1935*

By reviewing Dale's contributions it becomes apparent that while Loewi had been able to bring the "lady" of neurochemical transmission theory into public view, it was Sir Henry Dale who succeeded in making her a respectable companion acceptable to the general scientific community.

—*G. R. Fick, 1987*

Waves and Sparks

At the end of the nineteenth century, brain scientists were pondering how one neuron can communicate with another and how a neuron can make a muscle contract. The prevailing idea was that the nerve impulse must be transmitted in virtually the same way as an electrical charge shoots down a nerve—by an unbroken electrical "wave." Even if bridges or physical connections did not exist between cells, most investigators were still willing to assume that tiny electrical sparks from nerve endings will jump any small gap separating neurons from each other or from muscles.

A few scientists, however, were not entirely comfortable with these notions. These individuals wondered about the imagined electrical waves and if there really had to be an unbroken electrical event for one cell to communicate with another. One of the biggest problems with the electrical wave theory, they repeatedly pointed out, is that it cannot explain why neural transmission occurs in only one direction.[1]

In the opening decades of the twentieth century, there would be new ways of thinking about how neurons communicate with each other and how they stimulate muscles. The earlier idea of a continuous, unbroken electrical wave shooting from one neuron to another would be challenged by the notion of chemicals, called neurotransmitters being released at synapses.

What led scientists to conclude that axon endings liberate chemicals to stimulate or inhibit other cells? And what were some of the early ramifications of this important discovery? The answers to these questions can be appreciated only by looking at the lives of two very different twentieth-century scientists, Otto Loewi and Henry Dale.

The Hypothesis of Chemical Transmission

In 1877, twenty years before Charles Sherrington even coined the term *synapse,* the German physiologist Emil Du Bois-Reymond (Fig. 16.1) hypothesized that nerves can excite muscles either electrically, which was the prevailing opinion, or chemically, a radically different concept. Specifically, he proposed that chemicals released from nerve endings might make muscles contract:

> Of known natural processes that might pass on excitation, only two are, in my opinion, worth talking about. Either there exists at the boundary of the contractile substance a stimulative secretion in the form of a thin layer of ammonia, lactic acid, or some other powerful stimulatory substance, or the phenomenon is electrical in nature.[2]

Scientists, however, failed to respond to what Du Bois-Reymond had written. Most physiologists were completely unaware of his chemical idea because it was buried in two massive volumes of German text. Others read the passage but did not give it a second thought because it was not backed by solid evidence. Du Bois-Reymond did not help matters by presenting alternatives and not following up on his new idea.

As a result, the revolutionary idea of chemical transmission gathered dust for almost three decades. When the theory finally surfaced again, Du Bois-Reymond's name was not even associated with it. In fact, the British researchers who were now entertaining the chemical transmission idea were completely unaware that they were actually resurrecting a concept proposed years earlier by one of the leading neurophysiologists of the time.

FIGURE 16.1.

Emil Du Bois-Reymond (1818–1896). In 1877 he suggested that nerves might communicate by releasing chemical agents.

Adrenalin and the Autonomic Nervous System

Many of the scientists who figured prominently in the discovery of chemical transmission were first interested in the physiology of the autonomic nervous system. They knew the autonomic nervous system regulates the internal functions of the body, including heart rate, breathing, and digestion. Thanks largely to Walter H. Gaskell and John Newport Langley, Sherrington's mentors who were still teaching and doing research at Cambridge, they also knew that the autonomic nervous system could be divided into two parts, the sympathetic and parasympathetic divisions.

Considerable research indicated that electrical stimulation of the sympathetic nerves can speed up the heart, increase respiration, and divert oxygen-rich blood to the skeletal muscles. The sympathetic nervous system responds as if designed for flight-or-fight situations. In contrast, the literature was also showing that the parasympathetic nervous system plays a key role in digestion and aids in other activities that restore the vital resources of the body.

Another agreed-upon observation was that when both sympathetic and parasympathetic nerves go to the same organ, their actions oppose one another. For example, electrically stimulating the vagus nerves of the parasympathetic system slows the actions of the heart and increases movements in the gastrointestinal tract. In contrast, electrically stimulating the sympathetic accelerator nerve causes the heart to beat faster, while slowing digestive actions.

In the final years of the nineteenth century, many drug studies began to be conducted on the autonomic nervous system. Most were carried out by pharmacologists and physiologists alone or in small groups at universities. Some, however, were performed by practicing physicians and inquisitive amateurs.

One of the most significant discoveries began with some personal observations by George Oliver, an English physician. Oliver had a penchant for inventing simple instruments and testing them on himself and family members.[3] He tried to invent an instrument for measuring the diameter of an artery under the skin. To test the sensitivity of his new device, he administered extracts from various animal glands to his young son and recorded changes in his arteries. To his surprise, injecting adrenal gland extract caused a large artery from which he was recording to narrow dramatically, raising his blood pressure.

Oliver was so excited by the adrenal extract reaction that he went to London to tell Professor Edward Schäfer about it. He found him hard at work recording the blood pressure of a dog. Oliver urged Schäfer to inject some of the adrenal extract that he had brought along into the dog's veins. As soon as his own experiment was finished, Schäfer complied. Within minutes, he was amazed to see the mercury shoot to the top of the instrument he was using to measure blood pressure.

Oliver and Schäfer continued to study the actions of adrenal extract on dogs and frogs. They found many similarities between the effects of injecting the adrenal substance and stimulating the sympathetic nerves with electricity. Their thoughts and observations were published in the *Journal of Physiology* in 1894 and 1895.[4]

Oliver and Schäfer's findings were soon expanded by Russian neuropsychiatrist Max Lewandowsky and by Langley himself.[5] They showed that injections of adrenal extracts, like stimulation of the sympathetic nerves, can have many different actions. For example, both can cause salivation and widening of the pupil of the eye.

Neither the Russian nor the Cambridge physiologist, however, was ready to declare that the endings of the sympathetic nerves actually release an adrenal-like substance to regulate the smooth muscles of the body.

Langley, however, was intrigued enough to ask one of his students to try to learn more about the effects of the adrenal substance.[6] Thomas Renton Elliott published some of his findings in 1905.[7] He wrote that adrenaline stimulates the sympathetic nerves to the smooth muscles in a manner directly comparable to electrical stimulation. He added that a positive response to adrenaline can be used as a biological assay for identifying the nerves of the sympathetic nervous system, since other nerves are not activated by it.

Elliot said nothing about chemical transmission at the synapse in his 1905 paper. Nevertheless, he did address this issue in a talk before the Physiological Society in London in 1904.[8] He stated that the sympathetic nerves could well be producing their effects by releasing adrenaline or an adrenaline-like substance. The last sentence of the published abstract of his presentation said it all: "Adrenalin might then be the substance liberated when the nervous stimulus arrives at the periphery."[9]

A Second Transmitter?

Soon after Elliott placed his adrenaline hypothesis on the table, a parallel statement was made for the nerve endings of the opposing parasympathetic system.[10] In 1907 Walter Dixon, a pharmacologist at Cambridge, electrically stimulated the vagus nerves in a frog. He found that vagal activation inhibited the heart, confirming more than 150 years' worth of earlier experimentation. Dixon then removed the heart and partially purified an extract taken from it. When the extract was applied to another frog's heart, its beating slowed substantially. Dixon further observed that this inhibitory effect could be blocked with the drug atropine.

Dixon found considerably more of a substance resembling muscarine, an extract from the mushroom *Amanita muscaria,* in the fluid from the inhibited heart than in the fluid from a control heart. Since pharmacologists had known for years that administering muscarine can inhibit the heart, and that muscarine itself can be blocked by atropine, Dixon quickly put two and two together. He went well beyond Elliott when he postulated that a muscarinelike transmitter substance not only exists, but is safely stored in certain nerve endings until set free by neural activation. Drugs that slow the heart, Dixon thought, probably act by causing the vagus nerves to release more of this muscarinic agent.

Dixon's and Elliott's publications on adrenaline and muscarine were not particularly well received. Many scientists found their reports lacking in quantitative detail, unconvincing, and in conflict with their own pet theories. Dixon, in particular, was discouraged by the widespread skepticism. He became hesitant about conducting other experiments, even though he was remarkably close to the truth.

One person who was impressed by these experiments was Henry Hallett Dale (Fig. 16.2). He discussed the possibility of chemical transmission with Elliott and Dixon, both friends of his at Cambridge. More than half a century later, in his eulogy to Elliott, he would write:

> To me, however, it always seemed quite clear, that the earliest public and
> specific suggestion of a transmission from nerve-endings of excitatory or in-

FIGURE 16.2.

Sir Henry Hallett Dale (1875–1961). (From W. Haymaker and F. Schiller (Eds.), Founders of Neurology. *Springfield, IL: Charles C Thomas, 1970; courtesy of Charles C Thomas.)*

hibitory stimuli, by the liberation there of a chemical transmitter natural to the animal body, was that made and published by Elliott in 1904. And, for my own part, I am convinced that any contribution which I have been able to make in later years to the general development of that theme, owed its initial stimulus, and a continuing inspiration, to Elliott's published researches and ideas, and to my personal contacts with him in those early days.[11]

Henry Dale

Henry Dale was born in London in 1875 and began his schooling in Cambridge in 1894.[12] With a talent for the natural sciences, he turned to physiology and was introduced to the challenges of the autonomic nervous system by Walter Gaskell and John Newport Langley, the same two scientists who figured so prominently in Sherrington's life. Dale enjoyed Gaskell's lectures but, in his autobiography pointed out that it was the opportunity to work under Langley that really whetted his appetite for physiology and pharmacology. By studying physiology at the right place, at the right time, and with the right person, Dale was exposed to many groundbreaking discoveries and a plethora of new ideas.

In 1900 Dale left Cambridge for St. Bartholomew's Hospital in London to complete his medical training. Two years later he faced a major decision. Should he become a practicing physician or should he accept a stipend to do physiological research? After pondering the alternatives, he opted for the research position at University College, London. There he was given a small room and minimal equipment. But once again he had the opportunity to work with outstanding scientists, among whom were Ernest Starling and William Bayliss, whose studies on internal secretions were hailed far and wide.

After spending a summer at University College burdened with a heavy teaching load, Dale began to question his career choice. Fortunately, Henry Wellcome stepped into his life in 1904 and offered him a job at the Wellcome Physiological Research Laboratories. Dale was assured that he could pursue an active career in pharmacology without having to worry about the finances required to support a family. To understand more fully what Henry Wellcome was offering, a few words are in order about Burroughs Wellcome and Co.[13]

Henry Wellcome and Silas K. Burroughs, two American pharmacologists who wanted to introduce "compressed medicines" (tablets) in Europe, founded Burroughs Wellcome and Co. in London in 1880. Fourteen years later, a private research laboratory was set up as a somewhat separate entity to produce diphtheria antitoxin on a not-for-profit basis. Wellcome expected that the laboratory would also become involved in other research ventures. Within a few years, and after Burroughs died, this increasingly independent unit of Burroughs Wellcome and Co. became know as the Wellcome Physiological Research Laboratories.

In the opening years of the twentieth century, Wellcome decided to upgrade the staff at his well-equipped physiological laboratories. He searched for scientists with the vision to develop world-class research programs. It was in this context that he went to Dale, who was receptive to the idea of working for a pioneer such as Wellcome. In order to entice Dale to come on board, Wellcome told him that he would not be under pressure to produce results with immediate commercial appeal and that he would have job security. In addition, there would not be any administrative functions to bog him down.

As sweet as Wellcome's offer was, some of Dale's friends urged him not to sell his soul to the devil, meaning industry. University life was not perfect, that much was sure, but it did offer certain freedoms and a degree of stimulation that could not be matched by the money-driven corporate world. Dale listened to what his well-meaning friends had to say about the negatives of the private sector but, after evaluating the pros and the cons, still accepted Henry Wellcome's generous offer. Many years later Dale would write: "I never had serious or lasting reason to regret the change which I had made."[14]

From Ergot to Noradrenaline

Dale, now a married man, quickly settled down in his new job. He was determined to show that the word *Research* in the name Wellcome Physiological Research Laboratories could, for the first time, be justified. Fortunately, he took to heart a gentle suggestion made by Wellcome, who thought that it might be worthwhile to do some studies on the physiological properties of ergot, a fungus that grows on rye and which caused epidemics of St. Anthony's fire, a disorder associated with gangrene, convulsions, and intense burning pains in the extremities, during the Middle Ages (Fig. 16.3).

Ergot also causes uterine contractions and, for this reason, had been used in obstetrics for centuries. From Wellcome's perspective, an analysis of ergot not only would be interesting, but could result in better drugs for obstetricians. By initially trying to learn more about the physiology of ergot, which contains a number of chemical substances, and by a series of lucky accidents or coincidences, Dale was led into the most important discoveries of his scientific life.

Dale's interest in ergot stimulated him to publish a review of the effects of the

FIGURE 16.3.

Rye and wheat (far right) infected with horns of ergot fungus (Claviceps purpura). Dale's research on the physiological actions of ergot led him to chemical transmission in the nervous system.

drug in 1906.[15] He noted that an extract from the ergot fungus could block the stimulative effects of adrenaline. He also wrote that the ergot extract could block the effects of directly stimulating the sympathetic nerves.

A few years after writing his review, Dale and George Barger, a chemist who synthesized many of the compounds studied in the Wellcome Research Laboratories, showed that a drug closely related to adrenaline has even stronger stimulating effects than adrenaline itself.[16] This drug, which stood out from among some fifty chemically related amines, would become known as noradrenaline or norepinephrine.

Dale and Barger knew they were on to something important, but in 1910, the year of their paper, they were not quite ready to promote the idea that noradrenaline, and not adrenaline itself, may be the real chemical transmitter. The two investigators felt that their hands were tied because the drug was still only a synthetic laboratory curiosity. Until it could be isolated from the body, speculations about a role for noradrenaline in chemical transmission would best be avoided.

Acetylcholine

The synthetic drug that attracted Dale's attention even more than noradrenaline during his tenure at the Wellcome Physiological Research Laboratories, as well as afterward, was acetylcholine. First manufactured in 1867, little was known about its physiology until an American pharmacologist by the name of Reid Hunt and his assistant René Taveau drew attention to it in 1906.[17] They found that it was a hundred times more powerful in depressing blood pressure than adrenaline was in raising it, and a hundred thousand times more active than choline itself.

By coincidence, Hunt presented these findings at the same meetings in Toronto in which Dixon presented his vagus nerve studies. Nevertheless, the idea that Dixon's muscarinic transmitter could be Hunt's acetylcholine did not register on anyone's mind. It was as if two ships had passed in the night.

Dale's entree into the world of acetylcholine also came out of his interest in ergot. In 1913 he found that he could inhibit the heart of a cat by intravenous injection of another of the laboratory's ergot extracts. Additional tests showed that this ergot extract had powerful effects on other internal organs as well. All of this led Dale to hypothesize that the ergot fungus may perhaps contain muscarine or a muscarinelike compound, the chemical Dixon believed may be liberated when the vagus nerves are stimulated.

Dale changed his mind about the presence of muscarine in the ergot extract after the active ingredient was isolated by his colleague Arthur Ewins.[18] The isolated substance resembled muscarine, but it was less stable and not as long-lasting as muscarine itself. The investigators then recalled what Reid Hunt had said about the depressive cardiac actions of synthetic acetylcholine. Could their mysterious substance be natural acetylcholine? This thought prompted Ewins to synthesize some acetylcholine and to compare its actions to those of the powerful ergot extract. Physiologically and chemically they proved to be identical.

Dale next compared the effects of acetylcholine with the effects of related drug compounds.[19] In 1914 he reported that acetylcholine is by far the drug best able to mimic the natural activation of the parasympathetic nervous system. It could inhibit the heart, increase salivation, and cause intestinal contractions like those that take place during digestion.

In the same paper, Dale drew attention to two different actions of acetylcholine. On the one hand, he wrote, it causes parasympathetic changes like those produced with injections of muscarine. These muscarinic actions, including cardiac inhibition, can be blocked with the drug atropine. On the other hand, its actions on the ganglia of the autonomic nervous system and at nerve-skeletal muscle synapses closely resemble the effects of low doses of nicotine. These actions are unaffected by atropine but can be blocked with curare, also known as South American Indian arrow poison. Today, scientists still distinguish between the muscarinic and nicotinic actions of acetylcholine.

But Dale was troubled. He was still not dealing with a chemical actually isolated from an animal's body. Because he had found acetylcholine only in ergot, he concluded that it was not yet time to fight for it as a chemical transmitter. He went no further than to speculate that acetylcholine may eventually be found in the body, where it could be broken down very rapidly by another naturally occurring agent.

As we shall see, Dale could not have been more correct in his thinking about acetylcholine. Yet it would be years before strong evidence would be forthcoming to support his thoughts. One reason for the delay was the First World War, which pulled many scientists away from their research to help with the war effort.

Still, some significant milestones had already been reached. Indeed, two possible chemical transmitters were now attracting attention, thanks largely to Elliott, Dixon, and Dale. One was an adrenaline-like compound, which seemed to be associated with activation of the sympathetic nervous system. The second was acetylcholine, which can activate the parasympathetic nervous system and have effects on the skeletal muscles as well.

Was it finally time for the old electrical theory of "sparks" to give way to the theory of "soup" at the synapse? Although the Zeitgeist was changing, many scientists were not yet convinced that there was enough evidence to proclaim a victory for the forces of soup over those of sparks.[20] Stronger and better demonstrations were needed before the fence-sitters would change their minds—at least in print. As Dale himself put it: "Transmission by chemical mediators was like a lady with whom the neurophysiologist was willing to live and to consort in private, but with whom he was reluctant to be seen in public."[21]

Otto Loewi's Formative Years

Forty-four years after Emil Du Bois-Reymond had alluded to chemical transmission, and seventeen years after Elliott had written that chemical transmission must be seriously considered, Otto Loewi (Fig. 16.4) conducted the experiments that convinced many agnostics to support the soup position.[22] Born into a well-to-do Jewish family in Frankfurt am Main, Loewi had an enjoyable childhood in Germany. He attended an old-style Gymnasium for nine years, where he did especially well in the humanities.

Upon graduation in 1891, Loewi hoped to study art history. To his dismay, this upset family members, who encouraged him to be more practical and to consider a career in medicine. Family pressures eventually got the better of him, and Loewi signed up for medical studies at Strassburg University, which was then under German control.

At the university, Loewi learned from excellent teachers in anatomy and the other sciences. Nevertheless, he cut many of his science classes in order to attend lectures in philosophy, architecture, and the arts. In short, he thought like a humanist even though he was being fitted with physician's clothes.

For reasons that Loewi himself could not comprehend, he opted to do his thesis in pharmacology, an area in which he had little training. Notably, he decided to measure the effects of different drugs on the heart, using frogs as experimental subjects. His thesis project was well received, and after graduation he rewarded himself with

FIGURE 16.4.

*Otto Loewi
(1873–1961).*

a trip to Italy. This allowed him to visit outstanding art museums and gave him the time he needed to decide whether to compete for a research position at a university or to become a practicing physician. Before coming home, he made up his mind to try working in a hospital, to see what the life of a clinician would be like.

After a short stint at a Frankfurt hospital, during which he realized how dangerous and depressing it can be to deal with infected patients, he said goodbye to clinical medicine. Fortunately, there was an opening for an assistant pharmacologist in Marburg at the time. He applied and was offered the research position. He now began to study protein synthesis and metabolism.

By the turn of the century, Germany had forfeited some of its medical leadership to England. This was not lost on Loewi, who took the opportunity to visit some of England's leading institutions in 1902. His objectives were to pick up fresh ideas and to learn new methods. True to form, Loewi stopped in Holland for a week to see some of the great Dutch art collections before even crossing the English Channel.

Once in England, Loewi visited University College, where he met Henry Dale. The two men found that they had much in common and quickly became close friends. Loewi also visited Cambridge University, where he spoke with Langley and learned about some of the work that was being done on the reactions of the autonomic nervous system to drugs. Although he was introduced to Elliott, who was still a graduate student at the time, there is no evidence to show that the two men sat down and discussed the idea of chemical transmission at the synapse in 1902. Indeed, two more years would pass before Elliott would make his first public statement about adrenaline being the likely transmitter in the sympathetic nervous system.

Loewi returned to Marburg with admiration and warm feelings for the British scientists he had met. Two years later he moved to Vienna, the city of music and art. There, when not in the museums or concert halls, he began to devote himself to the study of the autonomic nervous system. Like his counterparts in London and Cambridge, he examined how salivation, blood pressure, and other internal functions are affected by various drugs.[23]

In 1909 he moved yet again. This time he accepted the chair of pharmacology in Graz, Austria's second-largest city. He now returned to his pharmacological research on metabolism and took on a heavier teaching load. And so the years passed until 1921, when Loewi conducted his landmark experiments on chemical transmission at the synapse.[24]

The Critical Experiments

The story of how Otto Loewi, then forty-eight years old, came up with the idea for his classic experiments on neurotransmitters has become folklore to researchers in the fields of pharmacology and neurophysiology. Loewi tells us in his autobiography that he was a restless sleeper, a man who suffered from insomnia and tended to awaken throughout the night.[25] One night he awoke with the idea for the critical experiments. Like Dixon, he reasoned that if the vagus nerves inhibit the heart by liberating a chemical, the substance might diffuse into a solution in contact with the heart. The presence of this substance then could be demonstrated by putting the fluid in contact with another heart. Loewi quickly scribbled down what he had to do the next morning and went back to sleep.

When he woke up at six o'clock, he was dumbfounded. He knew he had written something very important, but he was unable to read his own handwriting on the tiny piece of paper. As a result, he had absolutely no idea what to do! His memory of what he had written failed to return throughout the frantic day and into the night. He finally went to bed exhausted.

To his surprise, he awoke at three o'clock in the morning with the very same idea for new experiments. This time he took no chances. He got up and dashed off to his laboratory to perform the necessary research. By five o'clock in the morning, less than two hours after running out of the house, the experiments on chemical transmission were completed.

There are also other, less picturesque versions of the story. According to Dale, Loewi told him that he went back to bed after writing himself a more legible second note.[26] Early the next morning he supposedly picked up the paper, read it, and cheerfully went off to the laboratory. But whether Loewi ran in haste to the laboratory at three o'clock in the morning or first went back to sleep does not really matter. What is important is that Loewi suddenly realized that he had discovered a way to test the theory of chemical transmission. In his mind, he had formulated some experiments that could be conducted very quickly—simple studies that could topple the older theory that conduction across the neuromuscular synapse is due to an electrical wave spreading or a spark jumping from nerve to muscle.

Once in his laboratory, Loewi took a frog, removed its heart, and bathed it in neutral Ringer's solution. He then stimulated the vagus nerve that was still attached to the heart, slowing the frog's heartbeat. At the height of the inhibitory response, Loewi collected some of the Ringer's solution from the dissected frog's heart. He then transferred it to a chamber surrounding the isolated heart of the second frog, whose own vagus nerves were removed as an important control. To his delight, the second frog's heartbeat immediately became slower, just as if he had stimulated a vagus nerve. Because this effect did not occur when solution from an unstimulated frog's heart was used, Loewi deduced that a chemical liberated by the stimulated vagus nerve had to be responsible for the inhibition.

He then performed a second set of experiments to determine if the nerves of the sympathetic nervous system, which accelerate the heart, also function chemically. This time he isolated a frog's heart and stimulated its accelerator nerve. As expected, stimulating this nerve had the opposite effect of stimulating a vagus nerve. Again he collected some of the solution and, after cutting the sympathetic nerves to the heart in another frog, bathed the second frog's heart in it. As soon as the fluid was applied, the second frog's heartbeat accelerated (Fig. 16.5).

For the first time, Loewi understood how the nerve impulse, an excitatory event, can lead to inhibition. Before a chemical transmitter entered the equation, inhibition seemed impossible to explain—only excitation made sense. But now that he had convincing evidence that a chemical intermediary is involved, it was not hard to envision inhibitory events at the synapse. In his own words:

> There remains, to my mind, no possibility whatever of imagining how nerve stimulation can inhibit an organ otherwise than by humoral means. In other words, the humoral mechanism presents the only conceivable mechanism of peripheral inhibition.[27]

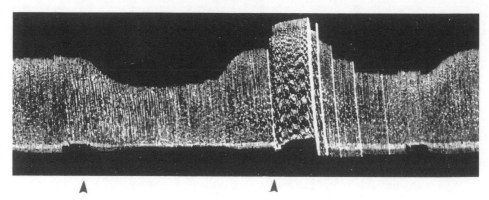

As far as many members of the scientific community were concerned, Loewi provided real proof for chemical transmission, which he called "neuro-humoral transmission," at the synapse. Before he published his experiments, as later stated by Dale, "none among those of us who had been taking a particular interest in such speculations, had been able to think of a method by which they could be put to the practical test of experiment."[28] As we have seen, Dixon came closest, but in Dale's mind it was Otto Loewi who really introduced the "shady lady" of neurotransmission into proper society.

Because Loewi was unsure about the identities of his two different humors, he simply called the inhibiting agent from the vagus nerves *Vagusstoff* (vagus substance) and the accelerating agent from the sympathetic nerves *Acceleransstoff* (acceleration substance). At the first opportunity, Dale asked Loewi why he did not just state that his *Vagusstoff* could, in fact, be acetylcholine, and that his *Acceleransstoff* could be adrenaline or at least an adrenaline-like compound.[29] Loewi responded that the substances probably are acetylcholine and adrenaline. But, he explained, he preferred to be cautious because he had recently been criticized for speculating too freely.

Some thirty years later, Loewi confessed that he never would have gone to the laboratory to conduct these experiments had he taken the time to consider what he intended to do. Instead, he said, he almost certainly would have concluded that not enough chemical substance could be released for it to be detected by applying fluid from one heart to a second heart. As far as he was concerned, fortune smiled upon him when he acted before thinking. "Sometimes," he sighed, "in order to become a discoverer one has to be a naïve ignoramus!"[30]

Loewi remained interested in dreaming and how ideas can be stored in the unconscious mind long after the experiments of 1921 brought him great fame. His side interest took him to Vienna in 1936, where he visited with Sigmund Freud. A few years later, when both men were in London, Loewi again visited Freud, not just to pay his respects to the dying father of psychoanalysis, but to talk to him about dreaming, the unconscious, and the psychological bases of discovery.

Even in one of his last public lectures, Loewi continued to address the role of dreaming, citing his own work and how the organic chemist Friedrich August Kekulé—after dreaming about a snake catching its tail—realized that benzene must have the structure of a ring.[31] As a cultured man with a sense of humor, Loewi might have thought about Shakespeare and changing Prospero's famous words in the last act of *The Tempest* to read: "*Science* is such stuff as dreams are made on."

Loewi Forges Ahead

Loewi continued to work on neurotransmitters after publishing his monumental findings in 1921 and 1922 and then replicating his experiments in front of several groups. Some of his new work concerned the drug atropine, which can block the effects of acetylcholine (his *Vagusstoff*). Prior to this time, it was believed that atropine worked by blocking nerve endings. Loewi, however, showed that the vagus nerve does not become paralyzed when atropine is administered. In fact, the vagus releases just as much *Vagusstoff* after atropine as it does in the absence of atropine. This important observation led him to suggest correctly that atropine affects only the released transmitter substance, not the release process itself.

Loewi now made another significant contribution to neurophysiology. Knowing that atropine can block the substance normally liberated by the parasympathetic nerve endings, he wondered whether there might be a naturally occurring substance in the body that can do the same thing. Indeed, for years he was perplexed by the fact that transmitter actions are very short lived. An enzyme that could break down *Vagusstoff* at the synapse, he reasoned, could certainly explain this phenomenon. Dale had also briefly entertained this possibility back in 1914.

In 1926 Loewi and a coworker, working once again with frogs, found that there is, in fact, a natural enzyme that breaks down *Vagusstoff* after it is released into the synapse.[32] Some additional experiments strongly supported this conclusion.[33] Because *Vagusstoff* appeared to be an ester of choline, the esterase (an enzyme that acts on a specific ester) was dubbed *cholinesterase*.

Loewi made yet another important discovery. He showed that acetylcholine can be made longer-lasting by introducing an agent that can block the destructive actions of cholinesterase—an *anticholinesterase*. In this phase of his work, he turned to physostigmine, an alkaloid extracted from the calabar bean, which grows on large, climbing vines in western Africa (Fig. 16.6).

The use of the calabar bean in rituals in parts of Africa had been known to Europeans for some time.[34] Explorers in precolonial and early colonial times reported that inhabitants of the Calabar region (now a part of Nigeria) used pulverized beans mixed with water for executions, as well as for gruesome witchcraft trials in which survival as a result of vomiting signified innocence. The word they used for the ritual swallowing of the mixture was *esere*, from which physostigmine's other name, *eserine*, was derived.

Neither the Africans nor the nineteenth-century scientists who began to cultivate the "ordeal beans" in Edinburgh, Scotland, knew the beans were stopping the heart by preventing the breakdown of acetylcholine from the vagus nerves. These scientists, however, succeeded in isolating the active ingredient from their beans. From trials on animals and testing on themselves, sometimes with near-fatal consequences, they discovered that even a tiny dose of the drug could stop the heart and the lungs.

To Loewi, the new experiments he performed on cholinesterase and anticholinesterase drugs seemed every bit as significant as his heart fluid transfer exper-

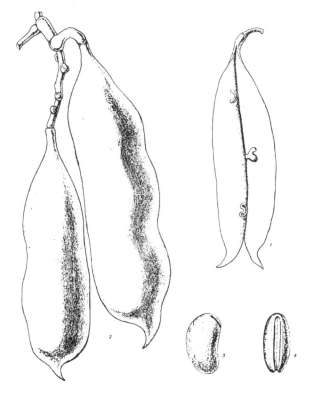

FIGURE 16.6.

*The calabar bean,
from which
physostigmine
(eserine),
a powerful
anticholinesterase
agent, was first
extracted. (From
Balfour, 1861.)*

iments from 1921. Nevertheless, he probably did not realize that by increasing the quantity of acetylcholine at the synapse, physostigmine would soon be used clinically. In addition, he probably did not envision how it would help scientists detect minute amounts of acetylcholine at skeletal muscle synapses, another important development.

Dale Returns to Acetylcholine

Loewi's discoveries were of more than passing interest to Henry Dale, who was now working at the newly constituted National Institute for Medical Research in Hampstead, a suburb of London. Even during the tumultuous war years, he never lost his interest in the possibility of chemical transmission at the synapse.

Yet Dale felt that something was still missing from the picture his cultured friend was painting. Specifically, he was bothered by the fact that acetylcholine still had not been isolated from animal tissue. Equally disheartening, no cholinesterase of any sort had been isolated from animal tissue. Thus, in the wake of Loewi's exciting new findings, he could do no more than speculate about Loewi's *Vagusstoff* being his own acetylcholine.

The situation changed for the better in 1929.[35] Dale and chemist Harold Dudley went to a local slaughterhouse and collected spleens from horses and oxen that had just been killed. They then minced the spleens, soaked the material in alcohol, filtered it, and manipulated it in other ways. Although seventy-one pounds of minced horse spleen yielded just one-third of a gram of acetylcholine, Dale and Dudley were thrilled. For the very first time, acetylcholine had actually been obtained from an animal organ.

Dale's new work on acetylcholine led Otto Loewi to abandon his cautious position about *Vagusstoff*. It took a while, but Loewi finally felt comfortable concluding that his English friend had had it right all along—*Vagusstoff* must be acetylcholine after all.

Loewi, however, still remained wary about calling the sympathetic transmitter adrenaline. In his own words:

> In spite of all analogies, however, and although personally I am convinced of the identity, I do not feel justified as yet in assuming that the sympathetic transmitter is adrenaline, and I will therefore [continue to] call it "the adrenaline-like substance.[36]

In the case of adrenaline, Loewi's caution was justified, especially when generalizing from frogs to mammals. During the 1930s, Walter Cannon and his associates at Harvard studied the sympathetic transmitter in cats and other mammals.[37] They found that the effects of injecting adrenaline and the effects of directly stimulating the sympathetic nerves are *not* perfectly identical. In 1946 Ulf von Euler, a Swedish scientist, finally succeeded in extracting the elusive transmitter directly from the sympathetic nerves of mammals. He reported that the active chemical agent is, in fact noradrenaline (norepinephrine), a substance very closely related to adrenaline, but not the same as it.

Histamine

In the same publication in which they isolated acetylcholine from the spleen, Dale and Dudley reported that histamine could also be found in the spleen. The search for a bodily source of histamine, also extracted previously from ergot, was actually the stated purpose for conducting these experiments. One of Dale's objectives from the time of the First World War was to understand the physiology of shock,[38] and both Dale's group and others capitalized on the histamine breakthrough by showing that its natural release plays a major role in allergic phenomena ranging from the sneezing of hay fever to a far more serious condition, anaphylactic shock, involving hypotension and constriction of the bronchioles.[39] The subsequent development of antihistamines to treat or prevent allergic reactions was a natural outcome of this pioneering research.

The Skeletal Muscles

Once Dale felt sure about the presence of acetylcholine in animals, and specifically about its role as a transmitter in the parasympathetic nervous system, he asked his coworkers to help him find evidence for it where the spinal nerves synapse with the skeletal muscles. Interestingly, when Loewi gave a Harvey Lecture in New York in

1933, he told his audience: "Personally, I do not believe in a humoral mechanism existing in the case of striated muscle."[40] He explained that the nervous impulse may jump directly to the skeletal muscles, because the nerves seem to have such intimate connections with them.

Not everyone agreed with Loewi. Charles Sherrington, for example, thought it likely that all synapses work by the same principle.[41] Although Dale also suspected that Loewi was wrong, he knew the issue would remain unresolved until someone found a way to test the hypothesis.

Detecting acetylcholine in the skeletal muscles was a real challenge for investigators at this time. There were the difficulties of only minute amounts of the substance being produced and its rapid breakdown at the synapse. Furthermore, and in contrast to the nerves of the autonomic nervous system, the motor nerves stimulating the skeletal muscles spread out over wide territories, preventing the liberated agent from accumulating in pools or pockets for easier sampling. Additionally, synaptic transmission seemed to be much faster at the nerve-skeletal muscle junction than at the junctions between the autonomic nerves and the internal organs. Obviously, new and more sensitive procedures would be needed for investigators to explore chemical transmission at the skeletal muscle synapse.

The assay Dale wished for was based on Loewi's work on anticholinesterase drugs and was imported from Germany. In 1932 Dale attended a meeting of the German Pharmacological Society in Wiesbaden. There he heard talks by Wilhelm Feldberg and Bruno Minz, who described how administering physostigmine can increase the amount of acetylcholine for capture after nerve stimulation. Expanding on earlier findings by others, they also showed how the contractions of a sensitized leech muscle could serve as a very sensitive biological assay for detecting even low concentrations of acetylcholine.

In 1933, soon after Hitler seized power in Germany, Feldberg, a Jew, was dismissed from his position at the Institute of Physiology in Berlin.[42] When Dale heard about Feldberg's plight, he began a series of inquiries aimed at finding Feldberg and. bringing him to England. His efforts included contacting a representative of the Rockefeller Foundation, who was going to Berlin to help Jewish scientists ousted by the German government find homes elsewhere. Dale told the representative that should he find Feldberg, he must tell him to pack his bags and join him in England.

The rest of the story is best told by Feldberg himself:

A few weeks later, someone told me that the representative of the Rockefeller Foundation was staying in Berlin and that I should try to see him. I succeeded. He was most sympathetic, but said something like this: "You must understand, Feldberg, so many famous scientists have been dismissed whom we must help that it would not be fair to raise any hope of finding a position for a young person like you." Then, more to comfort me, "But at least let me take down your name. One never knows." And when I spelt out my name for him, he hesitated, and said, "I must have heard about you. Let me see." Turning back the pages of his diary, he suddenly said, delighted [in] himself: "Here it is. I have a message for you from Sir Henry Dale whom I met in London about a fortnight ago. Sir Henry told me, if by chance I should meet Feldberg in Berlin, and if he has been dismissed, tell him I want him to come to London to work with me. So you are all right," he said warmly. "There is at least one person I needn't worry about any more."[43]

Soon Feldberg and his family found themselves living in a two-room flat just a short walk from the National Institute for Medical Research in Hampstead. Both Feldberg and Dale were excited by what they could now do, Dale so much so that he ran to the laboratory to join the good-natured Feldberg whenever he could break away from his administrative duties.

By using eserine to prevent acetylcholine from breaking down after nerve stimulation, and by turning to the leech as a biological assay for the transmitter, the researchers soon came forth with strong evidence for the presence of acetylcholine at nerve-skeletal muscle synapses.[44] Their studies looked at the motor nerves to the legs of frogs, cats, and dogs, as well as the nerves to the tongue muscles of the cat.

Once Loewi was able to learn about the accomplishments of the group at Hampstead, he was quick to admit that he had been wrong about the nerves innervating the skeletal muscles. From all indications, these nerves behave just like other nerves. To no one's great surprise, the remarkably insightful Sherrington had been right once again.

The Need for New Words

Some of Feldberg's research in England also shed more light on the autonomic nervous system. Although it seemed clear that an adrenalinelike compound is released by the sympathetic nerves synapsing with the smooth muscles, there were nagging questions about the nerves ending one gap back, in the sympathetic ganglia. By conducting experiments on the nerves entering the sympathetic ganglia, where they synapse with nerves going to the internal organs of the body, Feldberg provided strong evidence for the "preganglionic" transmitter being acetylcholine.[45]

Thus there seemed to be a major biochemical difference between the nerves to the sympathetic ganglia and the nerves that leave these ganglia to innervate the smooth muscles of the body. Although both are part of the sympathetic nervous system, the preganglionic nerve terminals release acetylcholine, whereas the postganglionic terminals liberate an adrenaline-like substance. From this perspective, could scientists still say that the transmitter of the sympathetic nervous system is adrenaline or a closely related agent? Wouldn't this be misleading and confusing?

In addition to this revelation, another complication with the sympathetic nervous system now became clearer. Drug experiments dating back to the turn of the century suggested that a few postganglionic sympathetic nerves may actually liberate acetylcholine, not an adrenaline-like substance.[46] For example, drug studies on the sympathetic nerves to the sweat glands of the cat's foot pad and the human hand suggested that they release acetylcholine. Newer and more direct experiments by Dale, Feldberg, and others at the Hampstead site soon confirmed that stimulation of these atypical sympathetic nerves causes an immediate release of acetylcholine, not an adrenaline-like compound.[47]

For these reasons, Dale felt that new terms had to be introduced—terms that would signify the pharmacological actions of the nerves regardless of their anatomical origins or connections. In 1933, after giving the situation considerable thought, he coined the terms *adrenergic* and *cholinergic* to describe the nerves releasing the adrenaline-like transmitter and acetylcholine, respectively.[48] In his brief note, he explained:

> We seem to need new words which will briefly indicate action by two kinds of chemical transmission, due in the one case to some substance like adrenaline, in the other case to a substance like acetylcholine, so that we may distinguish between chemical function and anatomical origin. I suggest the words "adrenergic" and "cholinergic," respectively, for use in this sense. We can then say that postganglionic parasympathetic fibres are predominantly, and perhaps entirely, "cholinergic," and that postganglionic sympathetic fibres are predominantly, though not entirely, "adrenergic," while some, and probably all, of the preganglionic fibres of the whole autonomic nervous system are "cholinergic."[49]

Thus, the biochemical labels *adrenergic* and *cholinergic* were introduced into the literature. Although they were first used to describe the two types of fibers liberating chemicals in the autonomic nervous system, these terms soon began to be used to describe the axons releasing these two transmitters in other places as well. Dale's terms are still being used today by neuroscientists around the world.

From Laboratory to Clinic

Once the first two neurotransmitters were identified, various diseases of the nervous system began to be thought of in terms of excesses and deficiencies of these chemicals or their enzymes. One of the first disorders perceived in this way was myasthenia gravis, a disease characterized by rapid fatigue and diminished use of the skeletal muscles. A patient suffering from myasthenia gravis may display a drooping jaw, one or two eyelids that cannot be raised (ptosis), indistinct speech, difficulty swallowing, weakness in the limbs, and loss of the ability to hold objects of moderate weight, such as a parcel. As the disease progresses, these patients lose their mobility and eventually find themselves confined to bed.

The idea of increasing the amount of acetylcholine at the neuromuscular synapses of myasthenia gravis patients emerged during the 1930s. Mary Walker, who published two short papers on the subject in 1934 and 1935, is often cited as the first person to treat myasthenia gravis patients successfully with anticholinesterase drugs.[50]

In her first communication, Walker, a quiet senior medical officer at St. Alfege's Hospital in Greenwich, England, explained the logic behind her clinical trial. She reasoned that the response of a skeletal muscle in a myasthenia gravis patient is much like that seen when a neuromuscular junction is rendered inoperative by curare poisoning. Walker deduced from the literature that the paralyzing effects of the South American arrow poison are probably due to an increase in cholinesterase and therefore could be blocked by anticholinesterase agents. From this premise, she thought it would be worthwhile to see if physostigmine, an anticholinesterase drug, could help an advanced myasthenia gravis patient.

Walker's subject was a fifty-six-year-old woman who was no longer able to lift a grocery bag or even hold her head up. Now bedridden, she was injected with a low dose of physostigmine once a day. The results were encouraging, although short-

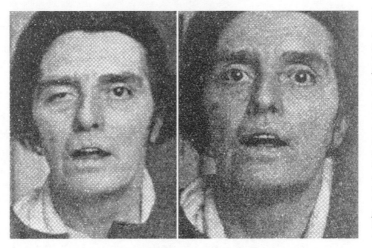

FIGURE 16.7.

A myasthenia gravis patient without physostigmine (left) and thirty minutes after being given the drug (right). These photographs were taken in 1934 by Mary Walker, a pioneer in the use of anticholinesterase drugs. (From Walker, 1934; with permission from Lancet.)

lived: "In from half an hour to an hour after the injection the left eyelid 'goes up,' arm movements are much stronger, the jaw drops rather less, swallowing is improved, and the patient feels 'less heavy.'"[51]

The woman responded even better to a moderate dose of physostigmine, and experienced six to seven hours of relief after a still larger dose. Documenting these changes with photographs (Fig. 16.7), the self-effacing Walker was encouraged. She suggested that anticholinesterase drugs may also be helpful in treating "botulism and cobra poisoning, in both of which a curare-like poisoning of the 'myoneural junctions' of the respiratory muscles has been stated to be the main cause of death."[52]

Walker's second paper was also a single case study. This time it involved a middle-aged woman with myasthenia gravis who was having difficulty keeping her eyelids up and swallowing food. Walker treated her with neostigmine (prostigmine), a synthetic drug like physostigmine, but one with fewer side effects. Within a few minutes she was able to get out of bed and walk a considerable distance without feeling tired. The improvement was greatest an hour after the drug was injected, and the beneficial effects of the drug lasted approximately six hours.

Walker's clinical achievements drew no immediate comments in the British medical journal *Lancet,* where her findings were published. They did, however, catch the attention of Henry Dale. He gave Walker full credit for being the first person to treat myasthenia gravis patients in a way that made biochemical sense.

Years later, scientists learned that a little-known German physician had tried to treat a myasthenia gravis patient with neostigmine two years before Walker performed her "miracle at St. Alfege's."[53] In 1932 Lazar Remen reported that a patient he had treated could stretch his hands, open his eyes better, and eat without choking after an injection of neostigmine. But because Remen was more interested in other things, his comments about the drug were brief (dosage was not even mentioned) and limited to a short paragraph in his thirteen-page paper on a different subject.

For attending physicians, the message from Walker and Remen was that myasthenia gravis patients can be helped, although not yet cured. For the laboratory scientists, their case studies showed that what was being learned about chemical transmission by working with animals had important, practical consequences in medicine. Neurotransmitter "therapy" would now find other applications, the best-known of which will be in the treatment of Parkinson's disease.

Honors and Terror

In 1936 Henry Dale and Otto Loewi, two old friends with common interests, shared the Nobel Prize "for their discoveries relating to chemical transmission of nerve impulses." Loewi, still a lover of the arts, wrote that he was twice moved during the award ceremony.[54] The first occasion was when the trumpeters announced the entrance of the Nobel laureates to the stage and the audience rose with eighty-year-old King Gustav to pay homage to the scientists. The second was after he and Dale received their prizes and the orchestra at the top of the gallery played Beethoven's "Egmont Overture," a call for freedom and an end to oppression.

Loewi, however, had little time to enjoy his award. Nazism was spreading, and he worried about his future because he was Jewish. He had good reason to be fearful. On March 12, 1938, the Nazis marched into Austria—this was their *Anschluss*, or annexation. At three o'clock in the morning Loewi was awakened by a dozen storm troopers and dragged from his home to the city jail. By the end of the day, two of his sons and hundreds of other Jewish males joined him behind bars. Fortunately, Loewi's two other children were out of the country at the time.

The guards did not let their prisoners read or write, and Loewi soon became obsessed with the thought that he would be murdered before he could publish his latest findings. Finally he persuaded a guard to give him a postcard and a pen. This minimal act of kindness enabled him to send a short communication about his latest discovery to a scientific journal.

Fortunately, Loewi was released by his captors two months later, albeit a hundred pounds lighter. His sons left the jail three weeks after he did. All were permitted to leave the country after Loewi agreed to give all of his Nobel Prize money and material possessions to the Nazis. Loewi knew that he had little choice, so he quietly surrendered everything he had and then boarded a ship bound for England.

While in England, Loewi received an offer from New York University. After weighing the alternatives, he decided to emigrate from an uncertain future in England to the safer confines of America. In order to enter the United States, however, he was required to apply for a special visa from the American consulate. He attended to the paperwork but ran into some trouble when the official in London asked him to prove he had been teaching, an indicator he could be employed. He showed the official the letter dismissing him from Graz, but was told that it did not prove that he had taught and could secure a paying job. He then asked the official to call Henry Dale. This was met with the reply that Sir Henry would not want to be bothered by some stranger.

Finally Loewi begged the man to open a copy of *Who's Who* and search for his name. The surprised official found Loewi's biography, read about him, and told him that he could now approve his papers. Loewi, who managed to maintain a sense of humor, asked the consul if he knew who had written the article. When the consul replied that he did not, Loewi informed him that he written it himself and then ran out of the building.

Loewi next took a physical examination. Upon its completion, he was given a sealed envelope by the doctor with the visa he needed for the trip to America. With his papers in order, he booked passage to New York. What happened next can only be told in his own words:

> Upon my arrival in New York harbor, a clerk prepared my papers for the immigration officer. While he was busy doing this, I glanced over the doctor's certificate—and almost fainted. I read: "Senility, not able to earn a living." I saw myself sent to Ellis Island and shipped back to Mr. Hitler. The immigration officer fortunately disregarded the certificate and welcomed me to this country. I arrived here June 1, 1940. On April 1, 1946, I became an American citizen.[55]

Otto Loewi and his wife, Guida, settled into New York remarkably well, given that he was now sixty-seven, had no material possessions, and did not have a close friend to greet them upon their arrival. These drawbacks, however, seemed to be of small consequence to the new immigrants. They felt blessed that every member of their immediate family was able to escape the Nazi terror and almost certain death.

From 1940 to 1955 Loewi continued to work at New York University, mostly on what he called "leftovers from earlier problems." During the summers, he and Guida went to the Marine Biological Station in Woods Hole, Massachusetts, where both enjoyed the scenery and he relished the opportunity to share his experiences and ideas with other scientists. Otto Loewi, who blended humanism and science to live life to its fullest, died in 1961, one year short of his ninetieth birthday.

As for Dale, he too was graced with practically all the honors a man in his profession could receive, including the presidency of the Royal Society from 1940 to 1945. Throughout his life, he remained charming, levelheaded, and helpful to others. To Dale, physiology was like treasure hunting. There could be false leads, but new adventures were guaranteed over each horizon. When he died in 1968, he was eulogized not just for his discoveries about neurotransmitters and their enzymes, but for his wonderful gift of conveying his contagious spirit of adventure to everyone around him.

17

Roger W. Sperry and Rita Levi-Montalcini

From Neural Growth to "Split Brains"

It was a spring day in 1951 when . . . it dawned on me that the tumor acted by releasing a growth factor of unknown nature.

—*Rita Levi-Montalcini, 1992*

Everything that we have seen so far indicates that the surgery has left these people with two separate minds, i.e. with two separate spheres of consciousness. What is experienced in the right hemisphere seems to be entirely outside the realm of awareness of the left hemisphere.

—*Roger W. Sperry, 1964*

Introduction

When Roger W. Sperry died in 1994, brain scientists knew that they had lost a heroic figure in the truest sense of the word. Sperry had questioned the convictions of the scientific establishment throughout his career. Although a rather shy and aloof person, he had a wonderful ability to design critical experiments that demanded looking at the growing and functioning nervous system in new ways.

As a cautious scientist, Sperry strove to make sure his findings were replicable and his theories airtight before making public statements. It was only after the evidence satisfied his own high standards that he took his famous stands, refuting even the theories of the authorities who had trained him. Guided by the firm belief that if the findings do not fit the theory, the theory must go, he devoted himself to the advancement of science.

Sperry's most important discoveries fall into two distinct domains. The first group began in the 1940s and concerned how axons grow to their proper places. Working with newts, frogs, fish, and rats, Sperry cut existing nerves, grafted muscles to unusual locations, damaged tracts, and even turned eyes upside down in their sockets. These studies showed that synapse formation is not haphazard, as others had assumed, but highly structured and incredibly regimented. His findings led him to propose that axons are guided to their targets by chemical attractants.

Early in the 1950s, with his reputation as an outstanding neurobiologist firmly established, he became interested in the role of the corpus callosum, the massive band of axons connecting the right and left cerebral hemispheres. He approached this subject by evaluating the effects of cutting the "great cerebral commissure," first in cats and then in monkeys. He and his students found that one cerebral hemisphere can be oblivious to the information going to the other hemisphere after such surgery. Sper-

FIGURE 17.1.

Roger Sperry
receiving the Nobel
Prize in Physiology
or Medicine
in 1981.
(Courtesy of
Ulf Norrsell,
Göteborg,
Sweden.)

ry and his associates then showed the same to be true for people who underwent this surgery to control severe epilepsy. For newspaper and television reporters hungry for a science story to stir the public imagination, his new findings with "split-brain" patients were a godsend.

In 1981, three years before he retired from the California Institute of Technology in Pasadena, Sperry was awarded the Nobel Prize in Physiology or Medicine (Fig. 17.1). He shared the prestigious award that December with two Harvard vision researchers, David Hubel and Torsten Wiesel. Hubel would later say of Sperry that his "contributions to neurobiology were titanic."[1] Let us now see why almost everyone familiar with Sperry's work knew that Hubel was not just being polite when eulogizing a fellow recipient of the Nobel Prize.

Sperry's Broad Education

Roger Wolcott Sperry was born in Hartford, Connecticut, in 1913.[2] His father was a banker who died young, leaving his widow with the responsibility of raising eleven-year-old Roger and his younger brother. Sperry's mother responded admirably to the tragedy. She secured a job as an assistant principal in a local high school and saw to it that her children were well cared for at home and had good educations.

Roger loved biology from an early age. In grade school he collected moths, and he raised wild animals that he caught in his junior high school and high school years. His other interests included dabbling in the fine arts and reading good literature. His pursuits, however, were not entirely intellectual. He was also gifted in athletics and won positions in three different sports on his high school varsity teams.

After completing his basic schooling, Sperry headed west to Ohio, where he attended Oberlin College on an academic scholarship. In 1935, after finishing a bachelor's degree in English, he remained at Oberlin to pursue a master's degree in experimental psychology. He received it in two years but chose to stay at Oberlin for yet another year in order to prepare for doctoral studies in zoology.

Sperry's new goal was to work under Paul Weiss, a highly regarded zoologist who had begun his career in Germany and was now at the University of Chicago. Weiss was well known for his research on the developing nervous system, and especially for his work on the effects of transplanting limb buds in young amphibians.

Weiss was convinced that nerves grow to muscles in a diffuse and nonselective manner, guided mainly by mechanical or structural factors. Like many other scientists in the 1930s and early 1940s, Weiss was a die-hard functionalist who did not see highly selective neural connections as a prerequisite for a finely tuned nervous system. Instead, the functionalists promoted the view that neurons possess a remarkable capacity to adjust to the characteristics of the individual muscles and sensory organs associated with them.

Sperry hoped to study with Weiss because he found his nerve and muscle transplantation experiments exciting. He also must have looked upon his functionalist theories as plausible. In addition, Weiss offered him the opportunity to continue to work on the neurobiology of movement, a subject that had aroused his interest at Oberlin.

Experiments in Graduate School

In contrast to his new mentor, who worked mostly on frogs, newts, and salamanders, Sperry opted to move higher up the phylogenetic scale for his own experiments at the University of Chicago. His selection of an animal for his research was altogether fitting and predictable for someone first trained in experimental psychology. He opted for the ubiquitous laboratory rat.

In 1940 Sperry published the results of a study he had begun two years earlier.[3] He switched the flexor and extensor muscles of the rat's hind limb to see if his animals could adjust to such a change. A year later he published a related study, one designed to see what would happen after he crossed (switched) the nerves serving the same two hind-limb muscles.[4] A third paper followed in 1942. This time he evaluated the effects of both crossing nerves and transposing muscles in rats.[5]

Would the rats in each of these experimental conditions show bizarre limb movements, or would they eventually adapt and perform normally? The functionalists held that the animals would quickly adapt to the new situations; the expectation was that they would soon be moving about perfectly normally. Sperry, however, was open-minded.

As would be typical of just about all of his experiments, the results of these studies were clear-cut. In every case, the rats exhibited reversed or extremely awkward limb movements. They often extended a hind limb when the situation called for contraction, and they contracted the same limb when extension was required. When these animals tried to move in a particular direction, the result was the opposite. The nerve-muscle changes were a disaster for the animals.

To Sperry's surprise, the rats did not overcome these aberrant behaviors. He found no evidence for functional reorganization, conditioning, or anything like thinking and planning to correct the bizarre movements. The maladaptive locomotion patterns persisted even when he tried to force corrective adjustments.

Postdoctoral Research

After completing his Ph.D. with Weiss in 1941, Sperry worked as a National Research Council fellow at Harvard University under Karl Lashley, the leading American psychobiologist of the era. Lashley was the world's most vocal opponent of pinpoint cortical localization, especially for higher functions such as thinking and memory. He was impressed by how well laboratory animals (usually rats) could still perceive, locomote, and remember how to negotiate mazes after specific lesions of the cerebral cortex. Although his experiments were not at all like the research of Weiss, he too was a functionalist who believed in the ability of the nervous system to adjust to change.

After a year at Harvard, Sperry accompanied Lashley to the Yerkes Laboratories of Primate Biology in Orange Park, Florida. Although the two men conducted some research on brain damage and the sense of smell, Sperry also continued to do experiments like those he had begun under Weiss. For example, he crossed two sensory nerves so that each now innervated the opposite hind leg.[6] When he then gave a mild electrical shock to such a rat's left hind leg, it lifted its right hind leg. As had been the case in the past, the rats did not adjust, even with lengthy recovery periods. Once again the nervous system seemed refractory to change.

Events outside the laboratory helped to persuade Sperry that these findings should be treated as more than laboratory curiosities. During the Second World War, many soldiers and civilians with peripheral nerve damage were undergoing nerve and muscle surgery to restore function. The surgeons usually assumed that pretty much any branch of a motor nerve could be attached to a muscle with beneficial effects. From his own work, coupled with a careful review of the clinical literature, Sperry now recognized that unusual surgical arrangements stood little chance of restoring normal function.

He now launched a crusade to convince neurosurgeons that attaching motor nerves to different muscles might restore some movement, but almost assuredly not proper function. In a lengthy review, he went over his own experiments, other laboratory animal studies, and what he considered the sadly disappointing human clinical literature.[7] He pointed out that poor methods, including superficial examinations, characterized the seemingly successful earlier reports on both animals and people.

Specificity in the Visual System

While working with Lashley early in the 1940s, Sperry read about a strange experimental finding with newts. Researchers had found that after they cut the optic nerves and even removed and replaced eyes in their sockets, new optic nerves formed to restore normal vision. What would happen, he wondered, if he took a newt, cut its optic nerves, and then rotated its eyes 180 degrees? After nerve regeneration, would such a newt see its world upside-down or right side up? And would its regenerating

optic nerve axons twist around to terminate in the same central nervous system locations as before, or would the axons end in a new, upside-down arrangement?

Sperry discovered that his newts acted as if everything were upside-down after he
cut their optic nerves and rotated their eyes.[8] When he presented food above their
eyes, they dipped down for it, and when a tempting morsel appeared below eye
level, they headed up for it.

When he traced the regenerated optic nerves into the brain, he observed that they
went back to their original locations in the optic tectum, the main visual center in
amphibians. The regenerating nerve fibers from the inverted eyes even twisted
around and bypassed vacant synaptic sites to reestablish the same connections as before. There was nothing random or disorderly about the arrangement, nor was it
something easily explained by mechanical guidance factors. Seemingly, and against
all odds, the original projection pattern from eye to brain was maintained, even
though such an end product would inevitably result in starvation.

Additional studies extended these observations to several species of toads and
frogs that were operated on as tadpoles.[9] In some experiments, the two eyes were
even removed and transplanted in the opposite eye sockets. Again and again, the regenerating optic nerves displayed an incredible ability to find their way back to their
original places of termination, producing predictable but highly maladaptive behavioral changes.

After the Second World War ended, Sperry continued to work on the visual sys-
tem.[10] Although he now held the position of assistant professor of anatomy at the
University of Chicago, some of his visual research took place in Bimini, a part of the
British West Indies. There he expanded his eye rotation experiments to fish, and he
also met his future wife, Norma Deupree. The two were married in 1949, when
Sperry was thirty-six years old.

The Chemoaffinity Hypothesis

In 1908, before Sperry was even born, Ross Granville Harrison spent hours in his
laboratory pondering the development of the nervous system.[11] Harrison, an American pioneer in developmental neurobiology and tissue culture work, stood back from
his microscope and asked:

> How can it possibly come about that these interlacing bundles always connect
> with their proper end stations? What are the factors that influence the laying
> down of the nerve paths during embryonic development?[12]

Harrison was never able to answer his own questions in a satisfactory way. Nor
was he alone in his frustration.

As the decades passed, newer generations of scientists never lost interest in Harrison's questions; all agreed that more had to be learned about how the nervous system
could organize itself so exquisitely and predictably. Now it was Sperry's turn to
think about Harrison's questions. His data repeatedly suggested a two-word answer—chemical guidance.

The idea that developing or regenerating axons may be guided by chemical gradients had been suggested before. It was mentioned in the 1880s by Santiago Ramón y
Cajal, the father of modern neuroanatomy (see Chapter 13).[13] He hypothesized that

developing axons start out following the path of least resistance (mechanical influences) and then are guided by chemical attractants released by target cells. But he was never able to support this theory with experiments. Nor was John Newport Langley, the Cambridge physiologist, any more successful.[14]

Although Ross Harrison himself found merit in chemical guidance, the theory languished. One reason was that investigators simply did not know how to test it. A second was that, between 1920 and 1945, when functionalism reigned supreme, few researchers felt any need to give chemical guidance theories serious thought.[15]

Looking back, it is clear that Sperry's initial studies with Weiss were not based on the belief that chemicals from muscles or other nerves must guide growing axons to specific locations. Nevertheless, the more he looked at his findings, the more he realized that prevailing functionalist theories could not explain the facts. Among other things, they could not account for why axons always head back into the same brain areas, even when they have to overcome barriers and have perfectly good vacated synaptic sites to occupy along the way.

When Sperry first began to bring up the idea of chemical guidance, he mentioned it as only one of several possibilities. He wrote about chemical guidance with increasing conviction after he began his work on the visual system. In 1943 he stated that optic nerve fibers leaving different parts of the retina might be distinguished from each other by "intrinsic physico-chemical properties."[16] A year later he used even stronger language, writing that orderly regeneration of the optic nerve "must therefore be regulated by growth factors."[17] On the same page he wrote that his work supported "the supposition that neurons of the central nervous system are specified biochemically in much greater degree than is evident from their morphological variations." And in 1949 he and a student wrote: "The patterning of synapses between sensory and central neurons is tentatively explained in terms of our chemoaffinity theory of synapse formation."[18]

During the 1950s and 1960s, Sperry continued to promote his chemoaffinity theory.[19] For example, when commenting on some of his newer work on optic nerve regeneration in goldfish, he wrote:

> The outgrowing fibers are guided by a kind of chemical touch that leads them along exact pathways in the enormously intricate guidance program that involves millions and perhaps billions of chemically different nerve cells. By selective chemical preferences, the respective nerve fibers are guided correctly to their separate channels at each of the numerous forks of decision points which they encounter as they travel to what is essentially a multiple Y-maze of possible channels. . . . Each fiber in the brain pathways has its own preference for particular prescribed trails by which it locates and connects with certain other neurons that have the appropriate cell flavor.[20]

Sperry never completely closed the door on the roles that may be played by environmental and mechanical factors. These influences, he acknowledged, could affect the structural and functional properties of neurons. But, he always added, their contributions have to be secondary to the more basic, genetically determined chemical matching programs that play the major role in organizing the growing nervous system.

Enter Rita Levi-Montalcini

Because Sperry's experiments were so elegant, he was able to convince many researchers that the time had come for them to abandon their functionalist views of the nervous system for ideas based on nerve specificity. Nevertheless, he did not conduct any biochemical experiments to isolate the growth attractants that he felt had to be present. Other mid-twentieth-century researchers, however, were soon able to prove that chemical growth and guidance factors do, in fact exist. The individual most responsible for this groundbreaking work was Rita Levi-Montalcini (Fig. 17.2).

Levi-Montalcini always had an interest in science. She attended medical school in Turin, Italy, the city in which she was born in 1909.[21] Her mentor in developmental neuroanatomy was a dedicated, ivory-tower academician by the name of Giuseppe Levi (no relation). Under his tutelage, she made a bold decision for a young woman in the 1930s—she would devote herself wholeheartedly to neurobiological research.

Everything was proceeding according to plan until 1938, when Mussolini issued a manifesto barring Jews from academic and professional careers. Because of government-sanctioned anti-Semitism, Levi-Montalcini, who was Jewish, decided to continue her studies at a neurological institute in Brussels. Soon, however, the Germans turned their guns toward Belgium. With dark clouds looming on the horizon, she went back to Italy, hoping the stormy situation would pass.

Upon her return, she continued her experiments, but now in the bedroom of her family home in Turin. After the Allies began to bomb the northern Italian cities, she and her family left for the Piedmont. By collecting chicken eggs and working on a table in their small mountain house, she diligently studied how nerves form, migrate, and differentiate during development.

FIGURE 17.2.

Rita Levi-Montalcini (b. 1909), whose research at Washington University led to the discovery of nerve growth factor, or NGF. (Courtesy of Washington University School of Medicine, St. Louis, MO.)

During the summer of 1943 the Mussolini government collapsed and "Il Duce" was placed under house arrest. The Germans responded by sending troops into Italy to rescue Mussolini and set him up as the titular head of the Republican Fascist Party. With the Nazis and their sympathizers now in control of northern Italy, Levi-Montalcini fled south to Florence. Using a false identity card, she served in the Italian health service for the duration of the war, which left her country in ruins.

Once the fighting ended, she returned to Turin. There she received a letter from Viktor Hamburger, a respected researcher at Washington University in Missouri. Hamburger, a German refugee and dedicated scientist, had read some of the young woman's publications. Because the two had different ideas, he proposed that she come to St. Louis for one or two years, during which time they could combine forces to study how embryonic structures, such as limb buds, influence the development of the nervous system. Levi-Montalcini was thrilled by the offer and sailed from Genoa to the United States. When she arrived in 1947, she had no idea that her next twenty-six years would be spent at Washington University, where Gasser and Erlanger had made their Nobel Prize–winning discoveries in neurophysiology during the 1920s (see Chapter 15).

Nerve Growth Factor

One day Viktor Hamburger walked into the laboratory and handed Levi-Montalcini a letter and a short article. The two were written by Elmer Bueker, who had been one of his students. Bueker explained that he had wanted to determine if developing nerve fibers will grow even faster in the presence of very rapidly growing peripheral tissue. Thus he replaced chick limb buds with mouse tumors and studied axonal growth. Amazingly, one of the tumors, a sarcoma, caused the sensory nerves in the region to grow unnaturally large.[22]

Intrigued by this finding, Levi-Montalcini and Hamburger decided to replicate Bueker's experiments. In 1951 their own paper appeared.[23] They reported that the sensory nerves are not the only structures affected by mouse sarcoma; parts of the sympathetic nervous system also swell in size.

Levi-Montalcini wondered whether these changes could have been triggered by chemical agents released by the tumor cells. She saw some support for this idea in the fact that some of the sympathetic nerves actually forced themselves inside blood vessels—a phenomenon that could be easily explained by tumor agent getting into the circulatory system. She then thought of a way to test this notion.

Her experiment involved implanting tumor cells onto a membrane overlying the chick embryo.[24] As she predicted, the nerve growth effects were comparable to those obtained when the tumor was implanted into the chick's body cavity itself. Some kind of growth factor was obviously making its way into the blood.

Identifying the mysterious agent now became a high priority at the laboratory. For this phase of her work, Levi-Montalcini needed a rapid assay. Therefore, she turned to the tissue culture techniques Ross Harrison had pioneered at the turn of the century. The plan was to expose nerve cells in culture dishes to agents extracted from tumors; within a day or so, she would know if a growth-promoting factor were present. The problem was that expert help was needed to do this sort of work. Remembering that an Italian friend of hers was doing tissue culture work in Rio de Janeiro, she flew to Brazil with a box of mice implanted with sarcomas.

With help, she learned how to place tumor extracts in culture dishes containing nerve cells. The results were spectacular. Sympathetic and sensory nerve cells exploded with a halo of extraordinary new growth, looking like a sun with a dense crown of rays within two days. The rapidly branching neurons in her tissue culture dishes also showed a directional effect, with more nerve growth toward the tumor substance than elsewhere. There was no longer any doubt about a growth factor existing. But could it be isolated? The year was 1954.

Upon her return to St. Louis she was greeted by Viktor Hamburger and introduced to Stanley Cohen, a biochemist hired to help identify the tumor factor. Cohen went to work and soon isolated a large protein substance responsible for the effect. It was given a most appropriate name, nerve growth factor or NGF.[25]

While Cohen was trying to determine the chemistry of the protein by exposing it to snake venom (which contains enzymes that break down nucleic acids), he made a surprising finding. Even minute amounts of the venom induced a denser halo in the nerves exposed to the tumor substance in the culture dishes. He and Levi-Montalcini immediately exposed sympathetic and sensory nerve cells to venom alone.[26] To their astonishment, snake venom was an even more potent source of NGF than mouse sarcoma cells. Since snake venom glands are modified salivary glands, Cohen decided to see if the same factor could be found in mouse salivary glands. The result was a third source of NGF.

These new findings were important for many reasons. But one that immediately registered on all of the people working on the projects was that it now seemed clear that NGF can be produced by healthy, nontumorous cells.[27]

More Experiments

Now using purified NGF, Levi-Montalcini and her coworkers found that the sympathetic and sensory ganglia of young chicks, mice, and rats grow extremely rapidly in its presence (Fig. 17.3).[28] They also observed that nerve cells from human fetuses respond to the purified growth factor. Everything they had done suggested that NGF plays a role in the growth and development of certain parts of the nervous system.

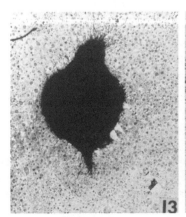 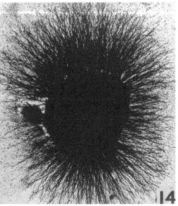

FIGURE 17.3.

Chick embryo sensory ganglia in a control medium (left) and after exposure to NGF (right). (From Levi-Montalcini and Angeletti, 1961; courtesy of the University of Chicago Press.)

Cohen now had a brilliant idea—why not try to make an antiserum to the growth factor to see if it would block sympathetic and sensory nerve development? His antiserum made its debut in 1959. The researchers were soon able to show that early exposure to the antiserum leads to fewer sympathetic and sensory nerve cells.[29]

These new findings had exciting medical implications. One was that inadequate amounts of NGF could account for some poorly understood but frightening disorders. One in particular is familial dysautonomia, also called Riley-Day syndrome, a genetic condition associated with a multitude of physiological problems, including unstable blood pressure, abnormal responses to stress, and decreased sensitivity to pain and temperature.[30] Anatomically, children with familial dysautonomia have a markedly reduced number of sympathetic nerves and fewer sensory nerves from the skin to the spinal cord.

In subsequent studies, including some by the present author, pregnant rats were injected with mouse NGF, causing them to produce anti-NGF antibodies that crossed the placenta. Their newborn pups displayed many of the same behavioral, physiological, and anatomical characteristics as the children with familial dysautonomia.[31] These findings suggested that familial dysautonomia is largely the clinical manifestation of having insufficient amounts of NGF to stimulate and guide nerve growth during development.

Levi-Montalcini continued to work on NGF during the 1960s. She then left St. Louis to help build a new medical research unit on Italian soil. In 1986 she became the fourth woman to receive the Nobel Prize in Physiology or Medicine.

Levi-Montalcini is important for our neuroscience voyage across time because she was able to demonstrate that Roger Sperry was headed in the right direction when he brought up the possibility of chemical guidance in the 1940s. But, as she herself emphasized, the pioneering research that she and Sperry had conducted can be looked upon only as a starting point. Among other things, NGF does not cause all neurons to grow, meaning other enterprising researchers will still have to look for additional growth factors. Further, it was unclear whether individual neurons within a structure respond to chemical gradients produced by one or perhaps two agents, or to a wealth of different chemical variations within a class of growth factors.[32]

The Corpus Callosum

Although Roger Sperry became well known in scientific circles for his research on the establishment of neural connections and for his theory of chemoaffinity, which would be confirmed by Levi-Montalcini, he achieved even broader name recognition for his experiments on the behavioral effects of severing the corpus callosum. The ancients knew about the massive bridge of axons uniting the two hemispheres, but they had no idea what it did.[33]

Later, during the 1700s, there were debates about the effects of damaging the great cerebral commissure. Some philosophically-minded scientists claimed that it may be the seat of the soul, because injury to it causes rapid death. Others seemed equally convinced that damage to it causes nothing beyond what should be expected after damage to any other part of the brain.

In the 1800s reports began to appear on individuals with the corpus callosum damaged and on an occasional man or woman who was born without this structure (Fig. 17.4). Again, there were two schools of thought. Some maintained that the corpus callosum has nothing to do with intellect,[34] whereas others proclaimed it critical for normal intelligence.[35] Because the subjects usually suffered from additional brain damage or other abnormalities, and because testing methods were poor, this issue could not be resolved at this time.

By the twentieth century, people were wondering whether damage to the corpus callosum might account for cases of dual personality. William McDougall, an Oxford University psychologist, was one scientist who predicted that consciousness will still remain unitary in an individual with a severed corpus callosum.[36] A colleague who knew McDougall even wrote:

> I remember him [McDougall] saying more than once that he tried to bargain with Sherrington . . . that if ever he should be smitten with an incurable disease, Sherrington should cut through his corpus callosum. "If the physiologists are right"—and by physiologists I suppose he meant Sherrington himself— "the result should be a split personality." "If I am right," he said, "my consciousness will remain a unitary consciousness." And he seemed to regard that as the most convincing proof of the existence of something like a soul.[37]

New findings continued to emerge in the 1930s, when Walter Dandy, a neurosurgeon at Johns Hopkins University, severed the corpus callosum in several patients to get to third-ventricle tumors.[38] He did not observe any major mental changes in his

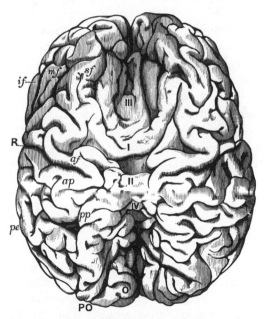

Vertex view of Cerebrum.   R, fissure of Rolando ; PO, parieto-occipital fissure ; O, occipital lobe ; *pe*, convolution of parietal eminence ; *pp*, postero-parietal convolution ; *ap*, ascending parietal convolution ; *af*, ascending frontal convolution ; *sf*, *mf*, *if*, superior, middle, and inferior frontal convolutions ; I., anterior, and II., posterior transverse convolutions ; III., grey matter in front of anterior transverse convolutions ; IV., mesial convolutions behind posterior transverse convolution.

FIGURE 17.4.

The brain of an institutionalized man born without a corpus callosum but possessing abnormal formations (I–IV) between the two cerebral hemispheres. (From Turner, 1877–1878.)

cases, nor did the operation produce multiple personalities. Instead, as had been predicted by McDougall, Dandy's patients seemed very much in command of their senses and their unified minds.

Dandy sided with those who were maintaining that the corpus callosum does not have anything to do with intellectual functions. He felt confident that the behavioral changes reported in some of the earlier studies were the result of associated brain damage. His behavioral evaluations, however, were not sophisticated in the least. His immediate concern was only with how his patients looked, felt, and interacted with others when they seemed ready to leave the hospital.

The most-cited studies on severing the corpus callosum in humans in the pre-Sperry era were conducted early in the 1940s by Andrew Akelaitis of the University of Rochester Medical School.[39] Akelaitis drew from a patient population that had been operated upon by William Van Wagenen to control severe epilepsy. It had been discovered that such surgery could prevent the spread of epileptic seizures from one hemisphere to the other and could also reduce the severity of the seizures in the diseased hemisphere. Most of the men and women Akelaitis studied had their corpus callosums only partially severed, but about a third had complete transections.

Akelaitis looked at a multitude of functions: recognition of objects by touch, block design, depth perception, picture completion, visual recognition, sound localization, body image, orientation, time relationships, praxis (executing movement on command), spoken and receptive language, reading, and writing. When possible, he used standardized tests developed by psychologists. In addition, he tried to evaluate the two sides of the body separately. He would ask a patient to write a sentence using his left hand, to identify an object in his right hand, to deal cards with his left hand, and so forth.

His findings appeared in a series of articles in leading American journals of neurology and psychiatry. He reported that his patients did not show postoperative personality changes, psychotic episodes, or serious abnormalities in general behavior. Given what Dandy had observed, there were no surprises here. The real eye-opener was the absence of severe or permanent deficits on any of the cognitive tests he administered in a systematic way.

Akelaitis offered a number of explanations for his negative findings. They included representation of some higher functions in both hemispheres and bilateral pathways to and from each hemisphere. The fact that many of his patients had incomplete sections of the corpus callosum and intact lower commissures, he recognized, could also have accounted for his decidedly negative findings.

Still, to seasoned brain researchers, such as Yale University's Warren McCullough, the Akelaitis findings were positively amazing. A structure with over two hundred million nerve fibers had been severed, yet these patients still read, talked, solved puzzles, and performed normally on tests of higher mental functions. These findings, combined with the beneficial effects of this surgery on seizure disorders, led McCullough to joke that the corpus callosum must have evolved for only one reason—"to aid in the transmission of epileptic seizures from one to the other side of the body!"[40]

Another tongue-in-cheek remark came from none other than one of Sperry's own mentors, Karl Lashley. He jested that the only known function of the corpus callosum "must be mechanical . . . i.e., to keep the hemispheres from sagging."[41] Lashley had conducted some corpus callosum surgeries of his own on laboratory animals. He too had failed to detect any disturbances of function.

The details of Lashley's experiments were not published, but his negative findings were a topic of conversation while Sperry was in residence at the Yerkes Laboratories. Thus, as one of the members of the Yerkes group later pointed out, "when Sperry left Orange Park [Yerkes] and went to Chicago, he carried with him knowledge of the apparently negative results of corpus callosum section in man and subhuman animals."[42]

The Myers and Sperry Split-Brain Experiments

Ronald Myers had been a student at the University of Chicago since 1947. It was not until 1950, however, that he got to know Sperry, who was then in the anatomy department. Myers entered Sperry's laboratory as a young man in need of money for schooling, and Sperry hired him, only to have him resign a few months later when classes began at the medical school. After Myers was accepted into a combined M.D.-Ph.D. program, he returned to Sperry's laboratory as a graduate student.

Early in 1952 Sperry approached Myers and asked him to think about evaluating information transfer between the hemispheres for his thesis project. Sperry had been interested in cerebral information processing for a number of years and explained that much could be learned by cutting the corpus callosum alone and in combination with the optic chiasm. Lying at the base of the brain, the optic chiasm allows each optic nerve to send some visual information to the opposite hemispheres, while the rest heads back to the hemisphere on the same side. By severing the optic chiasm, each eye would be able to send its information only to the hemisphere on the same side.

If a cat with a severed optic chiasm were fitted with a blinder that would allow it to see with only one eye, would the visual information from this eye still be accessible to the opposite hemisphere? Even more interesting, what would transpire if the experimenter also cut the corpus callosum? With the callosum and the optic chiasm both cut, would one hemisphere have even the vaguest idea about what the other was seeing? These questions would not be all that hard to answer. The blinder would just have to be switched from the initially covered eye to the opened eye for additional testing.

Myers figured out how to do the surgeries and was soon hard at work on his thesis experiments. He found that cutting the optic chiasm alone did not prevent the opposite side of the brain from continuing to discriminate circles versus squares and other visual stimuli, just like the trained side. The corpus callosum obviously permitted both sides of the brain to know about the stimuli that were presented to just one eye at a time.

The results were very different when the corpus callosum and optic chiasm were cut before any training took place. When the first eye was trained on a visual discrimination and then covered, the cats now forced to use the other eye acted as if they had never seen the visual problem before. In fact, it took just as long to learn the task with the second eye as it had taken the cats to learn it with the first eye.

This experiment pointed to two distinct roles for the corpus callosum. Under conditions in which the hemispheres receive different information, the corpus callosum allows each hemisphere to know what the other is experiencing. In this respect, the corpus callosum can be thought of as a conduit for the exchange of information between the worlds of the two hemispheres. In addition, it must allow both hemispheres to lay down memory traces of the stimuli, or one hemisphere to tap the memory traces stored in the other.

Myers and Sperry published their findings in 1953.[43] Their cat experiments ushered in a new era of corpus callosum research—experimentation that shed light on the functional differences between the hemispheres and the nature of consciousness itself. So much for the old jokes about the great cerebral commissure; Myers and Sperry now knew better.

But a Russian Was First

Interestingly, Konstantin Bykov, a physiologist who worked with Ivan Pavlov in Russia, had reached the same general conclusion as Myers and Sperry, but in the 1920s. He sectioned the corpus callosum in dogs that had been trained to salivate in response to a touch or a thermal stimulus on one side of the body, or to a sound to one ear. Although normal dogs tended to respond identically to tactile stimuli presented to the mirror site on the other side of the body, dogs with sectioned corpora callosa did not exhibit any such transference. In fact, it was easy to condition different responses, one on the right and another on the left, with them, although this seemed impossible to do with unoperated dogs. From all indications, Bykov was the first laboratory scientist to show by controlled experiment that severing the corpus callosum can prevent the spread of information from one hemisphere to the other.

A full report of Bykov's experiments appeared in the papers of the Pavlov Laboratory, which, unfortunately, were virtually unknown outside of the Soviet Union. Nevertheless, a short synopsis of some of his work was published in a leading German neurological journal and in a fairly well-known book of Pavlov's lectures, which appeared in English in 1928.[44]

Myers came across Bykov's work only after his thesis experiments had been completed. Afterward he cited Bykov's studies in a few of his papers and even referred to them as "elegant." But Sperry, for reasons unknown, chose not to do the same.

Thus, although Bykoff's classical conditioning experiments were published in the 1920s, they had no perceptible impact on the development of the brain sciences in the West. The elegant studies on the corpus callosum that came from Sperry's laboratory, in contrast, were another matter. This work, especially when expanded to humans, dramatically impacted the thinking of brain scientists around the world.

The Move from Chicago to Caltech

Sperry received some very disturbing news at this time. The chairman of his anatomy department decided to terminate his contract. The reason for this sudden decision had nothing to do with Sperry's intellect, productivity, or reputation. The issue, although almost hard to believe today, was solely one of health.[45]

Sperry's chairman had learned that he was fighting tuberculosis. This was one reason why Sperry was going to Bimini in the winter and traveling to Saranac Lake in New York in the summer. The chairman felt sure that Sperry would be forced to run off for fresh air and sunshine even more with each passing year. Why, he asked, would any chairman in his right mind hand a man destined to become an invalid a tenured university slot, a contract with unlimited job security?

Sperry did not take the news well. But before he could pack and leave Chicago, the head of the psychology department approached him. He told him that he was willing to offer him a position at a higher level (associate professor) with tenure. Sperry was appreciative of the offer and graciously accepted it. Nevertheless, he did not remain in Chicago much longer. In 1954 he accepted a special professorship at the California Institute of Technology and left the Windy City for the gentler, sunnier climate of Pasadena.

Sperry's health improved markedly in southern California. Soon he and his students were working on additional studies with split-brain cats.[46] In other experiments, they turned to more suitable models for humans, namely, monkeys (Fig. 17.5). The primates turned out to be especially interesting because they could quickly learn to respond in one way to a problem presented to the left hemisphere and in a different way when it was presented to the right hemisphere. Amazingly, with a special training box and eye filters, the two disconnected hemispheres could learn opposing responses at the same time.[47]

The Paratrooper with a Split Brain

In 1960 Joseph Bogen, a neurosurgeon in Los Angeles, called on Sperry. He told him he had a patient in his forties who suffered from severe epileptic attacks that could not be controlled by medication. The man's problems resulted from head injuries he had sustained as a paratrooper during the Second World War. He had been thirty years old when his parachute did not open during a jump into occupied Holland.

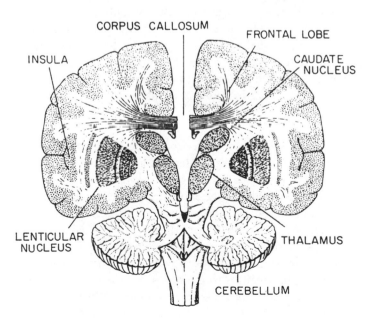

FIGURE 17.5.

A drawing showing the corpus callosum (top middle) severed in a primate brain. (From Sperry, 1968; with permission of Academic Press.)

When he hit the ground he suffered broken bones and remained unconscious for an unknown period of time. After being captured by the Nazis, he also endured a damaging blow from a rifle butt to the left side of his head.[48]

Bogen explained that the man's seizures were now threatening his life. He thought that cutting the corpus callosum could help, since the operation had been shown to reduce the frequency and severity of grand mal seizures many times in the past. But he wanted to know what Sperry thought. He also hoped to entice Sperry to study this man, should he be willing to undergo the surgery.

In 1962, after all other options were exhausted, the patient gave his permission to operate. The operation was performed by Philip Vogel, the senior neurosurgeon at the Los Angeles hospital, with Bogen assisting. To the relief of all concerned, the corpus callosum and the smaller, underlying anterior commissure were cut without serious complication.

There was more good news. The once-tormented man became almost seizure-free and displayed no radical changes in perception, temperament, manners, or intellect. He watched television, read books, remained coordinated, and communicated clearly with others. He never complained that his perception was distorted, irregular, or peculiar. In short, although he had some short-term memory problems (which would diminish), there was no outward reason for family members, acquaintances, or even observant physicians to suspect that the cables connecting his two hemispheres to each other had been severed.[49]

The way in which the patient looked and acted was consistent with the earlier clinical reports of Dandy and Akelaitis. But Sperry was anxious to draw on what he had learned from testing animals. Central to his plan was the idea of trying to present information to just one hemisphere and then testing the other hemisphere under highly controlled conditions. To help him with the behavioral testing, he turned to a new graduate student, Michael Gazzaniga.

Since the ex-paratrooper did not have his optic chiasm sectioned, the experimenters decided to present their visual stimuli either far to the right or far to the left side of the nose. They knew that objects flashed well to the right of the nose would be picked up by the nerves projecting to the left hemisphere, whereas objects flashed well to the left of the nose would go to the right hemisphere. With flash presentations of the stimuli precluding the possibility of head or eye movements, it was soon apparent that, like the cats and monkeys tested earlier, neither hemisphere knew what the other was seeing (Fig. 17.6).[50]

The experimenters now looked more carefully at language functions. In this domain, a wealth of new discoveries emerged. The paratrooper did well when it came to giving verbal labels to stimuli presented in the right visual field or the right hand, both of which project to the left or "verbal" hemisphere. The right hemisphere, in contrast, was unable to give correct responses using words. When forced to read or make verbal responses, what came out made little sense, and the patient verbally denied having any knowledge of the material flashed on the screen.

Parallel findings were obtained when answers had to be written down with the left and right hands—only the right hand (left hemisphere) showed any real capacity for writing the answers to questions. Yet, in many other respects, the mute right hemisphere was not retarded or unconscious. When allowed to respond to visual information nonverbally, such as by pointing or by selecting an object with the left

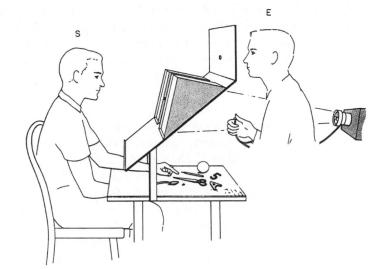

FIGURE 17.6.

The test apparatus used by Sperry and Gazzaniga for testing human patients. The patient looks straight ahead and is shown images on the right and/or the left screen. Objects on the table can be manipulated but not seen. (From Gazzaniga and Sperry, 1967; courtesy of Oxford University Press.)

hand, the "minor" hemisphere performed admirably. It demonstrated a high degree of comprehension, even understanding fairly detailed instructions.

These findings were exciting, to say the least. But there was a bothersome interpretative problem. Since the soldier had suffered serious brain injuries during the war, the experimenters knew they had to be very careful about jumping to conclusions about hemispheric differences. Patients without serious, additional brain damage would be needed for clarification. Fortunately, several "cleaner" split-brain subjects were soon available for testing.

More Subjects, More Findings

After 1967, Sperry and his team turned increasingly to the functions of the so-called silent hemisphere. By this time they had a number of additional split-brain patients available for study—men and women with complete sections of the corpus callosum and the smaller underlying commissures. Their second patient was a thirty-year-old woman without a history of traumatic brain injury who, after twelve years of seizures, underwent the same surgery as the ex-paratrooper.[51] An affable twelve-year-old boy who had an IQ of 115 and no history of acute head injury was their third patient. Soon there were ten subjects to work with, most without complications from acute brain injuries.

It now became apparent that, because of his additional brain damage, the ex-paratrooper was somewhat more impaired on certain functions than the newer split-brain patients. Nevertheless, many of the basic findings obtained with him still held up.[52] A good example has to do with the inability of the right hemisphere to label objects with spoken language. Witness what was written about one of the women who now underwent testing:

> The left hand is allowed to feel and to manipulate, say, a toothbrush under the table or out of sight behind a screen. Then a series of five to 10 cards are laid

out with names on them such as "ring," "key," "fork." When asked, the subject may tell you that what she felt in the left hand was a "ring." However, when instructed to point with the left hand, the speechless hemisphere deliberately ignores the erroneous opinions of its better half and goes ahead independently to point out the correct answer, in this case the card with the word "toothbrush."[53]

Jerre Levy-Agresti was one of the new students who chose to do her doctoral thesis under Sperry. She showed that spatial relationships are perceived more accurately by the right side of the brain.[54] The right hemisphere also came out better than the left in imagining how three-dimensional objects would look if unfolded into two-dimensional forms. When asked to identify faces, the right hemisphere was again superior to the left, except when the name of the person's face had to be spoken.

As studies like these continued to shed light on hemispheric differences, Sperry thought it reasonable to make some cautious generalizations about hemispheric specialization (Fig. 17.7).[55] He chose his words carefully when he concluded that the left hemisphere is more verbal than the right, being better programmed to think in words and communicate with speech or gestures. It is also more analytic, rational, and logical than the right. In addition, it is superior when confronted with mathematical problems.

The right hemisphere, in contrast, was characterized as more holistic, emotional, impulsive, and artistic, and superior to the left in dealing with geometric principles

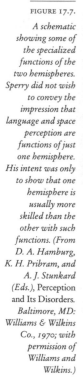

FIGURE 17.7.

A schematic showing some of the specialized functions of the two hemispheres. Sperry did not wish to convey the impression that language and space perception are functions of just one hemisphere. His intent was only to show that one hemisphere is usually more skilled than the other with such functions. (From D. A. Hamburg, K. H. Pribram, and A. J. Stunkard (Eds.), Perception and Its Disorders. Baltimore, MD: Williams & Wilkins Co., 1970; with permission of Williams and Wilkins.)

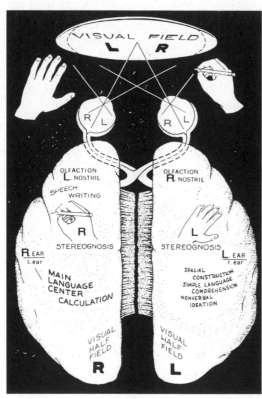

and shapeless designs, nonspeech sounds, and identifying objects by touch. Nevertheless, its own specializations may remain hidden unless it has a chance to respond nonverbally, such as by pointing.

In opposition to what had been thought previously, Sperry also noted that the right hemisphere can comprehend spoken language, make cognitive decisions, act volitionally, show self-awareness, and even understand consequences. To quote:

> The nonvocal hemisphere appears to be cognizant of the person's daily and weekly schedules, the calendar, seasons, and important dates of the year. The right hemisphere also makes appropriate discriminations that show concern with regard to the thought of possible future accidents and personal or family losses. The need for life, fire, and theft insurance, for example, seems to be properly appreciated by the extensively tested mute hemisphere of these patients.[56]

Clearly, concluded Sperry, the right hemisphere is not word-blind, word-deaf, or mute, as was being maintained by many neurologists. It may not be as skilled in language as the left, but it is not totally lacking in higher intellectual abilities. No longer would the right hemisphere be looked upon as the uneducated, bestial hemisphere, the one reformers in the Victorian era tried to educate. No longer would it be the Mr. Hyde to the cultured left hemisphere's Dr. Jekyll (see Chapter 10).

The Nature of Consciousness

Sperry's findings forced him to think long and hard about the nature of conscious awareness as defined by philosophers and psychologists. The more he pondered consciousness, the more he, like Sherrington before him, realized that it had to be viewed from an evolutionary perspective. The higher up the phylogenetic scale he went, the more organisms seemed to be aware of their actions and the need to control them.

Sperry eventually came to the conclusion that consciousness must be a higher property of the brain, one with the capability to modify other brain processes.[57] As such, he reasoned, consciousness should no longer be looked upon as just a passive property of brain physiology. Because of consciousness, we can override the basic wants and desires of our brainstem structures, make choices, and do what is morally right.

But why do we need two distinct consciousnesses rather than a single, indivisible one? Wouldn't a single set of controls be sufficient for managing emotion, language, intellect, perception, and memory?

Sperry concluded that many parts of the brain, especially in the brainstem, are unnecessarily duplicated. He looked upon this redundancy as an evolutionary mistake, commenting: "It is doubtful that all this redundancy has had any survival value."[58] Cerebral dominance, he continued, probably represents an attempt by Mother Nature to break away from these unnecessary duplications. In the case of speech, for example, there is no reason for both sides of the brain to have duplicate control centers.

From this premise, he argued that the species would best be served by having two different modes of cognitive processing, each controlled by a different hemisphere.

Not only will the organism be able to process a wider variety of information, but there will be less likelihood that the two hemispheres will do battle with each other.

To be maximally efficient, however, the two hemispheres must function as a unit. This is where the corpus callosum comes into play. Having two communicating minds working together—one verbal-analytic and the other holistic-spatial—would be superior to having just one type of mind or two separate minds that cannot communicate with each other.

As a result of Sperry's discoveries, the subject of consciousness took on a whole new look. He showed that consciousnesses can, in fact, be studied in the laboratory under controlled conditions. Once shunned by most experimental psychologists and neuroscientists, his work helped stimulate the "cognitive revolution"—a movement to study thinking, problem solving, and other higher mental processes with the tools of modern science.

Final Years

Sperry retired from Caltech in 1984 at age seventy-one. The idea that his tuberculosis would make him an invalid and cut his career short clearly had been wrong. When he left the laboratory, a form of lateral sclerosis was robbing him of his ability to speak clearly and work at full capacity. In fact, because his diction was already diminishing in 1981, the speech he had written for his Nobel Address had to be read by a member of Sweden's Karolinska Institute.

Sperry was as much a Renaissance man as any of the scientists we have met so far, including such greats as Charcot and Sherrington. He stepped into the scientific arena with formal training in English literature, experimental psychology, and zoology. He also had a plethora of side interests that he continued to cultivate. He collected fossils, enjoyed boating and camping, and was an artist of considerable skill who even decorated his home with his own drawings, watercolors, ceramics, and sculptures.

When Sperry died in 1994, he was eulogized as a man of strong personal convictions and exceptional clarity of thought—one who loved the challenge of research, the thrill of penetrating the unknown, and the excitement of discovery. With his gift for asking the right questions, designing critical experiments, and never losing sight of the big picture, he was instrumental in generating new ways of thinking about the now somewhat less mysterious organ of mind.

18

Pioneers and Discoveries in the Brain Sciences

Investigators do not march straight to their goal with ease and directness. . . .
Obstacles and difficulties are sure to be encountered. The search for understanding is
an adventure, or more commonly, a series of adventures.

—*Walter B. Cannon, 1945*

The examination of the steps by which our ancestors acquired our intellectual
estate . . . may teach us how to improve and increase our store . . . and afford us
some indication of the most promising mode of directing our future efforts.

—*William Whewell, 1837*

The Voyage

As this biographical voyage comes to an end, we would do well to think about how
little was known about the brain during the time of the unidentified but observant
physician from Egypt's Third Dynasty, when Galen began his experimentalism in
Rome, when Broca was examining Tan's brain in Paris, or even when Loewi came
forth with evidence for chemical transmission in Graz. By imagining ourselves in
these earlier times, we can begin to comprehend the scope of the forty-six hundred-
year trip we have just taken and better recognize how far we have come.

Of course, it is never possible to stop at every port of call on a voyage of this sort.
Indeed, the purpose of this book was not to look at the lives and discoveries of every
great contributor to the brain sciences. If it were, there would have been full chapters
on Hermann von Helmholtz, John Hughlings Jackson, Alan Hodgkin, John Carew
Eccles, and many others, especially from the twentieth century. Rather, my intent
was only to look at a selection of the minds who had a major impact on the brain sciences, even if what each had to say was not entirely correct when judged through
contemporary eyes (witness Greco-Roman humoral theory, Descartes' pineal physiology, or Gall's system of bumps). The idea was to show how these chosen individuals reasoned and to examine what each did to achieve great fame.

With this in mind, let us now turn back to our pioneers to see what we can determine about the discovery process itself. One issue worthy of discussion is the de-
pendence of these physicians and scientists on the work of others. A second relates
to the conditions that make the time ripe for discovery. And a third is the role that
may be played by chance encounters—at least for "prepared" minds. As this book
ends, I hope to show that the lessons that may be learned from a biographical history
of the brain sciences can be valuable to us all. But first let us begin with a question:

Did the leading players we have just met share many traits in common, or were their backgrounds, approaches to science, and even personalities very different?

Universal Traits?

Perhaps not surprisingly, those individuals who challenged others to see the brain in a new way share many common features. In terms of how they were raised, they typically had cultured parents who were interested in learning. Their fathers and mothers did everything they could to see to it that their children obtained quality educations—more likely than not, the best available at the time. Moreover, the learning process was not confined to the home or the classroom; nor did it stop with the awarding of some degree. Each person knew that formal education was merely a beginning or a base on which to build. Hence, after graduating, they visited other centers of learning and communicated with other scientists. They kept abreast of the written literature, attended lectures, and even formed groups to foster the exchange of ideas.

Coupled with this intense drive to learn and keep abreast of new developments, all settled in places where they had at least some like-minded colleagues—reasonably large cities and university towns where culture and learning were appreciated. Equally important, all hoped to see things with their own eyes and to reveal something new to those around them, as well as to the scientific community at large.

This driving force to learn and share information was not based on financial rewards, on the promise of material comforts, or even on the altruistic hope of immediately benefiting mankind. This love of discovery, although nurtured by enlightened parents and enhanced by stimulating teachers and colleagues, seemed to stem from some powerful inner force. But whatever the basis of this drive, all of the individuals described here exhibited what can almost be described as an addiction to, or intoxication with, learning and discovery.

Walter Cannon (Fig. 18.1), a Nobel Prize winner who spent years trying to understand the autonomic nervous system, admitted on many occasions to being happily obsessed with science. He once wrote:

> Primary among the rewards of the scientific explorer is the discovery of a new phenomenon. Only he who has had the experience knows the thrill of it.[1]

Another frequently encountered trait was an unwillingness to accept everything a teacher, or any other authority figure for that matter, had to say. All learned to be skeptical of established dogma and showed a healthy ability to break with the past and see things in a new, independent light. To a person, to paraphrase Mark Twain, they were the leaves that were capable of blowing against the wind.

In order to present the truth as they saw it, some even suffered verbal abuse and personal hardships. To cite but three examples: Vesalius was castigated by his teacher Sylvius, who even tried to have him imprisoned for questioning the authority of Galen; Gall was forced to leave everything that was dear to him in Austria, because he refused to stop presenting his "materialistic" philosophy; and Ferrier had to put up with the disruptive antics of the militant animal-rights activists in London, who went so far as to drag him into court.

Thinking in terms of the big picture was still another characteristic repeatedly

FIGURE 18.1.

Walter Bradford Cannon (1871–1945), the Nobel Prize–winning physiologist who compared his career in science to a series of wonderful adventures.

seen. Although some of our later pioneers could easily have drowned in a sea of details, they had wide horizons and never lost sight of larger goals, including what their discoveries were revealing about the brain as the organ of mind. They also recognized that even a wonderful new discovery with far-reaching ramifications would mean little if it could not be communicated clearly to others in a public forum. Broca, as we have seen, was effective at oral and written communication and clearly understood the rules of the game. In contrast, Emanuel Swedenborg and Marc Dax had no real impact on the brain sciences, not because they lacked great insights, but because they failed to make the scientific community aware of their discoveries.

In contrast to these traits and habits, some of the other characteristics common to these brain pioneers may be less expected by those not directly involved with science. One has to do with naïveté and impulsiveness occasionally dominating over reason. Otto Loewi, for example, said he would have calculated the odds of being successful as so low that he never would have conducted his experiments on neurotransmission had he taken the time to think about them.

Rita Levi-Montalcini also wrote about unwarranted optimism being a helpful trait. In her own words:

> Looking back now on the long path my life has followed, on the lives of my peers and colleagues, and on the briefer ones of the young recruits who have worked with us, I have become persuaded that, in scientific research, neither the degree of one's intelligence nor the ability to carry out one's tasks with thoroughness and precision are factors essential to personal success and fulfillment. More important . . . are total dedication and a tendency to underestimate difficulties.[2]

Another finding that may be unexpected has to do with the common stereotype of the dedicated scientist as a person who is so absorbed with work that he or she has no time for family or outside activities. True, Galen and Levi-Montalcini chose not to marry in order to dedicate themselves fully to their chosen careers. But, with the exception of Descartes (who still cared for a mistress and fathered an illegitimate daughter), all of the others about whom we know something were devoted to their wives and children and were also involved with the humanities.

Philosophy, languages, history, literature, the arts, and theology provided insight, comfort, inspiration, and perspective for Galen, Descartes, Broca, Charcot, Sher-rington, Loewi, Sperry, and just about everyone else. In this regard, it is interesting to come across Charcot saying that he had a better appreciation of human nature after reading Shakespeare; to see Sherrington being drawn to the poetry of Keats and writing his own poetry; and to find Loewi visibly moved during his own Nobel Prize ceremony by the playing of a piece of incidental music calling for freedom from oppression, Beethoven's invigorating "Egmont Overture."

Active engagement in the fine arts deserves special attention in this context. Per-haps reflecting their excellent visual-perceptual abilities, many of the people we have met also enjoyed drawing, painting, and sculpting. Ramón y Cajal and Charcot stand out the most in this regard, because they seriously contemplated careers as artists. Sperry was also a talented artist. As for Otto Loewi, his eye was so good that he initially hoped to study art history, and he might well have spent as much time in the world's art museums as he did in his own laboratory.

Although our brain pioneers were obviously very bright and shared many fea-tures—to which we could easily add such admirable traits as curiosity, dedication, and an infectious spirit of adventure—their basic personalities were anything but cast from a single mold. Galen had one of the greatest egos of all time and did not hesitate to push, shove, and claw his way to the top; Hitzig was described as being "a stern and forbidding character of incorrigible conceit and vanity";[3] and Adrian was said to have been rather "cold" and "aloof." In contrast to these three men, Vesalius was not the kind of person to be drawn into an emotional war of words with his op-ponents, Willis lived an almost saintly life, and Sherrington and Levi-Montalcini were modest to a fault. In terms of their social skills and underlying personalities, the minds behind the brain—so much alike in their love of learning, willingness to stand apart from the pack, and side interests—were not cast from the same mold.

Giants and Dwarfs

Let us now raise a different question. Is each successive generation of luminaries see-ing further by "standing on the shoulders of giants"?[4] More specifically, are the "gi-ants" we have read about—to continue the metaphor popularized in the 1670s by Sir Isaac Newton—the entire story?

In reality, the idea that succeeding generations of scientists can see farther by standing on the shoulders of giants is a partial truth. Typically, giants emerge only when others are already beginning to question existing theories and ideas. Thanks to such stirrings from these others, progress is rarely at an absolute standstill when new heroes venture forth with new experiments, observations, or insights to topple time-honored notions and theories.

Moreover, even great giants rarely work alone. Typically they have co-workers, students, and followers. Think about the *Virtuosi* who worked alongside Willis in the seventeenth century, or about the many rising stars who went to England to conduct experiments with Sherrington and Adrian early in the twentieth century. In most in-stances, these were students, physicians, or academicians. But, as we saw with Father Marin Mersenne in the chapter on Descartes, outsiders can also play influential roles.

Thus, even though this book has concentrated on the giants themselves, one can-not for a moment forget that many others of varying stature and recognition have also made contributions to a better understanding of the nervous system. To think of these lesser-known contributors as insignificant players or as mere "dwarfs" (to use the original term for exactly who is "standing on the shoulders of giants" in early versions of Newton's chosen metaphor[5]) is both undeserved and unjust. To use another metaphor, each generation may have some decorated generals who can see much farther than their predecessors, but they do so because the way has been cleared by other highly visible leaders and also because there were countless foot soldiers and lesser officers who were there as well, all helping to pave the way for new, important discoveries.

The Ripe Time

Clearly, our pioneers lived when the Zeitgeist, or spirit of the times, was such that their ideas did not fall upon deaf ears. Stated in a more positive way, these men and women came forth when there was a growing willingness to change course and see the world of the nervous system in a different way.

Technological developments are among the factors that can make the time ripe for new ideas. Neuron doctrine really had no chance of emerging in viable form before the nineteenth century, in part because the tissue-hardening methods, cutting procedures, stains, and microscopes were inadequate to permit scientists to see nerve cells and their processes with sufficient clarity. Similarly, researchers were not in any position to write about neural coding before sensitive amplifiers and recording instruments were built for collecting data from single neurons. Edgar Adrian knew this, which is why he maintained that one must first know something about the history of technology before trying to understand the rise of electrophysiology.[6]

Sir Charles Sherrington pointed to the importance of some additional factors when discussing the "ripe time" in his biography of Renaissance physiologist Jean Fernel. To quote:

> Essential to a great discoverer, in any field of Nature, would seem an intuitive flair for raising the right question. To ask something which the time is not ripe to answer is of small avail. There must be the means for reply and enough collateral knowledge to make the answer worth while.[7]

What Sherrington wrote about his own personal hero could certainly be applied to himself. He lectured and wrote about reflexes, integration in the nervous system, and excitatory and inhibitory actions at the synapse at a time when neuron doctrine was growing in acceptance and many scientists were trying to decipher the functional organization of the nervous system. He not only had a flair for raising the right questions, but also helped to make the time ripe for his own work by developing new investigative techniques and bringing facts and theories together in exciting new ways.[8]

Resistance to New Ideas

Sociologists of science inform us that there are many sources of resistance to scientific discovery, such as religion and politics, which can affect how scientists may function at any given moment in time. Today it is generally agreed that other scientists are among the greatest sources of resistance to new ideas.[9] As a result, as was once contended by the physicist Max Planck, some good ideas may forever go unrecognized and others may be recognized immediately, while still others may be fully appreciated only after their staunchest opponents finally die.[10]

Opposition from the scientific establishment can be based on many things. One, of course, is blind adherence to older theories: Think about Sylvius, who became enraged by what Vesalius had to say because he could not imagine Galen making any mistakes. A second is from perceived limitations in tools, techniques, or methodology. Here we should remember those scientists who expressed disbelief when told that Adrian and Zotterman were able to record from single neurons; they did not have the foggiest notion about how this could be done. Still a third source of opposition has to do with one's stature in the scientific community. Galvani was forced to fight a rough uphill battle in defense of "animal electricity" in part because he, a mere neophyte, was challenged by Alessandro Volta, then a recognized authority in the physics of electricity.

Stephen Jay Gould of Harvard University has written in an illuminating way about the career of Richard Owen, the nineteenth-century British surgeon and anatomist who explored the brain surface and gave us the term *prefrontal cortex*.[11] At the start of his article dealing with Owen's controversial position on evolution (he also coined the word *dinosaur*), Gould wrote:

> Even if one thinker experiences an emotional and transforming eureka, he must still work out an elaborate argument and gather extensive empirical support to persuade a community of colleagues often stubbornly committed to opposite views. Science, after all, is as much a social enterprise as an intellectual adventure.[12]

Gould is correct. Moreover, if it is natural for the old to resist the young, the establishment to question the upstarts, and even the most open-minded scientists to be shackled by preconceived notions about the nature of things, it follows that there will always be at least some resistance to new ideas.

From this perspective, the "ripe time" can now be defined as a period of time in which there is less rather than more resistance to new ideas, mainly because of cultural changes, technological advances, and ground-tilling by other scientists. At such opportune moments in time, scientists feel decidedly freer and more willing to deviate from the status quo with their research programs, their interpretations, and what they say in public. It is at such times that new Newtonian "giants" with novel ideas are most likely to burst on the scene.

What Does Chance Have to Do with It?

Let us now turn to a closely related issue. Mainly, are great discoveries and insights about the brain more likely to result from experiments that test, confirm, and extend established theories, or do they usually begin with chance observations and unexpected findings?

Books and articles written by modern laboratory scientists frequently convey the impression that new discoveries almost always come from well-thought-out hypotheses and exquisitely controlled experiments. Occasionally the reader is led to believe that a scientist had so much faith in his or her pet theory that the outcome was all but certain well in advance of the experiment itself.

Without a doubt, many discoveries do result from well-planned experiments designed to test mainstream ideas and corollary hypotheses. Nevertheless, as philosopher of science Thomas Kuhn has emphasized, experimental work that confirms preconceived ideas is unlikely to result in the so-called quantum leaps (Gould's "eureka" experiences) that stand out as transforming in the history of any science.[13] The shift to a new paradigm, meaning a new conceptual framework for seeing and evaluating a wide range of things, he suggests, will probably be triggered by sudden insights and unexpected findings that existing theories simply cannot explain.

Consider Andreas Vesalius, who for years had been unable to understand why Galen's "human" anatomy was so laced with errors. It was not until he placed a human and an ape skeleton next to each other during a trip to Bologna that he saw the light. On that wonderful afternoon in 1540, solving the bothersome "Galen problem" was not even on his mind. But once he saw the two skeletons side by side, he suddenly realized that Galenic anatomy was based entirely on animals. With his insight and new understanding of the older literature, he proceeded to sound his clarion call for a new, autopsy-based human neuroanatomy.

In contrast to Vesalius' fortuitous observation, Galen's discovery of the nerves of voice represents a particularly good example of an unexpected discovery stemming from an experiment conducted for a different purpose. His real intention was to discover the nerves responsible for respiration when he cut some nerves in the throat of a restrained but squealing pig. As soon as the nerves were severed, the pig immediately stopped squealing. Following up on this unexpected finding, Galen discovered that cutting the same pair of nerves had a comparable effect on a variety of other animals. And knowing that some people had abruptly lost their voices during neck operations for goiter, he astutely recognized that the recurrent laryngeal nerves must play a role in the production of human speech.

But how frequently do serendipitous findings and unexpected insights occur in the brain sciences? The answer to this question will never be known with certainty, but Rita Levi-Montalcini stands among those who are convinced that chance probably plays a greater role in significant discoveries than most people think. She contends that fresh new leads and ideas, such as those that culminated in her own group's discovery and isolation of nerve growth factor, typically result from unexpected or accidental findings. In her own words: "Of all sciences, biological sciences still remain the most dependent upon such fortuitous leads to unravel some of the intricate mechanisms of life."[14]

If what Levi-Montalcini contends is true, or even only half true, the roads traversed by the scientists we have met must have been considerably bumpier and fraught with more dead ends than is suggested by the final printed record. The problem for historians is that the role played by good fortune is not often discussed in the older literature. Even today, it is fairly unusual to find the role of serendipity mentioned in a public forum, except perhaps by an occasional retiring researcher who is asked to look back and comment on a successful career.[15] Of course, when only successes appear in the published record, it is hard to avoid the erroneous impression that the sciences always move firmly and irresistibly forward—from darkness, into shadow, and finally into daylight.

Sadly, some scientists have even gone out of their way to remove blemishes, errors, and misunderstandings in order to present the history of a field in a favorable way. Joseph Priestley, in a book on the status of electricity published in 1775, was one writer who made up his mind to present a history based solely on successes. The famed chemist who discovered oxygen even admitted:

> I made it a rule to myself, and I think I have constantly adhered to it, to take no notice of the mistakes, misapprehensions, and altercations. . . . All the disputes which have in no way contributed to the discovery of truth, I would gladly consign to eternal oblivion.[16]

In contrast to Priestley, Hermann Helmholtz, the first person to correctly estimate the speed of nerve conduction and a major contributor to sensory physiology, publicly acknowledged the bumpy and unpredictable nature of the road he had chosen. In 1891, on the occasion of his seventieth birthday, he gave a speech in Berlin. In it, he tried to correct the mistaken impression that he, for one, always knew exactly what to do. He explained:

> I would compare myself to a mountain climber who, not knowing the way, ascends slowly and toilsomely and is often compelled to retrace his steps because his progress is blocked; who, sometimes by reasoning and sometimes by accident, hits upon signs of a fresh path, which leads him a little farther; and who finally, when he has reached his goal, discovers to his annoyance a royal road on which he might have ridden up if he had been clever enough to find the right starting point at the beginning. In my papers and memoirs I have not, of course, given the reader an account of my wanderings, but have only described the beaten path along which one may reach the summit without trouble.[17]

The difference between what Priestley and Helmholtz were willing to make public is profound. In addition, to showing how misleading the published record can be—a point often made by historiographers—they make us wonder about a related unknown.[18] How can we even begin to guess how many significant observations and well-executed experiments went unpublished because they failed to confirm an established theory?

The Prepared Mind

If, as experimental psychologist B. F. Skinner contends, "science is a continuous and often a disorderly and accidental process," we must now ask why one scientist on the path of discovery may immediately recognize the importance of an anomaly, while another may stumble upon something unexpected and promptly sweep it under the rug.[19] I will suggest that being lucky enough to be in the right place at the right time and even having the right tools is not enough, although these factors certainly can improve the chances of making a significant breakthrough. My point, though hardly a new one, is that the truly great scientists, the men and women about whom volumes are written, are observant opportunists whose minds are more open than most to anomalies and new ideas. As Walter Cannon once wrote: "To persons who live according to pattern, adventures in ideas are impossible . . . wisdom counsels keeping our minds open."[20]

Without question, the giants we have met were inquisitive individuals—puzzle solvers, if you choose—who were ready to deal with the challenges of anomalies. They rank among the chosen who stand out for having the wisdom to follow up on an unexpected discovery or a novel idea and the background to put it into proper perspective. For such men and women, a serendipitous observation is but a lucky first step in a process that demands an open mind and, as Sir Henry Dale once put it, "a great deal of hard, systematic, and conscientious work."[21] Also very helpful in the discovery process, wrote Edgar Adrian, is the ability to know which unexpected leads are really worth pursuing and which are likely to be a waste of time.[22]

Swedish neurophysiologist Ragnar Granit once observed that "luck in science has a tendency to favour the well-prepared mind."[23] The subject of Granit's essay was none other than Sir Charles Sherrington, one of his teachers. Although Granit himself won a Nobel Prize, he was convinced that Sherrington was in a class of his own when it came to knowing if, when, and how to proceed when something unexpected turned up.

Walter Cannon, who delved into the history of the term "prepared mind," tells us that one of the first scientists to emphasize the importance of being intellectually ready was the Frenchman Louis Pasteur. In his inaugural lecture at the University of Lille on December 7, 1854, Pasteur stated: "*Dans les champs de l'observation, le hasard ne favorise que les espirits préparés*" (In the fields of observation, chance favors only the prepared mind).[24] But even before Pasteur, Cannon notes, Joseph Henry, the nineteenth-century genius in electrical science from America, had latched onto the same idea when he so beautifully explained: "The seeds of great discoveries are constantly floating around us, but they only take root in minds well prepared to receive them."[25]

The Past and the Future

Walter Cannon could well have gone on in his own inspirational book *The Way of an Investigator* to emphasize some sort of innate gift for seeing things from different perspectives. But, after thinking things through, he chose to stress how important learning can be for preparing minds for new, "inconvenient," and unusual findings. In a very real way, Cannon justifies books about the history of medicine and science such as this one. After all, he informs his readers, "only when we know what has been done by earlier contributors can we judge the present scene."[26]

Cannon's assertion is without question correct. But on further analysis it is just one of many reasons for those involved with science and medicine to look at the history of a discipline in biographical form. Some other reasons are given by Danish historiographer Helge Kragh. They include the benefits of inspiration and enlightenment that may be bestowed upon the researchers of today, the spillover to other fields (such as philosophy or sociology), the importance of such work for debunking myths about scientists, methods, and discoveries, and its use as an effective teaching tool.[27]

But is a history of brain science in biography in any way relevant to those who are not researchers or teachers? In closing, I would like to argue that it is. By knowing more about the past, all of us who read the daily papers or watch the news should now be in a better position to put modern scientific developments into proper perspective. By seeing how our chosen pioneers approached their work, we should also be able to realize just how important it is to keep an open mind in all walks of life. And by further preparing our own minds, we should be able to gain perspective and look upon unexpected events not as distractions or impediments, but as wonderful opportunities for seeing the exciting and mentally challenging world around us in fresh new ways.

Notes

Introduction: A Voyage Across Time

1. F. Schiller. 1967. The vicissitudes of the basal ganglia (Further landmarks in cerebral nomenclature). *Bulletin of the History of Medicine*, 41, 515–538.
2. F. W. H. Myers. 1886–87. Multiplex personality. *Proceedings of the Society for Psychical Research*, 4, 496–514.
3. Myers, ref. 2, p. 503.
4. C. S. Sherrington. 1941. *Man on His Nature*. New York: Macmillan, p. 225.
5. R. W. Sperry. 1964. *Problems Outstanding in the Evolution of Brain Function* (James Arthur Lecture). New York: American Museum of Natural History, p. 2.
6. Sherrington, ref. 4, pp. 223–224.
7. N. Steno. 1668. *Discours de Monsieur Stenon sur l'Anatomie du Cerveau*. Paris: Chez Robert de Ninville. Translated by E. Godfredsen as *A Dissertation on the Anatomy of the Brain*. Copenhagen: Nyt Nordisk Forlag Arnold Busck, 1950, p. 4.
8. M. Merbaum and M. R. Lowe. 1982. Serendipity in clinical research. In P. C. Kendall and J. N. Butcher (Eds.), *Handbook of Research Methods in Clinical Psychology*. New York: John Wiley & Sons, pp. 95–123.
9. H. H. Dale. 1948. Accident and opportunism in medical research. *British Medical Journal*, 2, 451–455.
10. W. B. Cannon. 1945. *The Way of an Investigator*. New York: W. W. Norton.

An Ancient Egyptian Physician: The Dawn of Neurology

1. J. B. Hurry. 1926. *Imhotep*. Humphrey Milford, UK: Oxford University Press.
2. J. H. Breasted. 1930. *The Edwin Smith Surgical Papyrus*. Chicago: University of Chicago Press, pp. 9, 75.
3. Breasted, ref. 2, pp 1–29.
4. P. Mazzone, M. A. Banchero, and S. Esposito. 1987. Neurological sciences at their origin: Neurology and neurological surgery in the medicine of ancient Egypt. *Pathologica*, 79, 787–800.
   W. A. Pahl. 1993. *Altägyptische Schädelchirurgie*. Stuttgart: Gustav Fischer Verlag.
   S. Finger. *Origins of Neuroscience*. New York: Oxford University Press, pp. 3–6.
5. Breasted, ref. 2, p. 220.
6. H. E. Sigerist. 1951. *A History of Medicine*, Vol. 1: *Primitive and Archaic Medicine*. New York: Oxford University Press, pp. 217–356.
7. B. Ebbell. 1937. *The Papyrus Ebers: The Greatest Egyptian Medical Document*. Copenhagen: Levin and Munksgaard.
8. Herodotus. 1972. *The Histories*. Trans. A. de Sélincourt; notes by A. R. Burn. London: Penguin Books, pp. 160–161.
9. A. J. Spencer. 1982. *Death in Ancient Egypt*. Harmondsworth, UK: Penguin Books, p. 112.
10. D. H. M. Woollam. 1958. Concepts of the brain and its functions in classical antiquity. In M. W. Perrin (Chair), *The Brain and Its Functions*. Wellcome Foundation. Oxford: Blackwell, pp. 5–18.

11. A.-P Leca. 1981. *The Egyptian Way of Death*. Garden City, NY: Doubleday.

12. R. David. 1978. *Mysteries of the Mummies: The Story of the Manchester University Investigation*. London: Book Club Associates.

Hippocrates: The Brain as the Organ of Mind

1. B. Snell. 1953. *The Discovery of the Mind*. Trans. T. G. Rosenmeyer. Oxford: Blackwell, p. 1–22.

2. G. Majno. 1975. *The Healing Hand*. Cambridge, MA: Harvard University Press, pp. 141–206.

    H. E. Sigerist. 1961. *A History of Medicine*, vol. 2: *Early Greek, Hindu, and Persian Medicine*. New York: Oxford University Press, pp. 44–79.

3. O. Bettmann. 1979. *A Pictorial History of Medicine*. Springfield, IL: Charles C Thomas, pp. 18–19.

4. Majno, ref. 2, pp. 201–205.

5. Snell, ref. 1, pp. 42–70.

6. Sigerist, ref. 2, pp. 85–115.

7. Snell, ref. 1, pp. 90–112.

8. L. Edelstein. 1967. *Ancient Medicine*. Baltimore, MD: Johns Hopkins University Press, pp. 201–246.

9. O. Temkin. 1991. *Hippocrates in a World of Pagans and Christians*. Baltimore, MD: Johns Hopkins University Press, pp. 39–78.

10. Majno, ref. 2, pp. 148, 192–200.

11. L. Canfora. 1990. *The Vanished Library*. Berkeley: University of California Press.

12. Sigerist, ref. 2, pp. 260–295.

13. Edelstein, ref. 8, pp. 319–348.

14. Edelstein, ref. 8, pp. 3–64.

15. Hippocrates. 1952. In *Great Books of the Western World*, vol. 10: *Hippocrates. Galen*. Chicago: Wm. Benton, p. 69.

16. G.-J. C. Lockhorst. 1982. An ancient Greek theory of hemispheric specialization. *Clio Medica*, 17, 33–38.

    G.-J. C. Lockhorst. 1996. The first theory about hemispheric specialization: Fresh light on an old codex. *Journal of the History of Medicine and Allied Sciences*, 51, 293–312.

17. Lockhorst 1982, ref. 16, pp. 34–35.

18. Lockhorst 1982, ref. 16, p. 35.

19. Hippocrates. 1952. *On Injuries of the Head*. Ref. 15, pp. 63–70.

20. Hippocrates, ref. 15, p. 69.

21. Hippocrates, ref. 15, p. 154.

22. J. O'Leary and S. Goldring. 1976. Epilepsy over the millennia. In *Science and Epilepsy*. New York: Raven Press, pp. 15–34.

23. E. Clarke. 1963. Apoplexy in the Hippocratic writings. *Bulletin of the History of Medicine*, 37, 301–314.

24. Clarke, ref. 23, p. 313.

25. Majno, ref. 1, pp. 167–169.

26. C. J. S. Thompson. 1938. The evolution and development of surgical instruments. *British Journal of Surgery*, 25, 726–734.

27. Canfora, ref. 11.

28. Edelstein, ref. 8, pp. 247–301.

29. J. F. Dobson. 1925. Herophilus of Alexandria. *Proceedings of the Royal Society of Medicine*, 18, 19–32.

J. Finlayson. 1893. Herophilus and Erasistratus. *Glasgow Medical Journal*, 39, 321–352.

H. von Staden. 1989. *Herophilus and the Art of Medicine in Early Alexandria*. Cambridge, UK: Cambridge University Press.

30. J. F. Dobson. 1927. Erasistratus. *Proceedings of the Royal Society of Medicine*, 20, 825–832.

31. A. C. Celsus. 1935–1938. *De medicina* (3 vols.). Ed. W. G. Spencer. Cambridge, MA: Loeb Classical Library.

32. Celsus, ref. 31, vol. 3, p. 15.

33. Finlayson, ref. 29, p. 326.

34. Aristotle. 1911. *Parts of Animals*. Trans A. L. Peck. London: Heinemann.

E. Clarke. 1963. Aristotelian concepts of the form and function of the brain. *Bulletin of the History of Medicine*, 37, 1–14.

C. G. Gross. 1995. Aristotle on the Brain. *The Neuroscientist*, 1, 245–250.

35. A. Karenberg and I. Hort. 1998. Medieval descriptions and doctrines of stroke: Preliminary analysis of selective sources. Parts I, II, and III. *Journal of the History of the Neurosciences*, 7, 162–200.

Galen: The Birth of Experimentation

1. G. Sarton. 1954. *Galen of Pergamon*. Lawrence, KS: University of Kansas Press, pp. 15–24.

J. S. Pendergast. 1930. The background of Galen's life and activities, and its influence on his achievements. *Proceedings of the Royal Society of Medicine*, 23, 53–70.

2. Sarton, ref. 1, pp. 30–38.

J. Walsh. 1934. Galen's writings and influences inspiring them. *Annals of Medical History*, 6 n.s., 1–30, 143–149.

3. J. Walsh. 1927. Galen's studies at the Alexandrian school. *Annals of the History of Medicine*, 9, 132–143.

4. J. Walsh. 1928. Galen clashes with the medical sects of Rome. *Medical Life*, 35, 408–443.

Walsh, ref. 3, pp. 413–414.

5. Galen. 1944. *Galen on Medical Experience*. Trans. R. Walzer. New York: Oxford University Press, p. 152.

6. C. Singer. 1949. Galen as a modern. *Proceedings of the Royal Society of Medicine*, 42, 563–570.

E. E. Smith. 1971. Galen's account of the cranial nerves and the autonomic nervous system. *Clio Medica*, 6, 79–80.

7. Galen. 1956. *Galen on Anatomical Procedures*. Trans. C. Singer. London: Oxford University Press, p. 227–236.

8. J. Walsh. 1926. Galen's discovery and promulgation of the function of the recurrent laryngeal nerve. *Annals of Medical History*, 8, 176–184.

9. Pendergast, ref. 1, p. 60.

10. Galen. 1917. De locis affectibus. *Annals of Medical History*, 1, 367.

11. O. Temkin. 1991. *Hippocrates in a World of Pagans and Christians*. Baltimore, MD: Johns Hopkins University Press, p. 72.

12. Galen. 1968. *On the Usefulness of the Parts of the Body*, vol. 1. Trans. M. T. May. Ithaca, NY: Cornell University Press, pp. 387–394.

13. Smith, ref. 6, pp. 77–98, 173–194.

R. E. Siegel. 1970. *Galen on Sense Perception*. Basel, Switz.: S. Karger.

14. Siegel, ref. 13, pp. 10–126.

15. Galen, ref. 12, vol. 2, pp. 463–503.

16. Siegel, ref. 13, pp. 127–139, 158–193.

17. Smith, ref. 6, pp. 173–181.

18. O. Temkin. 1951–1952. On Galen's pneumatology. *Gesnerus,* 8, 180–189.

19. Galen, ref. 12, p. 431.

20. C. G. de Gutiérrez-Mahoney and M. M. Schechter. 1972. The myth of the *rete mirabile* in man. *Neuroradiology,* 4, 141–158.

21. J. Rocca. 1997. Galen and the ventricular system. *Journal of the History of the Neurosciences,* 6, 227–239.

22. A. Karenberg and I. Hort. 1998. Medieval descriptions and doctrines of stroke: Preliminary analysis of selective sources. Part I: The struggle for terms and theories—Late antiquity and early middle ages (300–800). *Journal of the History of the Neurosciences,* 7, 162–173.

23. Walsh, ref. 2, p. 7.

24. P. Kellaway. 1946. The part played by electric fish in the early history of bioelectricity and electrotherapy. *Bulletin of the History of Medicine,* 20, 112–137.

    J. F. Blumenbach. 1795. *Elements of Physiology.* Philadelphia: Thomas Dobson at the Stone-House, pp. 217–218.

25. Scribonius Largus. 1529. *De compositionibus medicamentorum. Liber unus.* Ed. J. Ruel. Paris: Wechel.

26. A. Duggan. 1938. *He Died Old: Mithradates Eupator, King of Pontus.* London: Faber and Faber.

27. J. Walsh. 1930. Galen's second sojourn in Italy and his treatment of the family of Marcus Aurelius. *Medical Life,* 37, 473–505.

28. Walsh, ref. 27, p. 491.

29. C. S. Sherrington. 1946. *The Endeavour of Jean Fernel.* Cambridge: Cambridge University Press.

30. G. Majno. 1975. *The Healing Hand.* Cambridge, MA: Harvard University Press, pp. 415.

31. V. Nutton. 1995. Galen ad multos annos. *Dynamis,* 15, 25–39.

32. H. E. Sigerist. 1933. *The Great Doctors.* New York: W. W. Norton, p. 76.

Andreas Vesalius: The New "Human" Neuroanatomy

1. A. Karenberg and I. Hort. 1998. Medieval descriptions and doctrines of stroke: Preliminary analysis of selective sources. Parts I, II, and III. *Journal of the History of the Neurosciences,* 7, 162–200.

2. O. Temkin. 1991. *Hippocrates in a World of Pagans and Christians.* Baltimore, MD: Johns Hopkins University Press.

3. W. Telfer. 1955. *Cyril of Jerusalem and Nemesius of Emesa.* Philadelphia: The Westminster Press, pp. 341–342.

4. E. Clarke and K. Dewhurst. 1974. *An Illustrated History of Brain Function.* Berkeley: University of California Press, 10–48.

5. C. D. O'Malley. 1964. *Andreas Vesalius of Brussels, 1514–1564.* Berkeley: University of California Press, p. 10.

6. O'Malley, ref. 5, pp. 11–16.

7. Mondino de' Luzzi. 1495. *Anatomia corporis humani.* Venice: Bernardinus Venetus de Vitalibus, for Hieronymus de Durantibus. (First printed c. 1316.)

8. O'Malley, ref. 5, p. 12.

9. D. W. Amundsen. 1972. Medieval canon law on medical and surgical practice by the clergy. *Bulletin of the History of Medicine,* 52, 22–44.

10. C. D. O'Malley and J. B. de C. M. Saunders. 1952. *Leonardo da Vinci on the Human Body.* New York: Henry Schuman.

11. J. Berengario da Carpi. 1518. *Tractatus . . . de fractura calve sive cranei.* Bologna: de Benedictis.

12. O'Malley, ref. 5, p. 28.

13. O'Malley, ref. 5, pp. 35–61.

14. W. Stirling. 1902. Andreas Vesalius. In *Some Apostles of Physiology.* London: Waterlow and Sons Ltd., p. 2.

15. F. Baker. 1909. The two Sylviuses. An historical study. *Bulletin of the Johns Hopkins Hospital,* 20, 329–339.

16. O'Malley, ref. 5, pp. 62–72.

17. O'Malley, ref. 5, pp. 73–74.

18. A. Vesalius. 1538. *Tabulae anatomicae.* Venetiis: D. Bernardini.

19. R. Eriksson. 1959. *Andreas Vesalius' First Public Anatomy at Bologna 1540: An Eyewitness Report.* Uppsala: Almqvist and Wiksells Boktryckeri.

20. A. Vesalius. 1543. *De humani corporis fabrica.* Basilae: Joannis Oporini.

21. O'Malley, ref. 5, pp. 100–101, 111.

22. O'Malley, ref. 5, pp. 124–129.

23. A. Vesalius. 1543. *Epitome.* Basilae: Joannis Oporini.

24. O'Malley, ref. 5, p. 116.

25. C. Singer. 1952. *Vesalius on the Human Brain.* London: Oxford University Press, p. 40.

26. Singer, ref. 25, p. 57.

27. Singer, ref. 25, p. 4.

28. Singer, ref. 25, p. 7.

29. O'Malley, ref. 5, pp. 214–222, 238–240, 247–251.

30. O'Malley, ref. 5, p. 239.

31. J. Sylvius. 1551. *Vaesani cuiusdam calumniarum in Hippocratis Galenique rem anatomicam depulsio.* Parrhisiis: Apus Catharinam Barbé.

32. O'Malley, ref. 5, p. 239.

33. O'Malley, ref. 5, p. 250.

34. L. Fuchs. 1551. *Humani corporis fabrica epitome.* Lugduni: Apud Antonium Vincentium.

35. O'Malley, ref. 5, p. 246.

36. O'Malley, ref. 5, pp. 187–224.

37. A. Vesalius. 1555. *De humani corporis fabrica*, 2nd ed. Basilae: Joannis Oporini.

38. O'Malley, ref. 5, pp. 283–314, 407–419.

39. Paracelsus. 1941. *Four Treatises of Theophrastus von Hohenheim Called Paracelsus.* Ed. H. Sigerist. Baltimore, MD: Johns Hopkins University Press.

   Paracelsus. 1951. *Paracelsus: Selected Writings.* Trans. N. Guterman. New York: Pantheon.

   W. Pagel. 1958. *Paracelsus: An Introduction to Philosophical Medicine in the Era of the Renaissance.* Basel: S. Karger.

40. L. Bakay. 1987. *Neurosurgeons of the Past.* Springfield, IL: Charles C Thomas.

41. Berengario da Carpi, ref. 11 (presented in Bakay, ref. 40, pp. 22–26).

42. H. Peyligk. 1516. *Compendiosa capitis physici declaratio . . .* Leipzig: Impressit Wolfgangus Monacensis.

René Descartes: The Mind-Body Problem

1. J. R. Vrooman. 1970. *René Descartes: A Biography.* New York: G. P. Putnam's Sons, pp. 19–44.

R. Descartes. 1927. *Selections*. Ed. R. M. Eaton New York: Charles Scribner's Sons, pp. v–xl.

2. R. Descartes. 1970. *Philosophical Letters*. Ed. and trans. A. Kenny. London: Oxford University Press, p. 163.

3. Vrooman, ref. 1, pp. 45–67.

4. R. Fishman. 1971. Descartes' dream. *Archives of Ophthalmology*, 86, 446–448.
    Vrooman, ref. 1, pp. 68–134.

5. R. Descartes. 1972. *Treatise of Man*. Trans. T. S. Hall. Cambridge, MA: Harvard University Press pp. xxiii–xxv.

6. Vrooman, ref. 1, pp. 135–166.

7. R. Descartes. 1649. *Les Passions de l'âme*. Amsterdam: L. Elzevir. Translated by S. Voss as *The Passions of the Soul*. Indianapolis: Hackett Publishing Co., 1989, Art. XXXI, p. 36.

8. R. Descartes. 1662. *De homine figuris et latinitate donatus a Florentio Schuyl*. Leyden: Franciscum Moyardum and Petrum Leffen. Translated by Hall, ref. 5.

9. Vrooman, ref. 1, pp. 212–261.

10. G. Jefferson. 1949. René Descartes on the localisation of the soul. *Irish Journal of Medical Science*, 285, 691–706.

11. Descartes, ref. 1, p. xi.

12. F. Schiller, 1996. Pineal gland, perennial puzzle. *Journal of the History of the Neurosciences*, 4, 155–165.

13. J. A. Kappers and P. Pévet (Eds.). 1979. *The Pineal Gland of Vertebrates Including Man*. Progress in Brain Research, vol. 52. Amsterdam: Elsevier.
    L. Tamarkin, C. J. Baird and O. F. X. Almeida. 1985. Melatonin: A coordinating signal for mammalian reproduction? *Science*, 227, 714–720.
    M. R. Rosenzweig, A. L. Leiman, and S. M. Breedlove. 1996. *Biological Psychology*. Sunderland, MA: Sinauer Associates, pp. 235–236.

14. Descartes, ref. 5, pp. 21–22.

15. R. Descartes. 1637. *Discours de la méthode*. Reprinted in *Discourse on Method, and Meditations*. Trans L. J. Lafleur. New York: Library of Liberal Arts, 1960, p. 24.

16. Descartes, ref. 1, p. 358.

17. S. Finger. 1995. Descartes and the pineal gland in animals: A frequent misinterpretation. *Journal of the History of the Neurosciences*, 4, 166–182.

18. T. Willis. 1664. *Cerebri anatome, cui accessit nervorum descriptio et usus*. London: J. Martyn and J. Allestry. Reprinted as *The Anatomy of the Brain and Nerves*. Ed. W. Feindel. Montreal: McGill University Press, 1965, pp. 106–107.

19. Finger, ref. 17.

20. Descartes, ref. 2, p. 72.

21. Descartes, ref. 2, p. 70.

22. R. A. Watson. 1981. Cartesianism compounded: Louis de la Forge. *Studia Cartesiana*, 2, 165–171.

23. P. S. Regis. 1690. *Système de philosophie, contenant la logique, la métaphysique, la physique, & la morale* (3 vols.). Paris: Thierry.
    R. A. Watson. 1970. Introduction. In P. S. Regis, *Cours entier de philosophie*. New York: Johnson Reprint Co., pp. v–xx.

24. L. C. Rosenfield. 1940. *From Beast-Machine to Man-Machine*. New York: Oxford University Press, pp. 27–63.

25. Rosenfield, ref. 24, pp. 73–107.

26. Rosenfield, ref. 24.

27. S. Voss. 1993. *Essays on the Philosophy of Science of René Descartes*. New York: Oxford University Press, pp. 135–136.

28. C. Singer. 1952. *Vesalius on the Human Brain*. London: Oxford University Press, p. 67.

29. T. Bartholin. 1674. *Anatomia*. Lugduni Batavorum: Ex officina Hackiana.

30. Galen. 1968. *De usu partium*. Reprinted as *On the Usefulness of the Parts of the Body*. Trans. M. T. May. Ithaca, NY: Cornell University Press, pp. 419–420.

31. N. Steno. 1669. *Discours de Monsieur Stenon sur l'anatomie du cerveau*. Paris: R. de Ninville. Reprinted as *A Dissertation on the Anatomy of the Brain by Nicolaus Steno*. Trans. G. Douglas. Copenhagen: Nyt Nordisk Forlag, 1950. (Paper presented orally in 1665).

32. G. Scherz. 1968. *Steno and Brain Research in the Seventeenth Century*. Oxford: Pergamon Press, p. 83.

33. H. W. Smith. 1959. The biology of consciousness. In C. McC. Brooks and P. F. Cranefield (Eds.), *The Historical Development of Physiological Thought*. New York: Hafner, pp. 109–136.

34. Rosenfeld, ref. 24, p. 70.

35. Descartes, ref. 2, p. 208.

36. N. Malebranche. *Oeuvres complètes*, vol. 2. Ed. G. Rodis-Lewis. Paris: J. Vrin, p. 394.

37. J. O. de La Mettrie. 1748. *L'Homme-machine*. Leyden: E. Luzak Fils.

38. Rosenfield, ref. 24, pp. 111–114.

39. Rosenfield, ref. 24, pp. 154–179.

Thomas Willis: The Functional Organization of the Brain

1. T. Willis. 1664. *Cerebri anatome, cui accessit nervorum descriptio et usus*. London: J. Martyn and J. Allestry. Reprinted as *The Anatomy of the Brain and Nerves*. Ed. W. Feindel. Montreal: McGill University Press, 1965, p. 55.

2. H. Isler. 1968. *Thomas Willis (1621–1675): Doctor and Scientist*. New York: Hafner.

3. K. Dewhurst. 1982. Thomas Willis and the foundations of British Neurology. In F. C. Rose and W. F. Bynum (Eds.), *Historical Aspects of the Neural Sciences*. New York: Raven Press, pp. 327–346.

4. T. Willis. 1659. *Diatribae duae medico-philosophicae quarum prior agit de fermentation sive de motu intestino particulum in quovis corpore, altera de febribus sive de motu earundum in sanguine anamium; his accessit dissertatio episticola de urinis*. Londini: typis Tho. Roycroft; Impensis J. Martyn, J. Allesrty, & T. Dicas. Reprinted as *Dr. Willis's Practice of Physick*. Trans. S. Pordage. London: Dring, Harper, and Leigh, 1684.

5. A. G. Debus. 1966. *The English Paracelsians*. New York: Franklin Watts, pp. 13–48.

6. Isler, ref. 2, pp. 13–23.

7. Dewhurst, ref. 3, p. 330.

8. Willis, ref. 1 (1664).
    W. Feindel. 1965. The origins and significance of *Cerebri anatome*. In ref. 1 (trans.), pp. 3–9.

9. T. Willis. 1672. *De anima brutorum*. Oxonii: R. Davis. Reprinted as "Two Discourses Concerning the Soul of Brutes." in *Practice of Physick* (ref. 4).

10. Isler, ref. 2, pp. 19–20.

11. R. Gloclenius. 1594. *Psychologia: hoc es, de hominis perfectione* . . . Ex officina typographica Pauli Egenolphi.

12. M. Neuburger. 1981. *The Historical Development of Experimental Brain and Spinal Cord Physiology before Flourens*. Trans. and ed. with additional material by E. Clarke. Baltimore, MD: Johns Hopkins University Press, pp. 21–45.

13. Willis, ref. 1 (trans.), p. 92.

14. Willis, ref. 1 (trans.), Figure 4 and legend (p. 70) and p. 162.

15. A. Meyer and R. Hierons. 1964. A note on Thomas Willis' views on the corpus striatum and the internal capsule. *Journal of the Neurological Sciences*, 1, 547–554.

16. Willis, ref. 1 (trans.), p. 102.

17. Willis, ref. 1 (trans.), p. 102.

18. Willis, ref. 1 (trans.), p. 110–113.

19. S. Finger and D. Stein. 1982. *Recovery from Brain Damage: Research and Clinical Perspectives*. Orlando, FL: Academic Press, pp. 257–270.

20. Willis, ref. 9 (trans.), vol. 1, p. 188.

21. P. F. Cranefield. 1961. A seventeenth century view of mental deficiency and schizophrenia: Thomas Willis on "stupidity or foolishness." *Bulletin of the History of Medicine*, 35, 291–316.

22. Willis, ref. 9 (trans.), vol. 2, p. 134.

23. J. D. Spillane. 1981. *The Doctrine of the Nerves*. Oxford: Oxford University Press, pp. 65–67.

24. Spillane, ref. 23, pp. 72–73.
    Isler, ref. 2, pp. 127–135.

25. L. C. McHenry Jr. 1969. *Garrison's History of Neurology*. Springfield, IL: Charles C Thomas, p. 55.

26. N. Steno. 1669. *Discours de Monsieur Stenon sur l'anatomie du cerveau*. Paris: R. de Ninville. Reprinted as "A Dissertation on the Anatomy of the Brain by Nicolaus Steno." Trans. G. Douglas. Copenhagen: Nyt Nordisk Forlag, 1950. (Paper presented orally in 1665.)
    K. Dewhurst. 1968. Willis and Steno. In G. Scherz (Ed.), *Steno and Brain Research in the Seventeenth Century*. Oxford: Pergamon Press, pp. 43–48.

27. Steno, ref. 26 (trans.), p. 17.

28. Steno, ref. 26 (trans.), pp. 8–9.

29. Isler, ref. 2, p. 193.

30. G. Keynes. 1961. The history of myasthenia gravis. *Medical History*, 5, 313–326.

31. Isler, ref. 2, pp. 161, 176–182.

32. C. S. Sherrington. 1941. *Man on his Nature*. Cambridge, UK: Cambridge University Press, p. 245.

Luigi Galvani: Electricity and the Nerves

1. M. Brazier. 1958. The evolution of concepts relating to the electrical activity of the nervous system. In M. W. Perrin (Chair), *The Brain and Its Functions*. Oxford: Blackwell, pp. 191–222.
    R. W. Home. 1970. Electricity and the nervous fluid. *Journal of the History of Biology*, 3, 235–251.

2. I. Newton, 1704. *Opticks*. London: Smith and Walford. Reprint New York: Dover Publications, 1952.

3. A. von Haller. 1769. *Elementa physiologiae corporis humani. Tomus IV: Sensus externii internii*. Lausanne: Sumptibus Franciscis Grasset et Sociorum, p. 381.
    A. von Haller. 1786. *First Lines of Physiology*. Reprint. New York: Johnson Reprint Company, 1966.

4. H. Boerhaave. 1743. *Anatomical Lectures on the Theory of Physic*. London: Innys, vol. 2, p. 310.

5. A. Van Leeuwenhoek. 1674. More observations from Mr. Van Leeuwenhook, in a letter of Sept. 7. 1674. sent to the publisher. *Philosophical Transactions of the Royal Society*, 9, 178–182.

6. N. Stensen. 1667. *Elementorum myologiae specimen*. Florence: Stella, p. 5.

7. M. Brazier. 1984. *A History of Neurophysiology in the Seventeenth and Eighteenth Centuries*. New York: Raven, p. 191.

8. P. Kellaway. 1946. The part played by electric fish in the early history of bioelectricity and electrotherapy. *Bulletin of the History of Medicine*, 20, 112–137.

9. W. D. Hackmann. 1978. *Electricity from Glass*. Alphen aan den Rijn, The Netherlands: Sijthoff and Noordhoff.

10. Hackmann, ref. 9, p. 75.

11. J. G. Krüger. 1744. *Zuschrift an seine Zuhörer, worinnen er ihnen seive Gedanken von der Electricität mitteilet*. Halle: Carl Herrmann Hemmerde. Translated by J. S. Petrofsky in S. Finger, T. E. LeVere, C. R. Almli, and D. G. Stein (Eds.), *Brain Injury and Recovery: Theoretical and Controversial Issues*. New York: Plenum Press, 1988, p. 273.

12. C. Dorsman and C. A. Crommelin. 1957. The invention of the Leyden jar. *Janus*, 46, 275–280.

   Brazier, ref. 7, pp. 178–180.

   Hackmann, ref. 10, pp. 90–103.

13. J.-A. Nollet. 1746. *Essai sur l'electricité des corps*. Paris: Guerin.

14. F. Schiller, F. 1981. Reverend Wesley, Doctor Marat and their Electrical Fire. *Clio Medica*, 15, 159–176.

15. J. Wesley. 1747. *Primitive Physick, or, an Easy and Natural Way of Curing Most Diseases*. London: Thomas Trye.

16. J. Wesley. 1759. *The Desideratum: or, Electricity Made Plain and Useful. By a Lover of Mankind and of Common Sense*. London: Ballière, Tindall, and Cox.

17. A. H. Smith. 1905. *The Writings of Benjamin Franklin*, vol. 3. New York: Macmillan.

18. Smith, ref. 17, p. 426.

19. Schiller, ref. 14.

20. J. P. Marat. 1782. *Recherches physiques sur l'electricité*. Paris: Clousier.

   J. P. Marat. 1784. *Mémoires sur l'electricité médicale*. Paris: Méquignon.

21. S. Gray. 1731. A letter to Cromwell Mortimer, MD Sec. Roy. Soc. containing several experiments concerning electricity. *Philosophical Transactions of the Royal Society*, 37, 18–44.

   S. Hales. 1726–1732. *Statical Essays* (2 vols.). London: Innys Manby.

   A. Monro. 1781. *The Works of Alexander Monro (Collected by His Son)*. Edinburgh: C. Eliot.

22. F. Redi. 1675. *Experimenta circa generationem insectorum ad nobilissimum virum Carlo Dati*. Amstelodami: A Frisii.

   S. Lorenzini. 1678. *Observazioni interne alle Torpedini . . .* Florence: Per l'Onofri.

23. J. Davis. 1705. *The Curious and Accurate Observations . . . on the Dissection of the Cramp-Fish Done into the English from the Italian*, 4th ed.). London, p. 72.

24. J. Walsh. 1773. On the electric property of the torpedo. *Philosophical Transactions of the Royal Society*, 63, 461–477.

   J. Walsh. 1774. Of torpedos found on the coast of England. *Philosophical Transactions of the Royal Society*, 64, 464–473.

25. J. Hunter. 1773. Anatomical observations on the torpedo. *Philosophical Transactions of the Royal Society*, 63, 481–488.

   J. Hunter. 1775. An account of the *Gymnotus electricus*. *Philosophical Transactions of the Royal Society*, 65, 395–407.

26. T. Cavallo. 1795. *A Complete Treatise of Electricity in Theory and Practice with Original Experiments*, 4th ed., vol. 2. London: C. Dilly, p. 309.

27. I. B. Cohen, Introduction, in L. Galvani. 1791. De viribus electricitatis in motu muscu-lari commentarius. *De Bononiensi Scientiarum et Artium Instituto atque Academia Commentarii*, 7, 363–418. Reprinted as *Commentary on the Effects of Electricity on*

Muscular Motion. Trans. M. G. Foley. Norwalk, CT: Burndy Library, 1953, pp. 9–41.

28. G. C. Pupilli and E. Fadiga. 1963. The origins of electrophysiology. *Cahiers d'Histoire Mondiale*, 7, 547–588.

29. Galvani, ref. 27.

30. G. Aldini. 1804. *Essai théorique et expérimental sur le galvanisme.* Paris: Fournier.

31. G. Aldini. 1803. An account of the galvanic experiments performed by John Aldini, Professor of Experimental Philosophy in the University of Bologna, &c. on the body of a malefactor executed at Newgate, January 17, 1803. London: Cuthell Martin. Aldini, ref. 30, experiment 93, p. 81.

32. J. F. Blumenbach. 1795. *Elements of Physiology.* Philadelphia: Thomas Dobson at the Stone-House, pp. 217–218.

33. E. du Bois-Reymond. 1848. *Untersuchung über thierische Elektricität.* Berlin: G. Reimer, pp. 50–51.

34. A. Volta. 1793. Account of some discoveries made by Mr. Galvani from Mr. Alexander Volta to Mr. Tiberius Cavallo. *Philosophical Transactions of the Royal Society*, 83, 10–44.

35. G. Aldini. 1794. *De animali electricitate dissertationes duae.* Bologna: Emidio Dal-l'Olmo.

36. J. F. Fulton and H. Cushing. 1936. A bibliographical study of the Galvani and the Aldini writings on animal electricity. *Annals of Science*, 1, 239–268.
    Pupilli and Fadiga, ref. 28, pp. 567–569.

37. E. Clarke and L. S. Jacyna. 1987. *Nineteenth-Century Origins of Neuroscientific Concepts.* Berkeley: University of California Press, pp. 168–175.
    Pupilli and Fadiga, ref. 28, pp. 562–578.

38. F. A. Humboldt. 1797. *Versuche über die gereizte Muskel und Nervenfaser.* Posen, Decker, and Berlin: H. A. Rottmann.

39. A. Volta. 1800. Letter to Sir Joseph Banks, March 20th, 1800. On electricity excited by the mere contact of conducting substances of different kinds. *Philosophical Transactions of the Royal Society*, 90, 403–431.
    C. J. Pfeiffer. 1985. *The Art and Practice of Medicine in the Early Nineteenth Century.* Jefferson, NC: McFarland and Co.
    Dr. Bischoff. 1802. On Galvanism and its medical application. *Medical and Physical Journal*, 7, 529–540.

40. K. A. Weinhold. 1817. *Versuche über das Leben und seine, Grundkräfte, auf dem Wege der experimental-Physiologie.* Magdeburg: Creutz.
    S. Finger and M. B. Law. 1998. Karl August Weinhold and his "Science" in the Era of Mary Shelley's Frankenstein: Experiments on Electricity and the Restoration of Life. *Journal of the History of Medicine and Allied Sciences*, 53, 161–180.

41. Weinhold, ref. 40, pp. 35–36.

42. Finger and Law, ref. 40.

43. M. Shelley. 1818. *Frankenstein, or the Modern Prometheus.* London: Printed for Lackington, Hughes, Harding, Mavor and Jones.

44. M. Shelley. 1831. *Frankenstein, or the Modern Prometheus*, 3rd ed. London: Colburn and Bentley. Reprint. Oxford: Oxford University Press, 1990, pp. 8–9.

45. Shelley, ref. 44, p. 57.

Franz Joseph Gall: The Cerebral Organs of Mind

1. M. Ramström. 1910. *Emanuel Swedenborg's Investigations in Natural Science and the Basis for his Statements Concerning the Functions of the Brain.* Uppsala, Sweden: University of Uppsala.
    S. Toksvig. 1948. *Emanuel Swedenborg: Scientist and Mystic.* New Haven, CT: Yale University Press.

2. N. Steno. 1669. *Discours de Monsieur Stenon sur l'anatomie du cerveau*. Paris: R. de Ninville. Reprinted as "A Dissertation on the Anatomy of the Brain by Nicolaus Steno" Trans. G. Douglas. Copenhagen: Nyt Nordisk Forlag, 1950, p. 4. (Paper presented orally in 1665.)

3. G. Fritsch and E. Hitzig. 1870. Über die elektrische Erregbarkeit des Grosshirns. *Archiv für Anatomie und Physiologie*, 300–332.

4. E. Swedenborg. 1882. *The Brain, Considered Anatomically, Physiologically, and Philosophically* part I. Trans. and ed. R. L. Tafel. London: Speirs, p. 73.

5. E. Swedenborg. 1740–1741. *Oeconomia regni animalis*. Amsteldami: Franciscum Changuion.

6. E. Swedenborg. 1745. *Regnum animale*. Hagae: Adrianum Blyvenburgium.

7. Swedenborg, ref. 4.
    E. Swedenborg. 1940. *Three Transactions on the Cerebrum*. (3 vols.). Ed. A. Acton. Philadelphia: New Church Book Center.

8. K. Akert and M. P. Hammond. 1962. Emanuel Swedenborg (1688–1772) and his contribution to neurology. *Medical History*, 6, 255–266.

9. R. M. Young. 1970. *Mind, Brain and Adaptation in the Nineteenth Century*. Oxford: Clarendon Press, pp. 9–53.
    O. Temkin. 1947. Gall and the phrenological movement. *Bulletin of the History of Medicine*, 21, 275–321.
    E. H. Ackerknecht and H. V. Vallois. 1956. *Franz Joseph Gall, Inventor of Phrenology and His Collection*. Madison, WI: University of Wisconsin Press.

10. Ackerknecht and Vallois, ref. 9, pp. 8–9.

11. B. Hollander. 1901. *The Revival of Phrenology: The Mental Functions of the Brain*. New York: G. P. Putnam's Sons, p. 373.

12. Hollander, ref. 11, p. 377.

13. Tenon, Portal, Sabatier, Pinel and Cuvier. 1809. Report on a memoir of Drs. Gall and Spurzheim, relative to the anatomy of the brain, presented to, and adopted by the Class of Mathematical and Physical Sciences of the National Institute. (Given at the Institute, 15 April 1808) *Edinburgh Medical and Surgical Journal*, 5, 36–66.

14. F. J. Gall and J. Spurzheim. 1810–1819. *Anatomie et physiologie du système nerveux en général, et du cerveau en particulier* (4 vols.). Paris: F. Schoell. (Gall was the sole author of the first two volumes of the series.)

15. F. J. Gall. 1822–1826. *Sur les fonctions du cerveau* (6 vols.). Paris: Ballière. Reprinted as *On the Functions of the Brain and Each of Its Parts: with Observations on the Possibility of Determining the Instincts, Propensities, and Talents, or the Moral and Intellectual Dispositions of Men and Animals, by the Configuration of the Brain and Head*. Trans. W. Lewis Jr. Boston: Marsh, Capen and Lyon, 1935.

16. Ackerknecht and Vallois, ref. 9, p. 7.

17. Ackerknecht and Vallois, ref. 9, p. 37.

18. Ackerknecht and Vallois, ref. 9, pp. 37–86.

19. Gall, ref. 15, vol. 5, pp. 16–18.

20. N. Capen, *Reminiscences of Dr. Spurzheim and George Combe*. New York: Fowler and Wells, 1881.

21. J. G. Spurzheim. 1815. *The Physiognomical System of Drs. Gall and Spurzheim*. London: Baldwin, Cradock, and Joy.
    J. G. Spurzheim. 1832. *Outlines of Phrenology*. Boston: Marsh, Capen and Lyon.

22. E. Clarke and L. S. Jacyna. 1987. *Nineteenth-Century Origins of Neuroscientific Concepts*. Berkeley: University of California Press, p. 223.

23. T. I. M. Forster. 1815. Sketch for the new anatomy and physiology of the brain and nervous system of Drs. Gall and Spurzheim, considered as comprehending a complete sys-

tem of phrenology. *The Pamphleteer Respectfully Dedicated to Both Houses of Parliament*, 5, 219–243.

24. J. C. Spurzheim. 1818. *Observations sur la phrénologie*. Paris: Treuttel & Würtz.

25. Hollander, ref. 11, p. 384.

26. A. A. Walsh. 1971. George Combe: A portrait of a heretofore generally unknown behaviorist. *Journal of the History of Behavioral Sciences*, 7, 269–278.

27. Capen, ref. 20, p. 53.

28. G. Combe. 1827. *The Constitution of Man, Considered in Relation to External Objects*. Edinburgh: Neill.

29. R. M. Young. 1968. The functions of the brain: Gall to Ferrier (1808–1886). *Isis*, 59, 251–268.

30. C. Gibbon. 1878. *Life of George Combe* (2 vols.). London: Macmillan.
    Hollander, ref. 11, pp. 390–392.

31. F. R. Freemon. 1992. Phrenology as clinical neuroscience: How American academic physicians in the 1820s and 1830s used phrenological theory to understand neurological symptoms. *Journal of the History of the Neurosciences*, 1, 131–143.

32. J. D. Davies. 1955. *Phrenology: Fad and Science*. New Haven, CT: Yale University Press, p. 13.

33. C. Caldwell. 1824. *Elements of Phrenology*. Lexington, KY: T. T. Skillman.

34. A. A. Walsh. 1972. The American tour of Dr. Spurzheim. *Journal of the History of Medicine and Allied Sciences*, 27, 187–205.

35. Davies, ref. 32.

36. J. Jefferson. 1960. *Selected Papers of Sir. Geoffrey Jefferson*. Springfield, IL: Charles C Thomas, p. 96.

37. R. E. Riegel. 1930. Early phrenology in the United States. *Medical Life*, 37, 361–376.
    Davies, ref. 32.

38. J. M. D. Olmsted. 1953. Pierre Flourens. In E. A. Underwood (Ed.), *Science Medicine and History*, vol. 2. London: Oxford University Press, pp. 290–302.

39. M.-J.-P. Flourens. 1824. *Recherches expérimentales sur les propriétés et les fonctions du système nerveux dans les animaux vertébrés*. Paris: Ballière.

40. M.-J.-P. Flourens. 1846. *Phrenology Examined*. Trans. from the 2nd French ed. [1842] by C. de L. Meigs. Philadelphia, PA: Hogan and Thompson, p. 96.

41. Flourens. ref. 40, p. 102.

42. M.-J.-P. Flourens. 1864. *Psychologie Comparée*, 2nd ed. Paris: Garnier Frères, p. 234.

43. Anonymous. 1815. The doctrines of Gall and Spurzheim. *Edinburgh Review*, 25, 227.

44. Anonymous, ref. 43, p. 268.

45. Hollander, ref. 11, p. 406.

46. Anonymous. 1832. Boston Notions. *New England Magazine*, 3, 397–398.

47. Freemon, ref. 31.

48. G. E. Smith, 1923. Cited in Hollander, ref. 11, p. 380.

49. Anonymous. 1857. Phrenology in France. *Blackwood's Edinburgh Magazine*, 82, 669–671.

50. Anonymous, ref. 49, p. 672.

Paul Broca: Cortical Localization and Cerebral Dominance

1. B. Stookey. 1963. Jean-Baptiste Bouillaud and Ernest Aubertin. *Journal of the American Medical Association*, 184, 1024–1029.

2. J.-B. Bouillaud. 1825. Recherches cliniques propres à démontrer que la perte de la parole correspond à la lésion des lobules antérieurs du cerveau et à confirmer l'opinion de M. Gall sur le siège de l'organe du langage articulé. *Archives Générales de Médecine*, 8, 25–45.

J.-B. Bouillaud. 1825. *Traité clinique et physiologique de l'encéphalite ou inflammation du cerveau*. Paris: J. B. Ballière.

3. J.-B. Bouillaud. 1830. Recherches expérimentales sur les fonctions du cerveau (lobes cérébraux) en général, et sur celles de sa portion antérieure en particulier. *Journal Hebdomadaire de Médecine*, 6, 527–570.

J.-B. Bouillaud. 1839. Exposition de nouveaux faits à l'appui de l'opinion qui localise dans les lobules antérieurs du cerveau le principe législateur de la parole; examen préliminaire des objections dont cette opinion à été sujet. *Bulletin de l'Académie Royale de Médecine*, 4, 282–328.

J.-B. Bouillaud. 1848. *Recherches cliniques propres à démontrer que le sens du langage articulé et le principe coordinateur des mouvements de la parole résident dans les lobules antérieurs du cerveau*. Paris: J. B. Ballière.

4. A. L. Benton. 1984. Hemispheric dominance before Broca. *Neuropsychologia*, 22, 807–811.

5. G. Andral. 1840. *Clinique Médicale . . .* , 4th ed. Paris: Fortin, Masson et Cie.

6. F. Schiller. 1979. *Paul Broca: Founder of French Anthropology, Explorer of the Brain*. New York: Oxford University Press. Berkeley: University of California Press, pp. 174, 199.

Anonymous. 1865. The localization of speech. *British Medical Journal*, 1, 670.

A. Kussmaul. 1878. Disturbances of speech. In H. V. von Ziemssen (Ed.), *Cyclopedia of the Practice of Medicine*, vol. 14. Trans. J. A. McCreery. New York: Wood, pp. 581–875.

7. Schiller, ref. 6, p. 175.

8. Stookey, ref. 1.

9. E. Aubertin. 1861. Discussion of Gratiolet's paper, Sur le volume et la forme du cerveau. *Bulletins de la Société d'Anthropologie*, 2, 71–72, 74–75, 81, 209–220, 275–276, 278–279, 421–446.

10. Aubertin, ref. 9, pp. 217–218. Translated by E. Clarke and C. D. O'Malley in *The Human Brain and Spinal Cord*. Berkeley, CA: University of California Press, 1968, p. 492.

11. B. Stookey. 1954. A note on the early history of cerebral localization. *Bulletin of the N. Y. Academy of Medicine*, 30, 571.

12. P. Broca. 1861. Sur le volume et la forme du cerveau suivant les individus et suivant les races. *Bulletins de la Société d'Anthropologie*, 2, 139–207, 301–321, 441–446.

13. Schiller, ref. 6, pp. 7–89.

14. Schiller, ref. 6, pp. 90–119.

15. P. Broca. 1861. Remarques sur le siège de la faculté du langage articulé; suivies d'une observation d'aphémie (perte de la parole). *Bulletins de la Société Anatomique*, 6, 330–357, 398–407. Reprinted as "Remarks on the seat of the faculty of articulate language, followed by an observation of aphemia," in G. von Bonin, *Some Papers on the Cerebral Cortex*. Springfield, IL: Charles C Thomas, 1960, pp. 64–65.

16. P. Broca. 1861. Perte de la parole, ramollissement chronique et destruction partielle du lobe antérieur gauche du cerveau. *Bulletins de la Société d'Anthropologie*, 2, 235–238.

17. Broca, ref. 15.

18. P. Broca. 1873. Sur les crânes de la caverne de l'Homme Mort (Lozère). *Revue d'Anthropologie*, 2, 1–53.

19. S. Finger. 1994. *Origins of Neuroscience*. New York: Oxford University Press, pp. 316–331.

20. H. D. Rolleston. 1888. Description of the cerebral hemispheres of an adult Australian male. *Journal of the Anthropological Institute of Great Britain and Ireland*, 17, 32–42.

21. Rolleston, ref. 19, p. 33.

22. P. Broca. 1861. Nouvelle observation d'aphémie produite par un lésion de la moitié postérieure des deuxième et troisième circonvolutions frontales. *Bulletins de la Société Anatomique*, 6, 398–407.

23. P. Broca. 1863. Localisation des fonctions cérébrales. Siège du langage articulé. *Bulletins de la Société d'Anthropologie*, 4, 200–204.

    A. Trousseau. 1864. De l'aphasie, maladie décrite récemment sous le nom impropre d'aphémie. *Gazette des Hôpitaux, Paris*, 37, 13–14, 25–26, 37–39, 48–50.

24. P. Broca. 1865. Sur le siège de la faculté du langage articulé. *Bulletins de la Société d'Anthropologie*, 6, 337–393. Reprinted as Localization of speech in the third left frontal convolution. Trans. E. A. Berker, A. H. Berker, and A. Smith. *Archives of Neurology*, 43 (1986), 1065–1072.

25. G. Dax. 1865. Notes sur le même sujet. *Gazette Hebdomadaire de Médecine et de Chirurgie*, 2, 260–262.

    M. Dax. 1865. Lésions de la moitié gauche de l'encéphale coïncidant avec l'oubli des signes de la pensée (lu au Congrès méridional tenu à Montpellier en 1836). *Gazette Hebdomidaire de Médecine et de Chirurgie*, 2 (2nd ser.), 259–260.

26. A. Souques. 1928. Quelques cas d'anarthrie de Pierre Marie. *Revue Neurologique*, 2, 319–368.

    R. Cubelli and C. G. Montagna. 1994. A reappraisal of the controversy of Dax and Broca. *Journal of the History of the Neurosciences*, 3, 1–12.

    R. J. Joynt and A. L. Benton. 1964. The memoir of Marc Dax on aphasia. *Neurology*, 14, 851–854.

27. S. Finger and D. Roe. 1996. Gustave Dax and the early history of cerebral dominance. *Archives of Neurology*, 53, 806–813.

    D. Roe and S. Finger. 1996. Gustave Dax and his fight for recognition: An overlooked chapter in the early history of cerebral dominance. *Journal of the History of the Neurosciences*, 5, 228–240.

28. J. Osborne. 1833. On the loss of the faculty of speech depending on forgetfulness of the art of using the vocal organs. *Dublin Journal of Medical and Chemical Sciences*, 4, 157–170.

29. Osborne, ref. 28, pp. 161–162.

30. Osborne, ref. 28, p. 169.

31. T. Barlow. 1877. On a case of double cerebral hemiplegia, with cerebral symmetrical lesions. *British Medical Journal*, 2, 103–104.

32. J. Taylor. 1905. *Paralysis and Other Diseases of the Nervous System in Childhood and Early Life*. London: J. A. Churchill.

33. H. C. Bastian. 1898. *A Treatise on Aphasia and Other Speech Defects*. London: H. C. Lewis.

34. P. Eling. 1984. Broca on the relation between handedness and cerebral speech dominance. *Brain and Language*, 22, 158–159.

    L. J. Harris. 1991. Cerebral control for speech in right-handers and left-handers: An analysis of the views of Paul Broca, his contemporaries, and his successors. *Brain and Language*, 40, 1–50.

    L. J. Harris. 1993. Broca on cerebral control for speech in right-handers and left-handers: A note on translation and some further comments. *Brain and Language*, 45, 108–120.

35. P. Broca. 1869. Sur la siège de la faculté du langage articulé. *Tribune Médicale*, 74, 254–256; 75, 265–269.

36. J. H. Jackson. 1864. Hemiplegia on the right side, with loss of speech. *British Medical Journal*, 1, 572–573.

37. J. H. Jackson. 1868. Hemispheral coordination. *Medical Times and Gazette*, 2, 208–209.

38. J. H. Jackson. 1872. Case of disease of the brain—left hemiplegia—mental affection. *Medical Times and Gazette*, 1, 513–514.

39. J. H. Jackson. 1876. Case of large cerebral tumour without optic neuritis and with left hemiplegia and imperception. *Ophthalmic Hospital Reports*, 8, 434–444.

40. Jackson, ref. 39, p. 438.

41. J. H. Jackson. 1874. Remarks on systematic sensations in epilepsies. *British Medical Journal*, 1, 174.
     Jackson, ref. 39.

42. J. H. Jackson. 1874. On the nature of the duality of the brain. *Medical Press and Circular*, new ser., 17, 19–21, 41–44, 63–66.

43. C. Wernicke. 1874. *Der aphasische Symptomenkomplex: eine psychologische Studie auf anatomischer Basis.* Breslau: Cohn and Weigert.

44. A. Harrington. 1987. *Medicine, Mind and the Double Brain.* Princeton, NJ: Princeton University Press.
     A. Harrington. 1985. Nineteenth-century ideas on hemisphere differences and "duality of mind." *Behavioral and Brain Sciences*, 8, 617–660.

45. D. Daiches. 1973. *Robert Louis Stevenson and His World.* London: Thames and Hudson.

46. R. L. Stevenson and W. E. Henley. 1880. *Deacon Brodie; or the Double Life; a Melodrama, Founded on the Facts, in Four Acts and Ten Tableaux.* London.

47. R. L. Stevenson. 1886. *Strange Case of Dr. Jekyll and Mr. Hyde.* London: Longmans, Green, and Co. Reprint. Edinburgh: Cannongate Publishing, 1986.

48. L. C. Bruce. 1895. Notes of a case of dual brain action. *Brain*, 18, 54–65.

49. L. C. Bruce. 1897. Dual brain action and its relation to certain epileptic states. *Medico-Chirurgical Society of Edinburgh*, 16, 114–119.

50. Harrington, 1987, ref. 44, pp. 130–133.

51. C.-E. Brown-Séquard. 1874. Dual character of the brain. *Smithsonian Miscellaneous Collections*, 15, 1–21.

52. Brown-Séquard, ref. 51, p. 20.

53. J. Crichton-Browne. 1907. Dexterity and the bend sinister. *Proceedings of the Royal Institution of Great Britain*, 18, 624.

54. Schiller, ref. 6, pp. 136–164.

55. P. Broca. 1877. Sur la circonvolution limbique et la scissure limbique. *Bulletins de la Société d'Anthropologie*, 12, 646–657.
     P. Broca. 1878. Anatomie comparée des circonvolutions cérébrales. Le grand lobe limbique et la scissure limbique dans la série des mammifères. *Revue d'Anthropologie*, sér. 2, 1, 385–498.

David Ferrier and Eduard Hitzig: The Experimentalists Map the Cerebral Cortex

1. Anonymous. 1881. The festivities of the congress. *British Medical Journal*, 2, 303–304.
     J. D. Spillane. 1981. *The Doctrine of the Nerves.* Oxford: Oxford University Press, pp. 393–394.

2. Klein, Langley, and Schäfer. 1883. On the cortical areas removed from the brain of a dog and from the brain of a monkey. *Journal of Physiology*, 4, 231–247.

3. Spillane, ref. 1, p. 395.

4. Klein, Langley, and Schäfer, ref. 2.

5. Anonymous. 1881. Summons under the Vivisection Act. *British Medical Journal*, 2, 752.

6. M.-J.-P. Flourens. 1824. *Recherches expérimentales sur les propriétés et les fonctions du système nerveux dans les animaux vertébrés.* Paris: J. B. Ballière.

7. G. Fritsch and E. Hitzig. 1870. Über die elektrische Erregbarkeit des Grosshirns. *Archiv für Anatomie und Physiologie*, 300–332. Translated as "On the electrical excitability of the cerebrum" in G. von Bonin (Ed.), *Some Papers on the Cerebral Cortex*. Springfield, IL: Charles C Thomas, 1960, pp. 73–96.

8. G. Haller. 1927. Gustav Fritsch zum Gedächtnis. *Anatomischer Anzeiger*, 64, 257–269.
   A. Kuntz. 1953. Eduard Hitzig (1838–1907). In W. Haymaker (Ed.), *Founders of Neurology*. Springfield, IL: Charles C Thomas, pp. 138–142.

9. E. Hitzig. 1867. Über die Anwendung unpolarisirbarer Elektroden in Elektrotherapie. *Berliner klinische Wochenschrift*, 4, 404–406.
   E. Hitzig. 1869. Verhandlungen ärzlichen Gesellschaften. *Berliner klinische Wochenschrift*, 6, 420.

10. A. E. Walker. 1957. The development of the concept of cerebral localization in the nineteenth century. *Bulletin of the History of Medicine*, 31, 99–121.
   Kuntz, ref. 8.

11. Flourens, ref. 6, p. 274.
   L. Rolando. 1809. *Saggio sopra la vera struttura del cervello dell'uomo e degli animali e sopra le funzioni del sistema nervoso.* Sassari: Stamperìa da S. S. R. M. Privilegiata.

12. R. B. Todd. 1849. On the pathology and treatment of convulsive diseases. *London Medical Gazette*, 8, 661–671, 724–729, 766–772, 815–822, 837–846.

13. Fritsch and Hitzig, ref. 7 (trans.), p. 96.

14. H. R. Viets. 1938. West Riding, 1871–1876. *Bulletin of the Institute of the History of Medicine*, 6, 477–487.

15. E. Critchley. 1957. Sir David Ferrier: 1843–1928. *King's College Hospital Gazette*, 36, 243–250.
   T. G. Stewart. 1928. Sir David Ferrier, LL.D., Sc.D., M.D., F.R.C.P., F.R.S. 1843–1928. *Journal of Mental Science*, 74, 375–380.
   M. A. B. Brazier. 1988. *A History of Neurophysiology in the Nineteenth Century.* New York: Raven Press, pp. 165–170.

16. G. Holmes. 1954. *The National Hospital Queens Square: 1860–1948.* Edinburgh: E. & S. Livingstone, Ltd.

17. D. Ferrier. 1873. Experimental research in cerebral physiology and pathology. *West Riding Lunatic Asylum Medical Reports*, 3, 30–96.
   D. Ferrier. 1874. The localisation of function in the brain. *Proceedings of the Royal Society of London*, 22, 229–232.

18. D. Ferrier. 1874. Pathological illustrations of brain function. *West Riding Lunatic Asylum Medical Reports*, 4, 30–62.

19. D. Ferrier. 1874. On the localisation of the functions of the brain. *British Medical Journal*, 2, 766–767.

20. R. M. Young. 1970. *Mind, Brain and Adaptation in the Nineteenth Century.* Oxford: Clarendon Press, p. 237.

21. D. Ferrier. 1875. Experiments on the brains of monkeys—No. 1. *Proceedings of the Royal Society*, 23, 409–430.
   D. Ferrier. 1875. Experiments on the brain of monkeys: second series (Croonian Lecture). *Philosophical Transactions of the Royal Society of London*, 165, 433–488.

22. J. H. Jackson. 1861. Syphilitic affections of the nervous system. *Medical Times and Gazette*, 1, 648–652; 2, 59–60, 83–85, 133–135, 456, 502–503.

23. J. H. Jackson. 1863. Convulsive spasms of the right hand and arm preceding epileptic seizures. *Medical Times and Gazette*, 1, 110–111.

J. H. Jackson. 1863. Epileptiform convulsions (unilateral) after injury to the head. *Medical Times and Gazette*, 2, 65–66.

J. H. Jackson. 1863. Epileptiform seizures—Aura from the thumb—Attacks of coloured vision. *Medical Times and Gazette*, 1, 589.

24. J. H. Jackson. 1870. A study of convulsions. *Transactions of the St. Andrew's Medical Graduates Association*, 3, 162–204.

25. D. Ferrier. 1883. An address on the progress of knowledge in the physiology and pathology of the nervous system. *British Medical Journal*, 2, 805–808.

D. Ferrier. 1886. *The Functions of the Brain.* New York: G. P. Putnam's Sons. Reprint. Washington, DC: University Publications of America, 1978.

26. Ferrier 1874, ref. 17.

27. D. Ferrier. 1876. *The Functions of the Brain.* London: Smith, Elder and Company, p. 163.

28. S. Finger. 1994. *Origins of Neuroscience.* New York: Oxford University Press, p. 184.

29. Ferrier 1874, ref. 17.

Ferrier, ref. 21.

30. H. Munk. 1878. Weitere Mitteilungen zur Physiologie der Grosshirnrinde. *Verhandlungen der Physiologischen Gesellschaft zu Berlin*, 162–178.

H. Munk. 1881. *Über die Funktionen der Grosshirnrinde.* 3te Mitteilung, pp. 28–53. Berlin: A. Hirschwald. Translated as "On the Functions of the Cortex" in G. von Bonin (Ed.), *Some Papers on the Cerebral Cortex.* Springfield, IL: Charles C Thomas, 1960, pp. 97–117.

31. Finger, ref. 28, pp. 141–143.

32. Finger, ref. 28, pp. 172–173.

33. Ferrier in *Philosophical Transactions,* ref. 21.

34. Ferrier, ref. 27, pp. 283–288.

35. E. Hitzig. 1874. *Untersuchungen über das Gehirn.* Berlin: A. Hirschwald.

36. L. Bianchi. 1894. Über die Function der Stirnlappen. *Berliner klinische Wochenschrift,* 31, 309–310.

L. Bianchi. 1895. The functions of the frontal lobes. Trans. A. de Watteville. *Brain,* 18, 497–530.

37. J. M. Harlow. 1848. Passage of an iron rod through the head. *Boston Medical and Surgical Journal,* 39, 389–393.

H. J. Bigelow. 1850. Dr. Harlow's case of recovery from the passage of an iron bar through the head. *American Journal of Medical Sciences,* 19, 13–22.

J. M. Harlow. 1868. Recovery from the passage of an iron bar though the head. *Bulletin of the Massachusetts Medical Society,* 2, 3–20.

38. Harlow 1850, ref. 37 p. 13.

39. M. B. Macmillan. 1995. A wonderful journey through skull and brains: The travels of Mr. Gage's tamping iron. *Brain and Cognition,* 5, 67–107.

M. B. McMillan. 1995. Phineas Gage: A case for all reasons. In C. Code, C.-W. Wallesch, A.-R. Lecours, and Y. Joanette (Eds.), *Classic Cases in Neuropsychology.* London: Erlbaum, pp. 243–262.

40. D. Ferrier. 1878. *The Localization of Cerebral Disease.* London: Smith-Elder.

41. Ferrier, ref. 27.

42. G. H. Lewes. 1876. Ferrier on the brain. *Nature,* 15, 73–74, 93–95.

43. Ferrier, ref. 40.

44. Ferrier 1886, ref. 25.

45. L. Luciani and A. Tamburini. 1879. *Sulle funzioni del cervello.* Regio-Emilia: S. Calderini. Abstracted by A. Rabagliati in *Brain,* 1879, 2, 234–250.

Lewes, ref. 42.

46. Finger, ref. 28, pp. 126–129.

47. R. D. French. 1975. *Antivivisection and Medical Science in Victorian Society*. Princeton, NJ: Princeton University Press.

48. O. Wilde. 1894. *A Woman of No Importance*. London: John Lane.

49. F. P. Cobbe. 1894. *Life of Frances Power Cobbe*. Boston: Houghton, Mifflin, & Co.

50. Anonymous, 1881. Dr. Ferrier's localistions; For whose advantage? *British Medical Journal*, 2, 822–824.

51. Anonymous. 1881. The charge against Professor Ferrier under the Vivisection Act: Dismissal of the summons. *British Medical Journal*, 2, 836–842.

52. G. Jefferson. 1960. *Selected Papers of Sir Geoffrey Jefferson*. Springfield, IL: Charles C Thomas, pp. 132–149.

A. E. Walker, 1967. *A History of Neurological Surgery*. New York: Hafner, pp. 178–179.

53. W. Macewen. 1879. Tumour of the dura mater removed during life in a person affected with epilepsy. *Glasgow Medical Journal*, 12, 210–213.

54. W. Macewen. 1881. Intra-cranial lesions, illustrating some points in connexion with localisation of cerebral affections and the advantages of antiseptic trephining. *Lancet*, 2, 541–543.

W. Macewen. 1888. An address on the surgery of the brain and spinal cord. *British Medical Journal*, 2, 302–309.

55. A. H. Bennett and R. Godlee. 1884. Excision of a tumour from the brain. *Lancet*, 2, 1090–1091.

56. F. R. S. 1884. Brain surgery. *The Times* (London), Dec. 16, p. 5.

57. Bennett and Godlee, ref. 55.

58. W. Trotter. 1934. A landmark in modern neurology. *Lancet*, 2, 1207–1210.

59. A. H. Bennett and R. Godlee. 1885. Sequel to the case of excision of a tumour from the brain. *Lancet*, 1, 13.

60. A. H. Bennett and R. Godlee. 1885. Case of cerebral tumour. *British Medical Journal*, 1, 988–999.

61. C. S. Sherrington. 1928. Sir David Ferrier, 1843–1928. *Proceedings of the Royal Society of London*, ser. B, 103, vii–xvi.

Jean-Martin Charcot: Clinical Neurology Comes of Age

1. F. H. Garrison. 1925. Charcot. *International Clinics*, 35, 244–272. (Quote p. 245.)

2. C. G. Goetz, M. Bonduelle, and T. Gelfand. 1995. *Charcot: Constructing Neurology*. New York: Oxford University Press.

G. Guillain. 1959. *J.-M. Charcot (1825–1893): His Life—His Work*. Trans. P. Bailey. New York: Paul B. Hoeber.

A. Souques and H. Meige. 1939. Jean-Martin Charcot (1825–1893). *Les Biographies médicales*. Paris: J.-B. Baillière et Fils, vol. 4, pp. 321–336; vol. 5, pp. 337–352.

3. M. S. Micale. 1985. The Salpêtrière in the age of Charcot: An institutional perspective on medical history in the late nineteenth century. *Journal of Contemporary History*, 20, 703–731.

Guillain, ref. 2, pp. 35–49.

4. Guillain, ref. 2, p. 83.

5. Guillain, ref. 2, p. 52.

6. M. Bonduelle. 1994. Charcot intime. *Revue Neurologique*, 150, 524–528.

7. H. Meige. 1889. Charcot artiste. *Nouvelle Iconographie de la Salpêtrière*, 11, 489–516.

C. G. Goetz. 1991. Visual art in the neurologic career of Jean-Martin Charcot. *Archives of Neurology*, 48, 421–425.

8. C. G. Goetz. 1988. Shakespeare in Charcot's neurologic teaching. *Archives of Neurology*, 45, 920–921.

9. S. Finger. 1998. "A happy state of mind": The early history of mild elation, denial of disability, optimism, and laughing in multiple sclerosis. *Archives of Neurology*, 55, 241–250.

10. S. Fredrikson. 1991. Letter to the editor in response to Stenager and Jensen. *Perspectives in Biology and Medicine*, 32, 312.

    S. Fredrikson and S. Kam-Hansen. 1989. The 150th-year anniversary of multiple sclerosis: Does its early history give an etiological clue? *Perspectives in Biology and Medicine*, 32, 237–243.

11. D. Firth. 1948. *The Case of Augustus d'Esté*. Cambridge, UK: Cambridge University Press.

12. R. Carswell. 1838. *Pathological Anatomy: Illustrations of the Elementary Forms of Disease*. London: Longman, Orme, Brown, Green and Longman.

13. J. Cruveilhier. 1835–1842. *Anatomie pathologique du corps humain*. Paris: J. B. Baillière.

14. Cruveilhier, ref. 13, p. 22.

15. E. F. A. Vulpian. 1866. Note sur la sclérose en plaques de la moelle épinière. *L'Union Médicale* 30, 459–465, 475–482, 507–512, 541–548.

16. J.-M. Charcot. 1868. Histologie de la sclérose en plaques. *Gazette des Hôpitaux Civils et Militaires, 41,* 554–555, 557–558, 566.

    J.-M. Charcot. 1868. Séance du 14 mars: Un cas de sclérose en plaques généralisée du cerveau et de la moelle épinière. *Compte Rendus de la Société de Biologie 20,* 13–14.

17. J.-M. Charcot. 1879. Diagnostic des formes frustes de la sclérose en plaques. *Progrès Médical, 7,* 97–99.

18. J.-M. Charcot. 1872–1873. *Leçons sur les maladies du système nerveux*, vol. 1. Paris: A. Delahaye. Reprinted as *Lectures on the Diseases of the Nervous System*. Ed. and trans. G. Sigerson. London: New Sydenham Society, 1877, pp. 157–222.

19. S. Wilks. 1878. *Lectures on Diseases of the Nervous System*. London: Churchill.

20. M. Clymer. 1870. Notes on physiology and pathology of the nervous system with reference to clinical medicine. *New York Medical Journal*, 11, 225–262, 410–423.

21. Clymer, ref. 20, p. 231.

22. J. Parkinson. 1817. *An Essay on the Shaking Palsy*. London: Whittingham and Rowland, for Sherwood, Neely, and Jones. Reprinted in A. D. Morris, *James Parkinson: His Life and Times*. Boston: Birkhäuser, 1989, pp. 152–175.

23. A. D. Morris. 1989. A discussion of Parkinson's "Essay on the Shaking Palsy." In A. D. Morris, *James Parkinson: His Life and Times*. Boston: Birkhäuser, pp. 131–148.

    S. Finger. 1994. *Origins of Neuroscience*. New York: Oxford University Press, pp. 223–228.

24. Parkinson, ref. 22 (reprint), p. 152.

25. J.-M. Charcot and E. F. A. Vulpian. 1861–1862. De la paralyse agitante (A propos d'un cas tiré de la clinique du Professeur Oppolzer). *Gazette Hebdomadaire*, 8, 765–767, 816–820; 9, 54–59.

26. J.-M. Charcot. 1877. *Leçons sur les maladies du système nerveux*, vol. 2. Paris: A. Delahaye. Reprinted as *Lectures on the Diseases of the Nervous System*. Ed. and trans. G. Sigerson. London: New Sydenham Society, 1881.

27. Charcot, ref. 26 (trans.), p. 138.

28. F. Schiller. 1986. Parkinsonian rigidity: The first hundred-and-one years (1817–1918). *History and Philosophy of the Life Sciences*, 8, 221–236.

29. J.-M. Charcot. 1887. Lecture of December 13, 1887. In *Leçons du mardi à la Sal-*

pêtrière. Paris: Bureau du Progrès Médical. Translated by C. G. Goetz in *Charcot the Clinician: The Tuesday Lectures*. New York: Raven Press, 1987, pp. 63–65.

Goetz, Bonduelle, and Gelfand, ref. 2, pp. 157–162.

30. W. R. Gowers. 1893. *A Manual of Diseases of the Nervous System*, 2nd ed., vol. 2. Philadelphia: Blakiston, pp. 656–657.

31. E. Brissaud. 1895. *Leçons sur les maladies nerveuses*. Paris: Masson.

C. Trétiakoff. 1919. *Contribution à l'étude de l'anatomie pathologique du locus niger*. Thèse de Paris, no. 293.

32. M. Bonduelle. 1975. Amyotrophic lateral sclerosis. In P. J. Vinken and G. W. Bruyn (Eds.), *Handbook of Clinical Neurology*, vol. 22: *System Disorders and Atrophies*, part II. Amsterdam: North-Holland Publishing Co., pp. 281–338.

33. J.-M. Charcot and A. Joffroy. 1869. Deux cas d'atrophie musculaire progressive avec lésions de la substance grise et des faisceaux antéro-latéraux de la moelle épinière. *Archives de Physiologie Normale et Pathologique*, 1, 354–367; 2, 628–649; 3, 744–757.

34. Charcot, ref. 25, pp. 180–191.

35. Charcot, ref. 26 (trans.), p. 203.

36. J.-M. Charcot and P. Marie. 1886. Deux nouveaux cas de la sclérose latérale amyotrophie suivis d'autopsie. *Archives de Neurologie*, 10, 1–35, 168–186.

37. J.-M. Charcot. 1888. Lecture of February 28, 1888. In Goetz, ref. 29, pp. 164–186.

38. J.-M. Charcot and P. Marie. 1886. Sur une forme particulière d'atrophie musculaire progressive, souvent familiale, débutant par les pieds et les jambes et atteignant plus tard les mains. *Revue de Médecine*, 6, 97–138.

39. H. H. Tooth. 1886. *The Peroneal Type of Progressive Muscular Atrophy*. London: H. K. Lewis.

40. J.-B. Charcot. 1926. Charcot in the Franco-Prussian War. *Military Surgeon*, 37, 153–154.

41. Charcot, ref. 40, p. 153.

42. J. Gasser. 1995. *Aux origines du cerveau moderne: Localisations, langage, et mémoire dans l'ouevre de Charcot*. Paris: Fayard.

43. M. Jeannerod, M. 1994. Le contribution de J. M. Charcot à l'étude des localisations motrices chez l'homme. *Revue Neurologique*, 150, 536–542.

44. J.-M. Charcot and A. Pitres. 1877. Contribution à l'étude des localisations dans l'écorce des hémisphères du cerveau. Observations relatives aux paralyses at aux convulsions d'origine corticale. *Revue Mensuelle de Médecine et de Chirurgie*, 1, 1–18, 113–123, 180–195, 157–376, 437–457.

J.-M. Charcot and A. Pitres. 1878. Nouvelle contribution à l'étude des localisations dans l'écorce des hémisphères du cerveau. *Revue Mensuelle de Médecine et de Chirurgie*, 2, 801–815.

J.-M. Charcot and A. Pitres. 1878. Nouvelle contribution à l'étude des localisations dans l'écorce des hémisphères du cerveau. *Revue Mensuelle de Médecine et de Chirurgie*, 3, 127–156.

45. P. Guilly. 1982. Gilles de la Tourette. In F. C. Rose and W. F. Bynum (Eds.), *Historical Aspects of the Neurosciences*. New York: Raven Press, pp. 397–413.

A. J. Lees. 1986. Georges Gilles de la Tourette, the Man and His Times. *Revue Neurologique*, 142, 808–816.

46. G. M. Beard. 1880. Experiments with the "jumpers" or "jumping Frenchman" of Maine. *Journal of Nervous and Mental Disease*, 7, 487–490.

G. Gilles de la Tourette. 1881. Les sauteurs du Maine (Etats-Unis). *Archives de Neurologie*, 5, 146–150.

47. H. A. O'Brien. 1883. Latah. *Journal of the Straits Branch of the Royal Asiatic Society in Singapore*, 6, 145–158.

48. W. A. Hammond. 1884. Miryachit, a newly described disease of the nervous system, and its analogs. *New York Medical Journal,* 39, 191–192.

49. G. Gilles de la Tourette. 1884. Jumping, latah, myriachit. *Archives de Neurologie,* 8, 68–74.

C. Lajonchere, M. Nortz, and S. Finger. 1996. Gilles de la Tourette and the discovery of Tourette's syndrome (Includes a translation of his 1884 paper). *Archives of Neurology,* 53, 567–574.

50. G. Gilles de la Tourette. 1885. Étude sur une affection nerveuse caractèrisée par de l'incoordination motrice, accompagnée d'écholalie et de coprolalie (Jumping, latah, myriachit). *Archives de Neurologie,* 9, 19–42, 158–200.

51. J. M. G. Itard. 1825. Mémoire sur quelques fonctions involuntaires des appareils de la locomotion, de la préhension et de la voix. *Archives Générales de Médecine,* 8, 385–407.

52. Itard, ref. 51, pp. 404–405.

53. H. I. Kushner. 1995. Medical fictions: The case of the cursing marquise and the (re)construction of Gilles de la Tourette's syndrome. *Bulletin of the History of Medicine,* 69, 224–245.

54. J.-M. Charcot. 1885. Ospedale della Salpêtrière. Intorno ad alcuni casi di tic convulsivo con coprolalia ed echolalia. Lecone raccolta dal Dott. G. Melotti. *La Riforma Medica,* August 5, p. 2; August 19, p. 2; August 31, p. 2.

55. G. Guinon. 1886. Sur la maladie des tics convulsifs. *Revue de Médecine,* 6, 50–80.

G. Guinon. 1887. Tics convulsifs et hystérie. *Revue de Médecine,* 7, 509–518.

H. I. Kushner and L. S. Kiessling. 1996. The controversy over the classification of Gilles de la Tourette syndrome, 1800–1995. *Perspectives in Biology and Medicine,* 39, 409–435.

56. A. K. Shapiro and E. Shapiro. 1968. Treatment of Gilles de la Tourette's syndrome with haloperidol. *British Journal of Psychiatry,* 114, 345.

57. M. S. Micale. 1990. Charcot and the idea of hysteria in the male: Gender, mental science, and medical diagnosis in late-nineteenth-century France. *Medical History,* 34, 363–411.

58. M. S. Micale. 1994. Charcot and *les névroses traumatiques*: Historical and scientific reflections. *Revue Neurologique,* 150, 498–505.

J. Goldstein. 1987. *Console and Classify.* Cambridge, UK: Cambridge University Press, pp. 322–377.

59. S. Freud. 1886. Bericht über Meine mit Universitäts-jubiläums Reisestipendium unternommene Studienreise nach Paris und Berlin. Reprinted as "Report on my studies in Paris and Berlin" in *The Standard Edition of the Complete Psychological Works of Sigmund Freud,* vol. 1. Trans. J. Strachey. London: Hogarth Press, 1966, pp. 3–31.

60. J. Goldstein. 1982. The hysteria diagnosis and the politics of anticlericalism in late-nineteenth-century France. *Journal of Modern History,* 54, 209–239.

61. J.-M. Charcot and P. Richer. 1887. *Les Démoniaques dans l'art.* Paris: Delahaye et Lecrosnier.

J.-M. Charcot and P. Richer. 1889. *Les Difformes et les malades dans l'art.* Paris: Lecrosnier et Babé.

62. D. Widlöcher and N. Dantchev. 1994. Charcot et l'hystérie. *Revue Neurologique,* 150, 490–497.

H. F. Ellenberger. 1970. *The Discovery of the Unconscious.* New York: Basic Books, pp. 53–109.

R. E. Fancher. 1990. *Pioneers of Psychology,* 2nd ed. New York: W. W. Norton, pp. 320–389.

Santiago Ramón y Cajal: From Nerve Nets to Neuron Doctrine

1. A. Kölliker. 1896. *Handbuch der Gewebelehre des Menschen,* 6th ed. Leipzig: Engelmann.

W. His. 1889. Die Neuroblasten und deren Entstehung im embryonalen Marke. *Sächsischen Akademie der Wissenschaften zu Leipzig. Math-Phys. Cl., 15*, 313–372.

2. W. von Waldeyer-Hartz. 1891. Über einige neuere Forschungen im Gebiete der Anatomie des Centralnervensystems. *Deutsche medizinische Wochenschrift, 17*, 1213–1218, 1244–1246, 1267–1269, 1287–1289, 1331–1332, 1352–1356.

3. M. Foster. 1897. *A Text-Book of Physiology*, 7th ed. London: Macmillan, p. 929.

4. R. Hooke. 1667. *Micrographia: or some Physiological Descriptions of Minute Bodies Made by Magnifying Glasses*. London: J. Martyn and J. Allestry.

5. F. J. Cole. 1937. Leeuwenhoek's zoological researches. *Annals of Science, 2*, 1–46.

6. A. Van Leeuwenhoek. 1674. More observations from Mr. Van Leeuwenhoek, in a letter of Sept. 7. 1674. sent to the publisher. *Philosophical Transactions of the Royal Society of London, 9*, 178–182

   A. Van Leeuwenhoek. 1675. Microscopical observations of Mr. Van Leeuwenhoek concerning the optic nerve, communicated to the publisher in Dutch, and made by him in English. *Philosophical Transactions of the Royal Society of London, 10*, 378–380.

7. A. Van Leeuwenhoek. 1719. *Epistolae physiologicae super compluribus naturae arcanis*. Epistola XXXII. Delft: Beman.

8. F. T. Lewis. 1953. The introduction of biological stains: Employment of saffron by Vieussens and Leeuwenhoek. *Quarterly Journal of Microscopical Science, 54*, 229–253.
   Van Leeuwenhoek, ref. 7.

9. R. de Vieussens. 1684. *Neurographia universalis*. Lyons: Certe.

10. E. G. T. Liddell. 1960. *The Discovery of Reflexes*. Oxford: Clarendon Press, pp. 1–30.

11. J. C. Reil. 1809. Untersuchungen über den Bau des grossen im Menschen . . . Vierte Fortsetzung VIII. *Archiv für Physiologie, 9*, 136–146.

12. A. Hannover. 1840. Die Chromsäure, ein vorzügliches Mittel beim mikroscopischen Untersuchungen. *Archiv* für Anatomie und Physiologie, 549–558.

13. H. J. John. 1959. *Jan Evangelista Purkyně*. Philadelphia: American Philosophical Society.

14. J. E. Purkyně. 1838. Bericht über die Versammlung deutcher Naturforscher und Ärzte in Prag im September, 1837. *Bericht über die Versammlung deutcher Naturforscher und Ärzte, 15*, 174–175, 177–180.

15. M. J. Schleiden. 1838. Beiträge zur Phytogenesis. *Archiv für Anatomie, Physiologie und wissenschaftliche Medizin*, 137–176.

16. T. Schwann. 1839. *Mikroscopische Untersuchungen über die Übereinstimmung in der Struktur und dem Wachsthum der Thiere und Pflanzen*. Berlin: G. E. Reimer.

17. G. Shepherd. 1991. *Foundations of the Neuron Doctrine*. New York: Oxford University Press, pp. 25–47.

18. H. J. Conn. 1946. The evolution of histological staining. *Ciba Symposium, 7*, 270–300.

19. A. Corti. 1851. Recherches sur l'organe de l'ouîe des mammifères. *Zeitschrift für wissenschaftliche Zoologie, 3*, 109–169.

20. J. von Gerlach. 1858. *Mikroscopische Studien aus dem Gebiete der menschlichen Morphologie*. Enke: Erlangen.

   G. Clarke and F. K. Kasten. 1983. *History of Staining*. Baltimore, MD: Williams and Wilkins.

21. O. Deiters. 1865. *Untersuchungen über Gehirn und Rückenmark des Menschen und der Säugethiere*. Ed. M. Schultze., Braunschweig: Wieweg.

22. A. Kölliker. 1867. *Handbuch der Gewebelehre des Menschen*, 5th ed. Leipzig: Engelmann.

23. J. von Gerlach. 1872. Über die struktur der grauen Substanz des menschlichen Grosshirns. *Zentralblatt für die medizinischen Wissenschaften, 10*, 273–275.

24. C. Golgi. 1873. Sulla struttura della grigia del cervello. *Gazetta Medica Italiana*, 6, 244–246. Translated as "On the structure of the gray matter of the brain," in M. Santini (Ed.), *Golgi Centennial Symposium, Proceedings*. New York: Raven Press, 1975, pp. 647–650.

25. C. Da Fano. 1926. Camillo Golgi, 1843–1926. *Journal of Pathology and Bacteriology*, 29, 500–514.

    H. R. Viets. 1926. Camillo Golgi, 1843–1926. *Archives of Neurology and Psychiatry*, 15, 623–627.

    Shepherd, ref. 17, pp. 79–101.

26. E. Clarke and C. D. O'Malley. 1968. *The Human Brain and Spinal Cord*. Berkeley: University of California Press, p. 842.

    Liddell, ref. 10, p. 25.

    Shepherd, ref. 17, p. 84.

27. C. Golgi. 1886. *Sulla fina anatomia degli organi centrali del sistema nervoso*. Milano: Hoepli.

28. S. Finger. 1994. *Origins of Neuroscience*. New York: Oxford University Press, pp. 53–54.

29. W. His. 1886. Zur Geschichte des menschlichen Rückenmarkes und der Nerven-wurzeln. *Sächsische Akademie der Wissenschaften zu Leipzig. Math-Phys Cl.*, 13, 147–209, 477–513.

30. A. H. Forel. 1887. Einige hirnanatomische Betrachtungen und Ergebnisse. *Archiv für Psychiatrie*, 18, 162–198.

31. A. H. Forel. 1937. *Out of My Life and Work*. New York: Norton, p. 163.

32. D. F. Cannon. 1949. *Explorer of the Human Brain: The Life of Santiago Ramón y Cajal*. New York: Henry Schuman, pp. 264–265.

    Shepherd, ref. 17, pp. 127–258.

33. S. Ramón y Cajal. 1937. *Recollections of My Life*. Trans. E. H. Craigie with the assistance of J. Cano. Philadelphia: American Philosophical Society. First published as *Recuerdos de mi vida*, 1901–1917; republished in 1989 by MIT Press; second printing, 1991.

34. Cajal, ref. 33, p. 169.

35. Cajal, ref. 33, p. 280.

36. S. Ramón y Cajal. 1888. Estructura de los centros nerviosos de las aves. *Revista de Histologia Normal y Patologica*, 1, 1–10.

37. S. Ramón y Cajal. 1889. Conexión general de los elementos nerviosos. *La Medicina Práctica*, 2, 341–346.

38. Waldeyer, ref. 2.

39. E. G. Jones. 1994. The neuron doctrine, 1891. *Journal of the History of the Neurosciences*, 3, 3–20.

    Shepherd, ref. 17, pp. 186–193.

40. S. Ramón y Cajal. 1933. *¿Neuronismo o reticularismo?* Reprinted as *Neuron Theory or Reticular Theory?* Trans. M. U. Purkiss and C. A. Fox. Madrid: Instituto Ramón y Cajal, 1954, p. 3.

41. Cajal, ref. 33, pp. 382–392.

42. S. Ramón y Cajal. 1890. A quelle époque apparaissent les expansions des cellules nerveuses de la moëlle épinière du poulet? *Anatomischer Anzeiger*, 5, 631–639.

43. R. G. Harrison. 1907. Observations on the living developing nerve fiber. *Anatomical Record*, 1, 116–118.

44. H. Rabl-Rückhard. 1890. Sind die Ganglienzellen amöboid? Eine Hypothese zur Mechanik psychischer Vorgägne. *Neurologisches Centralblatt*, 9, 199.

    E. Tanzi. 1893. I fatti e la induzione nell'odierna istologia del sistema nervoso. *Revista*

Sperimentale di Freniatria e Medicina Legale delle Alienazioni Mentali, 19, 149.

M. Duval. 1895. Remarques à propos de la communication de M. Lépine. *Comptes Rendus Hebdomadaires des Séances et Mémoires de la Société de Biologie*, 47, 86–87.

M. Duval. 1895. Hypothèse sur la physiologie des centres nerveux; théorie histologique du sommeil. *Comptes Rendus Hebdomadaires des Séances et Mémoires de la Société de Biologie*, 47, 74–77.

45. A. Bain. 1872. *Mind and Body: The Theories of Their Relation.* London: Henry S. King.

46. S. Ramón y Cajal. 1895. *Algunas conjeturas sobre el mecanismo anatómico de la ideación y attención.* Madrid.

47. Kölliker, ref. 2.

48. S. Ramón y Cajal. 1894. La fine structure des centres nerveux. *Proceedings of the Royal Society of London*, Series B, 55, 444–467.

49. W. Greenough. 1975. Experiential modification of the developing brain. *American Scientist*, 63, 37–46.

50. Shepherd, ref. 17, pp. 259–270.

51. C. Golgi. 1906. The neuron doctrine—theory and facts. In *Nobel Lectures Physiology or Medicine 1901–1921.* New York: Elsevier. 1967, pp. 189–217.

52. Cajal, ref. 33, p. 553.

53. S. Ramón y Cajal. 1906. The structure and connexions of neurons. Nobel Lecture, December 12, 1906. Reprinted in *Nobel Lectures Physiology or Medicine 1901–1921.* New York: Elsevier, 1967, pp. 220–253.

54. Cajal, ref. 33, pp. 509–510.

55. C. S. Sherrington. 1949. Santiago Ramón y Cajal, 1852–1934. *Obituary Notices of the Royal Society*, 4, 425–441.

Charles Scott Sherrington: The Integrated Nervous System

1. J. P. Swazey. 1969. *Reflexes and Motor Integration: Sherrington's Concept of Integrative Action.* Cambridge, MA: Harvard University Press.

Lord Cohen of Birkenhead. 1958. *Sherrington: Physiologist, Philosopher and Poet.* Liverpool, UK: Liverpool University Press.

J. C. Eccles and W. C. Gibson. 1979. *Sherrington: His Life and Thought.* New York: Springer International.

R. Granit. 1967. *Charles Scott Sherrington: An Appraisal.* Garden City, NY: Doubleday.

D. Denny-Brown. 1957. The Sherrington School of Physiology. *Journal of Neurophysiology*, 20, 543–548.

2. Denny-Brown, ref. 1, p. 547.

3. J. N. Langley and C. S. Sherrington. 1884. Secondary degeneration of nerve tracts following removal of the cortex of the cerebrum in the dog. *Journal of Physiology*, 5, 49–65.

4. E. Pflüger. 1853. *Die sensoriellen Funktionen des Rükenmarkes der Wirbelthiere.* Berlin: Hirschwald.

E. G. T. Liddell. 1960. *The Discovery of Reflexes.* Oxford: Clarendon Press, pp. 84–86.

5. C. S. Sherrington. 1885. On secondary and tertiary degenerations in the spinal cord of the dog. *Journal of Physiology*, 6, 177–191.

6. M. Hall. 1850. *Synopsis of the Diastaltic Nervous System . . .* London: J. Mallett.

7. C. S. Sherrington. 1889. On nerve-tracts degenerating secondarily to lesions of the cortex cerebri. *Journal of Physiology*, 10, 429–432.

8. C. E. R. Sherrington. 1979. Memories (Beaumont Lecture, Yale University, November, 1957). Reprinted as appendix 17 in Eccles and Gibson, ref. 1, pp. 239–259.

9. R. Descartes. 1662. *De homine figuris et latinitate donatus a Florentio Schuyl.* Leyden: Franciscum Moyardum and Petrum Leffen. Translated in J. Cottingham, R. Stroothoff, D. Murdach, and A. Kenny, (Eds.), *The Philosophical Writings of Descartes,* vol. 1. Cambridge, UK: Cambridge University Press, 1985.

10. T. Willis. 1664. *Cerebri anatome, cui accessit nervorum descriptio et usus.* London: J. Martyn and J. Allestry. Reprinted as *The Anatomy of the Brain and Nerves.* Ed W. Feindel. Montreal: McGill University Press, 1965.

11. R. Whytt. 1751. *An Essay on the Vital and Other Involuntary Motions of the Animal.* Edinburgh, UK: Hamilton, Balfour, and Neill.

   R. Whytt. 1755. *Physiological Essays.* Edinburgh, UK: Hamilton, Balfour, and Neill.

12. Hall, ref. 6.

13. I. M. Sechenov. 1863. *Reflexes of the Brain.* Trans. S. Belsky. Cambridge, MA: MIT Press, 1965.

14. M. Foster. 1897. *A Text-Book of Physiology, Part III* 7th ed. London: Macmillan, p. 929.

15. C. S. Sherrington. 1937. Letter. In Swazey, ref. 1, p. 76.

16. W. H. Erb. 1875. Über Sehenreflexe bei Gesunden und Rückenmarkskranken. *Archiv für Psychiatrie und Nervenkrankheit,* 5, 792.

   C. I. O. Wesphal. 1875. Über einige Bewegungserscheinungen an gelähmten Gliedern. *Archiv für Psychiatrie und Nervenkrankheit,* 5, 803.

17. C. S. Sherrington. 1891. Note on the knee-jerk. *St. Thomas's Hospital Reports,* 21, 145–147.

   C. S. Sherrington. 1892. Note toward the localisation of the knee-jerk. *British Medical Journal,* 1, 545; addendum, 654.

18. C. S. Sherrington. 1900. The muscular sense. In E. A. Schäfer (Ed.), *Text-Book of Physiology,* vol. 2. Edinburgh: Young J. Pentland, pp. 1002–1025.

19. C. S. Sherrington. 1894. On the anatomical distribution of nerves of skeletal muscles; with remarks on recurrent fibres in the ventral spinal nerve-root. *Journal of Physiology,* 17, 211–258.

20. Sherrington, ref. 18.

21. C. S. Sherrington. 1892. Note on the arrangement of some motor fibres in the lumbo-sacral plexus. *Journal of Physiology,* 13, 621–772 (plates 20–23).

22. C. S. Sherrington. 1894. Experiments in examination of the peripheral distribution of the fibres of the posterior roots of some spinal nerves. *Philosophical Transactions of the Royal Society,* 184B, 641–763, plates 42–52.

   C. S. Sherrington. 1898. Experiments in examination of the peripheral distribution of the fibres of the posterior roots of some spinal nerves. *Philosophical Transactions of the Royal Society,* 190B, 45–186.

   C. S. Sherrington. 1899. On the spinal animal. *Royal Medical and Chirurgical Society of London,* 82, 449–475.

23. H. Head and A. W. Campbell. 1900. The pathology of *Herpes zoster* and its bearing on sensory localization. *Brain,* 23, 353–523.

24. Granit, ref. 1, p. 85.

25. Granit, ref. 1, p. 50.

26. Descartes, ref. 9.

   C. Bell. 1823. On the nerves of the orbit. *Philosophical Transactions of the Royal Society,* 113, 289–307.

   Hall, ref. 6.

27. Sherrington, ref. 22, p. 178.

28. C. S. Sherrington. 1898. Decerebrate rigidity, and reflex co-ordination of movements. *Journal of Physiology*, 12, 319–332.

29. C. S. Sherrington. 1910. Remarks on the reflex mechanism of the step. *Brain*, 33, 1–25.

30. G. Paton. 1846. On the perceptive power of the spinal cord as manifested by cold-blooded animals. *Edinburgh Medical and Surgical Journal*, 65, 251–269.

31. C. S. Sherrington. 1906a. Observations on the scratch reflex in the spinal dog. *Journal of Physiology*, 34, 1–50.

    C. S. Sherrington and E. E. Laslett. 1903. Observations on some spinal reflexes and the interconnection of spinal segments. *Journal of Physiology*, 29, 58–96.

32. W. T. Clower. 1998. Early contributions to the reflex chain hypothesis. *Journal of the History of the Neurosciences*, 7, 32–42.

33. A. Bain. 1855. *The Senses and the Intellect*. London: John Parker and Son, pp. 265–266. Sechenov, ref. 13.

34. C. E. Beevor and V. H. Horsley. 1890. A record of the results obtained by electrical excitation of the so-called motor cortex and internal capsule in an orang-outang *(Simia satyrus)*. *Philosophical Transactions of the Royal Society*, 181, 129–158.

35. A. S. F. Grünbaum and C. S. Sherrington. 1901. Observations on the physiology of the cerebral cortex of some of the higher apes. *Proceedings of the Royal Society of London*, 69, 206–209.

    A. S. F. Grünbaum and C. S. Sherrington. 1903. Observations on the physiology of the cerebral cortex of the anthropoid apes. *Proceedings of the Royal Society of London*, 72, 152–155.

    C. S. Sherrington and A. S. F. Grünbaum. 1901. Localisation in the "motor" cerebral cortex. *British Medical Journal*, 2, 1857–1859.

    H. E. Roaf and C. S. Sherrington. 1906. Experiments in examination of the "locked-jaw" induced by tetanus toxin. *Journal of Physiology*, 34, 315–331.

36. C. S. Sherrington. 1894. Experimental degeneration of the pyramidal tract. *Lancet*, 1, 571.

37. A. S. F. Leyton and C. S. Sherrington. 1917. Observations on the excitable cortex of the chimpanzee, orang-utan and gorilla. *Quarterly Journal of Experimental Physiology*, 11, 135–222.

38. W. Penfield and E. Boldrey. 1937. Somatic motor and sensory representation in the cerebral cortex of man as studied by electrical stimulation. *Brain*, 60, 389–443.

39. Lord Cohen, ref. 1, p. 15.

40. S. Finger and D. G. Stein. 1982. *Brain Damage and Recovery*. New York: Academic Press, pp. 287–302.

41. C. S. Sherrington. 1906. *The Integrative Nature of the Nervous System*. New Haven, CT: Yale University Press, p. 390.

42. Sherrington, ref. 41.

43. C. S. Sherrington. 1947. *The Integrative Nature of the Nervous System*, 2nd ed. New Haven, CT: Yale University Press.

44. C. E. R. Sherrington, ref. 8, p 250.

45. Eccles and Gibson, ref. 1, pp. 44–45.

46. Eccles and Gibson, ref. 1, p. 57.

47. R. S. Creed, D. Denny-Brown, E. G. T. Liddell, J. C. Eccles, and C. S. Sherrington. 1932. *Reflex Activity of the Spinal Cord*. Oxford: Clarendon Press.

48. C. S. Sherrington. 1940. *The Assaying of Brabantius and Other Verse*, 2nd ed. London: Humphrey Milford.

    C. S. Sherrington. 1925. *The Assaying of Brabantius and Other Verse*. Oxford: Oxford University Press.

49. Lord Cohen, ref. 1, p. 13.

50. C. S. Sherrington. 1941. *Man on his Nature*. Cambridge: Cambridge University Press.

51. Sherrington, ref. 50, p. 201.

52. C. S. Sherrington. 1933. *The Brain and Its Mechanism*. Cambridge, UK: Cambridge University Press, pp. 22–23.

53. J. Fernel. 1542. *De naturali parte medicinae*. Paris: Apus Simonem Colineum.

54. C. S. Sherrington. 1946. *The Endeavour of Jean Fernel*. Cambridge: Cambridge University Press.

55. Lord Cohen, ref. 1, p. 15.

56. R. Brain. 1948. Discussion of "Early developments of ideas relating the mind to the brain" by H. W. Magoun. In G. E. W. Wolstenholme and C. M. O'Connor (Eds.), *Ciba Foundation Symposium on the Neurological Bases of Behavior, in Commemoration of Sir Charles Scott Sherrington*. Boston: Little, Brown, p. 24.

57. L. G. Brock, J. S. Coombs, and J. C. Eccles. 1951. Action potentials with intracellular electrode. *Proceedings of the University of Otago Medical School*, 29, 14–15.

   L. G. Brock, J. S. Coombs, and J. C. Eccles. 1952. The recording of potentials from motoneurons with an intracellular electrode. *Journal of Physiology*, 117, 431–460.

   J. C. Eccles. 1959. Excitatory and inhibitory synapses. *Annals of the New York Academy of Sciences*, 81, 247–264.

58. H. Viets. 1952. Charles Scott Sherrington, 1857–1952: An appreciation. *New England Journal of Medicine*, 246, 481.

Edgar D. Adrian: Coding in the Nervous System

1. E. D. Adrian. 1932. *The Mechanism of Nervous Action. Electrical Studies of the Neurone*. Philadelphia: University of Pennsylvania Press, p. 2.

2. H. E. Hoff and L. A. Geddes. 1960. Ballistics and the instrumentation of physiology. The velocity of the projectile and the nerve impulse. *Journal of the History of Medicine*, 15, 133–146.

3. H. E. Hoff and L. A. Geddes. 1957. The rheotome and its prehistory: A study in the historical interrelation of electrophysiology and electromechanics. *Bulletin of the History of Medicine*, 31, 212–234, 327–347.

   T. Lenoir. 1986. Models and instruments in the development of electrophysiology. *Historical Studies in the Physical and Biological Sciences*, 17, 1–54.

4. H. E. Hoff and L. A. Geddes. 1961. The capillary electrometer: The first graphic recording of bioelectric signals. *Archives Internationales d'Histoire des Sciences*, 56–57, 275–290.

5. F. Gotch and V. Horsley. 1888. Observations upon the mammalian spinal cord following electrical stimulation of the cortex cerebri. *Proceedings of the Royal Society*, 45, 18–26.

6. F. Gotch and C. J. Burch. 1899. The electrical response of nerve to two stimuli. *Journal of Physiology*, 117, 410–426.

7. R. G. Frank Jr. 1988. The telltale heart: Physiological instruments, graphic methods, and clinical hopes, 1854–1914. In W. Coleman and F. L. Holmes (Eds.), *The Investigative Enterprise: Experimental Physiology in Nineteenth-Century Medicine*. Berkeley: University of California Press, pp. 211–290.

8. W. M. Bayliss. 1917–1919. Keith Lucas, 1879–1916. *Proceedings of the Royal Society of London*, ser. B, 90, xxxi–xlii.

9. K. Lucas. 1905. On the gradation of activity in a skeletal muscle-fibre. *Journal of Physiology*, 33, 124–137.

10. K. Lucas. 1909. The "all-or-none" contraction of the amphibian skeletal muscle-fibre. *Journal of Physiology*, 38, 113–133.

11. Lucas, ref. 10, p. 133.

12. Lucas, ref. 10, p. 132.

13. A. L. Hodgkin. 1979. Edgar Douglas Adrian: Baron Adrian of Cambridge. *Biographical Memoirs of Fellows of the Royal Society,* 25, 1–73.

14. E. D. Adrian. 1912. On the conduction of subnormal disturbances in normal nerve. *Journal of Physiology,* 45, 389–412.

15. E. D. Adrian. 1914. The all-or-none principle in nerve. *Journal of Physiology,* 47, 460–474.

16. K. Lucas. 1917. *The Conduction of the Nervous Impulse.* London: Longmans, Green.

17. J. C. Eccles. 1970. Alexander Forbes and his achievement in electrophysiology. *Perspectives in Biology and Medicine,* 13, 388–404.

   W. O. Fenn. 1969. Alexander Forbes: May 14, 1882–May 27, 1965. *Biographical Memoirs of the National Academy of Sciences,* 40, 113–141.

18. A. Forbes and A. Gregg. 1915. Electrical studies in mammalian reflexes. I. The flexion reflex. *American Journal of Physiology,* 37, 118–176.

   A. Forbes and A. Gregg. 1915. Electrical studies in mammalian reflexes. II. The correlation between the strength of stimuli and the direct and reflex nerve response. *American Journal of Physiology,* 39, 172–235.

19. A. Forbes and C. Thacher. 1920. Electron tube amplification with the string galvanometer. *American Journal of Physiology,* 51, 177–178.

   A. Forbes and C. Thacher. 1920. Amplification of action currents with the electron tube in recording with the string galvanometer. *American Journal of Physiology,* 52, 409–471.

   A. Forbes. 1922. The interpretation of spinal reflexes in terms of present knowledge of nerve conduction. *Physiological Review,* 2, 361–414.

20. R. G. Frank Jr. 1994. Instruments, nerve action, and the all-or-none principle. *Osiris,* 9, 208–235.

21. L. H. Marshall. 1983. The fecundity of aggregates: The axonologists at Washington University, 1922–1942. *Perspectives in Biology and Medicine,* 26, 613–636.

22. H. S. Gasser and H. S. Newcomer. 1921. Physiological action currents in the phrenic nerve: An application of the thermionic vacuum tube to nerve physiology. *American Journal of Physiology,* 57, 1–26.

23. H. S. Gasser and J. Erlanger. 1921. The cathode ray oscillograph as a means of recording nerve action currents and induction shocks. *American Journal of Physiology,* 59, 473–475.

24. H. S. Gasser and J. Erlanger. 1922. A study of the action current of nerve with the cathode ray oscillograph. *American Journal of Physiology,* 62, 496–524.

25. G. Bishop, 1966, quoted in J. O'Leary and S. Goldring, *Science and Epilepsy.* New York: Raven Press, 1976, p. 150.

26. J. Erlanger and H. S. Gasser ("with the collaboration, in some of the experiments, of G. H. Bishop"). 1924. The compound nature of the action current as disclosed by the cathode ray oscillograph. *American Journal of Physiology,* 70, 624–666.

   Gasser and Erlanger, ref. 24.

27. H. S. Gasser and J. Erlanger. 1927. The role played by the sizes of the constituent fibers of a nerve trunk in determining the form of its action potential. *American Journal of Physiology,* 80, 522–547.

28. Erlanger and Gasser, ref. 26.

   H. S. Gasser. 1937. The control of excitation in the nervous system. *Harvey Lectures,* 169–193.

   H. S. Gasser. 1945. Mammalian nerve fibers. In *Nobel Lectures: Physiology or Medicine 1942–1962.* Amsterdam: Elsevier, 1964, pp. 34–47.

   J. Erlanger. 1947. Some observations on the responses of single nerve fibers. In *Nobel Lectures: Physiology or Medicine 1942–1962.* Amsterdam: Elsevier, 1964, pp. 50–73.

29. E. D. Adrian. 1954. Memorable experiences in research. *Diabetes*, 3, 17–18.

30. Adrian, ref. 29, p. 18.

31. E. D. Adrian. 1926. The impulses produced by sensory nerve endings. *Journal of Physiology*, 61, 49–72.

32. Adrian, 1926, ref. 31, p. 72.

33. Y. Zotterman. 1969. *Touch, Tickle and Pain.* Oxford: Pergamon Press, p. 220.

34. E. D. Adrian and Y. Zotterman. 1926. The impulses produced by sensory nerve endings. Part 2: The response of a single end-organ. *Journal of Physiology*, 61, 151–171.

35. Zotterman, ref. 33, p. 230.

36. A. Forbes. 1926. Letter to Adrian. In Hodgkin, ref. 13, pp. 21–22.

37. E. D. Adrian and D. W. Bronk. 1928. The discharge of impulses in motor nerve fibres. Part I. Impulses in single fibres of the phrenic nerve. *Journal of Physiology*, 66, 81–101.
    E. D. Adrian and D. W. Bronk. 1929. The discharge of impulses in motor nerve fibres. Part II. The frequency of discharge in reflex and voluntary contractions. *Journal of Physiology*, 67, 119–151.

38. G. Liljestrand. 1932. Presentation speech for the Nobel Prize in Physiology or Medicine. In *Nobel Lectures: Physiology or Medicine, 1922–1941.* Amsterdam: Elsevier, 1965, p. 277.

39. E. D. Adrian. 1932. The activity of nerve fibres. In *Nobel Lectures: Physiology or Medicine, 1922–1941.* Amsterdam: Elsevier, 1965, pp. 293–300.

40. Adrian, ref. 39, p. 293.

41. Adrian, ref. 39, p. 295.

42. Adrian, ref. 39, p. 300.

43. F. A. Gibbs. 1953. Hans Berger. In W. Haymaker (Ed.), *Founders of Neurology.* Springfield, IL: Charles C Thomas, pp. 105–107.
    P. Gloor. 1969. The work of Hans Berger. *Electroencephalography and Clinical Neurophysiology*, 17, 649.

44. H. Berger. 1929. Über das Elektrenkephalogramm des Menschen. *Archiv für Psychiatrie*, 87, 527–570.

45. R. Caton. 1875. The electric currents of the brain. *British Medical Journal*, 2, 278.
    R. Caton. 1877. Interim report on investigation of the electric currents of the brain. *British Medical Journal*, 1, Suppl. L, 62.
    R. Caton. 1887. Researches on electrical phenomena of cerebral grey matter. *Transactions of the Ninth International Medical Congress*, 3, 246–249.

46. E. D. Adrian and B. H. C. Matthews. 1934. The interpretation of potential waves in the cortex. *Journal of Physiology*, 8, 440–471.

47. E. D. Adrian. 1934. Electrical activity of the nervous system. *Archives of Neurology and Psychiatry*, 32, 1125–1136.
    E. D. Adrian. 1936. The electrical activity of the cortex. *Proceedings of the Royal Society of Medicine*, 29, 197–200.
    E. D. Adrian and B. H. C. Matthews. 1934. The Berger rhythm: Potential changes from the occipital lobes in man. *Brain*, 57, 355–385.
    E. D. Adrian and K. Yamagiwa. 1935. The origin of the Berger rhythm. *Brain*, 58, 323–351.

48. Adrian and Matthews, ref. 47, p. 356.

49. O'Leary and Goldring, ref. 25, pp. 135–152.

50. Adrian 1936, ref. 47, p. 199.

51. R. Lemke and F. Schiller. 1956. Personal reminiscences of Hans Berger. *Electroencephalography and Clinical Neurophysiology*, 8, 708.

52. E. D. Adrian. 1943. Afferent areas in the cerebellum connected with the limbs. *Brain*, 66, 289–315.

53. S. Finger. 1994. *Origins of Neuroscience.* New York: Oxford University Press, pp. 141–145.

54. E. D. Adrian. 1940. Double representation of the feet in the sensory cortex of the cat. *Journal of Physiology,* 98, 16P–18P.

E. D. Adrian. 1941. Afferent discharges to the cerebral cortex from peripheral sense organs. *Journal of Physiology,* 100, 159–191.

55. C. N. Woolsey. 1943. "Second" somatic receiving areas in the cerebral cortex of cat, dog, and monkey. *Federation Proceedings,* 2, 55.

C. N. Woolsey and D. Fairman. 1946. Contralateral, ipsilateral and bilateral representation of cutaneous receptors in somatic areas I and II of the cerebral cortex of pig, sheep and other mammals. *Surgery,* 19, 684–702.

56. E. D. Adrian. 1943. Afferent areas in the brain of ungulates. *Brain,* 66, 89–103.

E. D. Adrian. 1943. Sensory areas of the brain. *Lancet,* 2, 33–35.

E. D. Adrian. 1946. The somatic receiving area in the brain of the Shetland pony. *Brain,* 69, 1–8.

E. D. Adrian. 1947. *The Physical Background of Perception.* Oxford: Clarendon Press.

57. E. D. Adrian. 1948. The sense of smell. *Advancement of Science,* 4, 287–292.

E. D. Adrian. 1953. Sensory messages and sensation: The response of the olfactory organ to different smells. *Acta Physiologica Scandinavica,* 29, 4–14.

E. D. Adrian. 1956. The action of the mammalian olfactory organ. *Journal of Laryngology and Otology,* 70, 1–14.

58. E. D. Adrian. 1963. Olfaction and taste. In Y. Zotterman (Ed.), *Wenner-Gren International Symposium,* vol. 1. Oxford, UK: Pergamon Press, pp. 1–4.

Otto Loewi and Henry Dale: The Discovery of Neurotransmitters

1. H. H. Dale. 1952. Transmission of effects from nerve-endings. Lecture at the University of St. Andrews Medical School, 13 March. In H. H. Dale (Ed.), *Autumn Gleanings.* London: Pergamon Press, 1954, pp. 212–225.

2. E. du Bois-Reymond. 1877. *Gesammelte Abhandlungen zur Allgemeinen Muskel- und Nervenphysik,* vol. 2. Leipsig: Veit & Co., p. 700.

3. H. H. Dale. 1938. Natural chemical stimulators. *Edinburgh Medical Journal,* 45, 461–480.

H. H. Dale. 1948. Accident and opportunism in medical research. *British Medical Journal,* 2, 451–455.

4. G. Oliver and E. O. Schäfer. 1894. On the physiological action of extract of the suprarenal capsule. *Journal of Physiology,* 16, i–iv.

G. Oliver and E. O. Schäfer. 1895. On the physiological action of extract of the suprarenal capsule. *Journal of Physiology,* 18, 230–276.

5. M. H. Lewandowsky. 1899. Über die Wirkung des Nebennierenextractes auf die glatten Muskeln im Besonderen des Auges. *Archiv für Physiologie,* 360–366.

J. N. Langley. 1901. Observations on the physiological action of extracts of the suprarenal bodies. *Journal of Physiology,* 27, 237–256.

6. H. H. Dale. 1961. Thomas Renton Elliott, 1877–1961. *Biographical Memoirs of Fellows of the Royal Society,* 7, 53–74.

7. T. R. Elliott. 1905. The action of adrenalin. *Journal of Physiology,* 32, 401–467.

8. T. R. Elliott. 1904. On the action of adrenalin. *Journal of Physiology,* 31, xx–xxi.

9. Elliott, ref. 8, p. xxi.

10. W. E. Dixon. 1907. The mode of action of drugs. *Medical Magazine,* 16, 454–457.

11. Dale, ref. 6, p. 64.

12. H. H. Dale. 1958. Autobiographical sketch. *Perspectives in Biology and Medicine*, 1, 125–137.

   C. D. Leake. 1970. Henry Dale (1875–1968). In W. Haymaker and F. Schiller (Eds.), *Founders of Neurology*. Springfield, IL: Charles C Thomas, pp. 282–285.

13. E. M. Tansey. 1989. The Wellcome Physiological Research Laboratories 1894–1904: The home office, pharmaceutical firms, and animal experiments. *Medical History*, 33, 1–41.

14. Dale, ref. 12, p. 129.

15. H. H. Dale. 1906. On some physiological actions of ergot. *Journal of Physiology*, 34, 163–206.

16. Barger, G., and H. H. Dale. 1910. Chemical structure and sympathomimetic action of amines. *Journal of Physiology*, 41, 19–59.

17. R. Hunt and R. de M. Taveau. 1906. On the physiological action of certain choline derivatives and new methods for detecting cholin. *British Medical Journal*, 2, 1788–1791.

18. A. Ewins. 1914. Acetyl-choline, a new active principle of ergot. *Biochemical Journal*, 8, 44–49.

19. H. H. Dale. 1914. The action of certain esters and ethers of choline, and their relation to muscarine. *Journal of Pharmacology and Experimental Therapeutics*, 6, 147–190.

20. H. H. Dale. 1954. The beginnings and prospects of neurohumoral transmission. *Pharmacological Reviews*, 6, 7–13.

21. Dale, ref. 20, p. 10.

22. H. H. Dale. 1962. Otto Loewi, 1873–1961. *Biographical Memoirs of Fellows of the Royal Society*, 8, 67–89.

   O. Loewi. 1953. *From the Workshop of Discoveries*. Lawrence: University of Kansas Press.

   O. Loewi. 1963. The excitement of a life in science. In D. J. Ingle (Ed.), *A Dozen Doctors*. Chicago: University of Chicago Press, pp. 109–126.

23. O. Loewi and A. Frölich. 1910. Über eine Steigerung der Adrenalinempfindlichkeit durch Cocaïn. *Archiv für experimentelle Pathologie und Pharmakologie*, 62, 159–169.

24. O. Loewi. 1921. Über humorale Übertragbarkeit der Herznervenwirkung. I Mitteilung: *Pflüger's Archiv für die gesamte Physiologie*, 189, 239–242.

   O. Loewi. 1922. Über humorale Übertragbarkeit der Herznervenwirkung. II. Mitteilung. *Pflüger's Archiv für die gesamte Physiologie*, 203, 201–213.

   O. Loewi. 1936. Chemical transmission of nerve action. In *Nobel Lectures: Physiology or Medicine, 1922–41*. Amsterdam, Elsevier, 1965.

25. Loewi, ref. 22 (1953), p. 33.

26. Dale, ref. 22, p. 76.

27. O. Loewi. 1935. Problems connected with the principle of humoral transmission of nervous impulses. *Proceedings of the Royal Society*, ser. B, 118, 306.

28. Dale, ref. 22, p. 77.

29. Dale, ref. 22, p. 78.

30. Loewi 1953, ref. 22, p. 34.

31. Loewi 1953, ref. 22, p. 27.

32. O. Loewi and E. Navratil. 1926. Über humorale Übertragbarkeit der Herznervenwirkung. X Mitteilung: Über das Schicksal des Vagusstoffs. *Pflüger's Archiv für die gesamte Physiologie*, 214, 678–688.

   O. Loewi and E. Navratil. 1926. Über humorale Übertragbarkeit der Herznervenwirkung. XI Mitteilung: Über den Mechanismus der Vaguswirkung von Physostigmin und Ergotamin. *Pflüger's Archiv für gesamte Physiologie*, 214, 689–696.

33. E. Engelhard and O. Loewi. 1930. Fermentative Azetylcholinspaltung im Blut und ihre

Hemmung durch Physostigmin. *Archiv für experimentelle Pathologie und Pharmakologie, 150,* 1–13.

34. J. H. Balfour. 1861. Description of the plant which produces the ordeal bean of Calabar. *Transactions of the Royal Society of Edinburgh, 22,* 305–312.

B. Holmstedt. 1972. The ordeal bean of old Calabar: The pageant of *Physostigma venenosum* in medicine. In T. Swain (Ed.), *Plants in the Development of Modern Medicine.* Cambridge, MA: Harvard University Press, pp. 305–360.

A. G. Karczmar. 1970. History of research with anticholinesterase agents. In A. G. Karczmar, E. Usdin, and J. H. Wills (Eds.), *Anticholinesterase Agents,* vol. 1. Oxford: Pergamon Press, pp. 1–44.

35. H. H. Dale and H. W. Dudley. 1929. The presence of histamine and acetylcholine in the spleen of the ox and the horse. *Journal of Physiology, 68,* 97–123.

36. Loewi, ref. 27, p. 300.

37. Z. M. Bacq. 1934. La pharmacologie du système nerveux autonome, et particulièrement du sympathetique, d'après la théorie neurohumorale. *Annales de Physiologie, 10,* 467–553.

W. B. Cannon. 1934. The story of the development of our ideas of chemical mediation of nerve impulses. *American Journal of the Medical Sciences, 188,* 145–159.

W. B. Cannon and A. Rosenblueth. 1933. Sympathin E and I. *American Journal of Physiology, 104,* 557–574.

38. J. H. Burn. 1955. Sir Henry Dale's contribution to therapeutics. *British Medical Journal, 1,* 1357–1359.

39. H. H. Dale. 1950. The pharmacology of histamine: With a brief survey of evidence for its occurrence, liberation, and participation in natural reactions. *Annals of the New York Academy of Science, 50,* 1017–1028.

H. H. Dale. 1950. The action and uses of the antihistamine drugs as applied to dermatology. *The British Journal of Dermatology and Syphilis, 62,* 151–158.

40. O. Loewi. 1933. The humoral transmission of nervous impulse. *Harvey Lectures,* 232.

41. H. H. Dale. 1948. Transmission of effects from the endings of nerve fibres. *Nature, 162,* 558.

42. W. Feldberg. 1977. The early history of synaptic and neuromuscular transmission by acetylcholine: Reminiscences of an eye witness. In A. L. Hodgkin, A. F. Huxley, W. Feldberg, W. A. H. Rushton, R. A. Gregory, and R. A. McCance (Eds.), *The Pursuit of Nature: Informal Essays on the History of Physiology.* Cambridge, UK: Cambridge University Press, pp. 65–83.

43. Feldberg, ref. 42, pp. 68–69.

44. G. L. Brown, H. H. Dale, and W. Feldberg. 1936. Reaction of the normal mammalian muscle to acetylcholine and to eserine. *Journal of Physiology, 87,* 394–424.

H. H. Dale, W. Feldberg, and M. Vogt. 1936. Release of acetylcholine at voluntary motor nerve endings. *Journal of Physiology, 86,* 353–380.

45. W. Feldberg and J. H. Gaddum. 1934. The chemical transmitter at synapses in a sympathetic ganglion. *Journal of Physiology, 80,* 12P–13P.

46. H. H. Dale. 1934. Pharmacology and nerve endings. *Proceedings of the Royal Society of Medicine, 28,* 319–332.

47. H. H. Dale and W. Feldberg. 1934. The chemical transmission of secretory impulses to the sweat glands of the cat. *Journal of Physiology, 82,* 121–128.

Feldberg and Gaddum, ref. 45.

48. H. H. Dale. 1933. Nomenclature of fibres in the autonomic nervous system and their effects. *Journal of Physiology, 80,* 10P–11P.

49. Dale, ref. 48, p. 11p.

50. M. B. Walker. 1934. Treatment of myasthenia gravis with physostigmine. *Lancet,* 1, 1200–1201.

M. B. Walker. 1935. Case showing effect of prostigmin on myasthenia gravis. *Proceedings of the Royal Society of Medicine, 28,* 759–761.

H. Viets. 1935. The miracle at St. Alfege's. *Medical History, 9,* 184–186.

51. Walker 1934, ref. 50, p. 1200.

52. Walker, 1934, ref. 50, p. 1201.

53. L. Remen. 1932. Zur Pathogenese und Therapie der Myasthenia gravis pseudoparalityka. *Deutsche Zeitschrift für Nervenheilkunde, 128,* 66–78.

54. Loewi 1963, ref. 22, p. 126.

55. Loewi 1963, ref. 22, p. 129.

Roger W. Sperry and Rita Levi-Montalcini: From Neural Growth to "Split Brains"

1. D. Hubel. 1994. Roger W. Sperry (1913–1994). *Nature, 369,* 186.

2. C. Trevarthen. 1990. Roger W. Sperry's lifework and our tribute. In C. Trevarthen (Ed.), *Brain Circuits and the Functions of the Mind: Essays in Honor of Roger W. Sperry.* Cambridge, UK: Cambridge University Press, pp. xxvii–xxxvii.

3. R. W. Sperry. 1940. The functional results of muscle transposition in the hind limb of the rat. *Journal of Comparative Neurology, 73,* 379–404.

4. R. W. Sperry. 1941. The effect of crossing nerves to antagonistic muscles in the hind limb of the rat. *Journal of Comparative Neurology, 75,* 1–19.

5. R. W. Sperry. 1942. Transplantation of motor nerves and muscles in the forelimb of the rat. *Journal of Comparative Neurology, 76,* 283–321.

6. R. W. Sperry. 1943. Functional results of crossing sensory nerves in the rat. *Journal of Comparative Neurology, 78,* 59–90.

7. R. W. Sperry. 1945. The problem of central nervous reorganization after nerve regeneration and muscle transposition. *Quarterly Review of Biology, 20,* 311–369.

8. R. W. Sperry. 1943. Visuomotor coordination in the newt *(Triturus viridescens)* after regeneration of the optic nerve. *Journal of Comparative Neurology, 79,* 33–55.

9. R. W. Sperry. 1944. Optic nerve regeneration with return of vision in anurans. *Journal of Neurophysiology, 7,* 57–69.

R. W. Sperry. 1945. Restoration of vision after crossing of optic nerves and after contralateral transplantation of eye. *Journal of Neurophysiology, 8,* 17–28.

10. R. W. Sperry. 1948. Orderly patterning of synaptic associations in regeneration of intracentral fiber tracts mediating visuomotor coordination. *Anatomical Record, 102,* 63–75.

R. W. Sperry. 1948. Patterning of central synapses in regeneration of the optic nerve in teleosts. *Physiological Zoology, 21,* 351–361.

11. R. G. Harrison. 1908. Embryonic transplantation and development of the nervous system. *Anatomical Record, 2,* 385–410.

12. Harrison, ref. 11, p. 389.

13. S. Ramón y Cajal, S. 1972. *The Structure of the Retina.* Translated by S. A. Thorpe and M. Glickstein. Springfield, IL: Charles C Thomas, pp. 148–151.

14. J. N. Langley. 1895. Note on regeneration of preganglionic fibres of the sympathetic. *Journal of Physiology, 18,* 280–284.

15. R. K. Hunt and W. M. Cowan. 1990. The chemoaffinity hypothesis: An appreciation of Roger W. Sperry's contributions to developmental biology. In Trevarthen, ref. 2, pp. 19–74.

16. Sperry, ref. 8, p. 45.

17. Sperry 1994, ref. 9, p. 67.

18. R. W. Sperry and N. Miner. 1949. Formation within sensory nucleus V of synaptic associations mediating cutaneous localization. *Journal of Comparative Neurology, 90,* 422.

19. R. W. Sperry. 1963. Chemoaffinity in the orderly growth of nerve fiber patterns and connections. *Proceedings of the National Academy of Sciences*, 50, 703–710.

R. W. Sperry. 1965. Embryogenesis of behavioral nerve nets. In R. L. DeHaan and H. Ursprung (Eds.), *Organogenesis*, vol. 6. New York: Holt, Rinehart and Winston, pp. 161–185.

20. Sperry 1965, ref. 19, pp. 170–171.

21. R. Levi-Montalcini. 1988. *In Praise of Imperfection*. New York: Basic Books.

R. Levi-Montalcini. 1992. NGF: An uncharted route. In F. G. Worden, J. P. Swazey, and G. Adelman (Eds.), *The Neurosciences: Paths of Discovery*, vol. 1 Boston: Berkhäuser, pp. 245–265.

22. E. D. Bueker. 1948. Implantation of tumors in the hind limb field of the embryonic chick and the developmental response of the lumbrosacral nervous system. *Anatomical Record*, 102, 369–389.

23. R. Levi-Montalcini and V. Hamburger. 1951. Selective growth stimulating effects of mouse sarcoma on the sensory and sympathetic nervous system of the chick embryo. *Journal of Experimental Zoology*, 116, 321–361.

24. R. Levi-Montalcini and V. Hamburger. 1953. A diffusable agent of mouse sarcoma, producing hyperplasia of sympathetic ganglia and hyperneurotization of viscera in the chick embryo. *Journal of Experimental Zoology*, 123, 233–287.

25. S. Cohen, R. Levi-Montalcini, and V. Hamburger. 1954. A nerve growth-stimulating factor isolated from sarcomas 37 and 180. *Proceedings of the National Academy of Sciences*, 40, 1014–1018.

26. S. Cohen and R. Levi-Montalcini. 1956. A nerve growth-stimulating factor isolated from snake venom. *Proceedings of the National Academy of Sciences*, 42, 571–574.

27. S. Cohen. 1958, A nerve growth-promoting protein. In W. D. McElroy and B. Glass (Eds.), *Chemical Basis of Development*. Baltimore, MD: Johns Hopkins University Press, pp. 665–676.

28. R. Levi-Montalcini and P. U. Angeletti. 1961. Growth control of the sympathetic system by a specific protein factor. *Quarterly Review of Biology*, 36, 99–108.

R. Levi-Montalcini and B. Booker. 1960. Excessive growth of the sympathetic ganglia evoked by a protein isolated from mouse salivary glands. *Proceedings of the National Academy of Sciences*, 46, 373–384.

29. R. Levi-Montalcini and B. Booker. 1960. Destruction of the sympathetic ganglia in mammals by an antiserum to a nerve-growth protein. *Proceedings of the National Academy of Sciences*, 46, 384–391.

30. F. B. Axelrod, R. Nachtigal, and J. Dancis. 1974. Familial dysautonomia: Diagnosis, pathogenesis and management. *Advances in Pediatrics*, 21, 75–96.

C. M. Riley and R. H. Moore. 1966. FD differentiated from related disorders: Case reports and current concepts. *Pediatrics*, 37, 435–446.

31. J. Bell, M. Gruenthal, S. Finger, P. Lundberg, and E. Johnson. 1982. Behavioral effects of early deprivation of nerve growth factor: Some similarities with familial dysautonomia. *Brain Research*, 234, 409–421.

32. R. Levi-Montalcini. 1990. Ontogenesis of neuronal nets: The chemoaffinity theory, 1963–1983. In C. Trevarthen (Ed.), *Brain Circuits and the Functions of the Mind*. Cambridge, UK: Cambridge University Press, pp. 3–18.

33. L. Harris. 1995. The corpus callosum and hemispheric communication: An historical survey of theory and research. In F. L. Kitterle (Ed.), *Hemispheric Communication: Mechanisms and Models*. Hillsdale, NJ: Lawrence Erlbaum.

R. J. Joynt. 1971. The corpus callosum: History of thought regarding its function. In M. Kinsbourne and W. L. Smith (Eds.), *Hemispheric Disconnection and Cerebral Function*. Springfield, IL: Charles C Thomas, pp. 117–125.

34. W. W. Ireland. 1886. *The Blot upon the Brain*. New York: G. P. Putnam's Sons, p. 318.

35. J. Turner. 1877–1878. A human cerebrum imperfectly divided into two hemispheres. *Journal of Anatomy and Physiology*, 12, 241–253.

    A. Bruce. 1889–1890. On the absence of the corpus callosum in the human brain, with the description of a new case. *Brain*, 12, 171–190.

36. W. McDougall. 1911. *Body and Mind*. London: Methuen.

37. C. Burt, quoted in O. L. Zangwill. 1974. Consciousness and the cerebral hemispheres. In S. J. Dimond and J. G. Beaumont (Eds.), *Hemisphere Function in the Human Brain*. New York: Halsted Press, p. 265.

38. W. E. Dandy. 1930. Changes in our conceptions of localization of certain functions in the brain. *American Journal of Physiology*, 93, 643.

    W. E. Dandy. 1936. Operative experience in cases of pineal tumor. *Archives of Surgery*, 33, 19–46.

39. A. J. Akelaitis. 1940. A study of gnosis, praxis and language following partial and complete section of the corpus callosum. *Transactions of the American Neurological Association*, 66, 182–185.

    A. J. Akelaitis. 1941. Psychobiological studies following section of the corpus callosum. *American Journal of Psychiatry*, 97, 1147–1157.

    A. J. Akelaitis. 1943. Studies on the corpus callosum. VII. Study of language functions (tactile and visual lexia and graphia) unilaterally following section of the corpus callosum. *Journal of Neuropathology and Experimental Neurology*, 2, 226–262.

40. R. W. Sperry. 1964. The great cerebral commissure. *Scientific American*, 210, 42.

41. Sperry, ref. 40, p. 42.

42. E. V. Evarts. 1990. Coordination of movement as a key to higher brain function: Roger W. Sperry's contributions from 1939–1952. In Trevarthen, ref. 2, pp. xx–xxi.

43. R. E. Myers and R. W. Sperry. 1953. Interocular transfer of a visual form discrimination habit in cats after section of the optic chiasma and corpus callosum. *Anatomical Record*, 115, 351–352.

44. K. Bykoff. 1924–1925. Versuche an Hunden mit Durchschneiden des Corpus callosum. *Zentralblatt für die gesammte Neurologie und Psychiatrie*, 39, 199.

    I. P. Pavlov. 1928. *Lectures on Conditioned Reflexes*. Trans. W. H. Gantt. New York: International Publishers, pp. 150–151.

    S. M. Kanne and S. Finger. 1999. Konstantin M. Bykov and the discovery of the role of the corpus callosum. *Journal of the History of Medicine and Allied Sciences*, in press.

45. J. G. Miller. 1994. Roger Wollcott Sperry. *Behavioral Science*, 39, 265–267.

46. J. S. Stamm and R. W. Sperry. 1957. Function of corpus callosum in contralateral transfer of somesthetic discrimination of cats. *Journal of Comparative and Physiological Psychology*, 50, 138–143.

47. C. Trevarthen. 1962. Double visual learning in split brain monkeys. *Science*, 136, 258–259.

48. J. E. Bogen and P. J. Vogel. 1962. Cerebral commissurotomy in man. *Bulletin of the Los Angeles Neurological Society*, 27, 169–172.

49. J. E. Bogen and P. J. Vogel. 1963. Treatment of generalized seizures by cerebral commissurotomy. *Surgical Forum*, 14, 431–433.

50. M. S. Gazzaniga, J. E. Bogen, and R. W. Sperry. 1962. Some functional effects of sectioning the cerebral commissures in man. *Proceedings of the National Academy of Sciences*, 48, 1765–1769.

    M. S. Gazzaniga, J. E. Bogen, and R. W. Sperry. 1965. Observations on visual perception after disconnection of the cerebral hemispheres in man. *Brain*, 88, 221–236.

    M. S. Gazzaniga and R. W. Sperry. 1967. Language after section of the cerebral commissures. *Brain*, 90, 131–148.

R. W. Sperry. 1968. Mental unity following surgical disconnection of the cerebral hemispheres. *Harvey Lectures,* 62, 293–323.

51. J. E. Bogen, E. D. Fisher, and P. J. Vogel. 1965. Cerebral commissurotomy: A second case report. *Journal of the American Medical Association,* 194, 160–161.

52. R. W. Sperry and M. S. Gazzaniga. 1967. Language following surgical disconnection of the hemispheres. In F. C. Darley (Ed.), *Brain Mechanisms Underlying Speech and Language.* New York: Grune and Stratton, pp. 108–121.

Gazzaniga, Bogen, and Sperry 1965, ref. 50.

53. R. W. Sperry. 1964. *Problems Outstanding in the Evolution of Brain Function.* New York: American Museum of Natural History, p. 17.

54. J. Levy, C. Trevarthen, and R. W. Sperry. 1972. Perception of bilateral chimeric figures following hemisphere disconnection. *Brain,* 95, 61–78.

J. Levy-Agresti and R. W. Sperry. 1968. Differential perceptual capacities in major and minor hemispheres. *Proceedings of the National Academy of Sciences,* 61, 1151.

55. R. W. Sperry. 1970. Perception in the absence of the neocortical commissures. In D. A. Hamburg, K. H. Pribram, and A. J. Stunkard (Eds.), *Perception and Its Disorders.* Baltimore, MD: Williams & Wilkins Co., pp. 123–138.

56. R. W. Sperry. 1982. Some effects of disconnecting the cerebral hemispheres (Nobel lecture). *Science,* 217, 1224.

57. R. W. Sperry. 1965. Mind, brain, and humanist values. In J. R. Platt (Ed.), *New Views of the Nature of Man.* Chicago: University of Chicago Press, pp. 71–92.

R. W. Sperry. 1969. A modified concept of consciousness. *Psychological Review,* 76, 532–536.

R. W. Sperry. 1977. Forebrain commissurotomy and conscious awareness. *Journal of Medicine and Philosophy,* 2, 101–126.

R. W. Sperry. 1984. Emergence. In P. Weintraub (Ed.), *The Omni Interviews.* New York: Omni Press, pp. 187–207.

Sperry, ref. 53.

58. Sperry, ref. 53, p. 12.

Pioneers and Discoveries in the Brain Sciences

1. W. B. Cannon. 1945. *The Way of an Investigator.* New York: W. W. Norton, p. 205.

2. R. Levi-Montalcini. 1988. *In Praise of Imperfection.* New York: Basic Books, p. 5.

3. G. von Bonin. 1960. *Some Papers on the Cerebral Cortex.* Springfield, IL: Charles C Thomas, p. xii.

4. I. Newton. Unpublished letter to Robert Hooke dated 5 February 1675/6. R. K. Merton. *On the Shoulders of Giants: A Shandean Postscript.* New York: Free Press.

5. Merton, ref. 4. (Section 281.25).

6. E. D. Adrian. 1932. *The Mechanism of Nervous Action. Electrical Studies of the Neurone.* Philadelphia: University of Pennsylvania Press, p. 2.

7. C. S. Sherrington. 1946. *The Endeavour of Jean Fernel.* Cambridge: Cambridge University Press, p. 142.

8. J. C. Eccles and W. C. Gibson. 1979. *Sherrington: His Life and Thought.* New York: Springer International.

R. Granit. 1967. *Charles Scott Sherrington: An Appraisal.* Garden City, NY: Doubleday & Co., p. 28.

9. B. Barber. 1961. Resistance by scientists to scientific discovery. *Science,* 134, 596–602.

10. M. Planck. 1936. *Philosophy of Physics.* New York: Norton, p. 97.

11. S. J. Gould. 1998. An awful, terrible, dinosaurian irony. *Natural History,* 107, 24–26, 61–68.

R. Owen. 1868. *On the Anatomy of Vertebrates.* London: Longmans, Green.

S. Finger. 1994. *Origins of Neuroscience.* New York: Oxford University Press, p. 317.

12. Gould, ref. 11, pp. 26–27.

13. T. Kuhn. 1970. *The Structure of Scientific Revolutions,* 2d ed. Chicago: University of Chicago Press.

14. Levi-Montalcini, ref. 2, p. 108.

15. M. Merbaum and M. R. Lowe. 1982. Serendipity in clinical research. In P. C. Kendall and J. N. Butcher (Eds.), *Handbook of Research Methods in Clinical Psychology.* New York: John Wiley & Sons, pp. 95–123.

16. J. Priestley. 1775. *The History and Present State of Electricity.* London: J. Dodsley, p. xi.

17. H. Helmholtz. 1891. An autobiographical sketch (Translation of an address delivered in Berlin on the occasion of his seventieth birthday). In R. Kahl (Ed.), *Selected Writings of Hermann von Helmholtz.* Middletown, CT: Wesleyan University Press, 1971, p. 474.

18. H. Kragh. 1987. *An Introduction to the Historiography of Science.* Cambridge, UK: Cambridge University Press, pp. 21–60.

19. B. F. Skinner. 1968. A case history in scientific method. In A. C. Catania (Ed.), *Contemporary Research in Operant Behavior.* Glenville, IL.: Scott Foresman, p. 38.

20. Cannon, ref. 1, p. 77.

21. H. H. Dale. 1948. Accident and opportunism in medical research. *British Medical Journal,* 2, 451–455. Reprinted in H. H. Dale, *An Autumn Gleaning.* New York: Interscience Publishers, 1954, p. 126.

22. A. Hodgkin. 1979. Edgar Douglas Adrian: Baron Adrian of Cambridge. *Biographical Memoirs of Fellows of the Royal Society,* 25, 49–50.

23. Granit, ref. 8, p. 28.

24. Cannon, ref. 1, pp. 75–76.

25. Cannon, ref. 1, p. 76.

26. Cannon, ref. 1, p. 76.

27. Kragh, ref. 18, pp. 32–40.

Index